# LASER BEAM SCANNING

## OPTICAL ENGINEERING

*Series Editor*

**Brian J. Thompson**
William F. May Professor of Engineering
*and* Provost, University of Rochester
Rochester, New York

*Laser Engineering Editor:* **Peter K. Cheo**
United Technologies Research Center
Hartford, Connecticut

*Laser Advances Editor:* **Leon J. Radziemski**
New Mexico State University
Las Cruces, New Mexico

*Optical Materials Editor:* **Solomon Musikant**
General Electric
Valley Forge, Pennsylvania

1. Electron and Ion Microscopy and Microanalysis: Principles and Applications, *by Lawrence E. Murr*
2. Acousto-Optic Signal Processing: Theory and Implementation, *edited by Norman J. Berg and John N. Lee*
3. Electro-Optic and Acousto-Optic Scanning and Deflection, *by Milton Gottlieb, Clive L. M. Ireland, and John Martin Ley*
4. Single-Mode Fiber Optics: Principles and Applications, *by Luc B. Jeunhomme*
5. Pulse Code Formats for Fiber Optical Data Communication: Basic Principles and Applications, *by David J. Morris*
6. Optical Materials: An Introduction to Selection and Application, *by Solomon Musikant*
7. Infrared Methods for Gaseous Measurements: Theory and Practice, *edited by Joda Wormhoudt*
8. Laser Beam Scanning: Opto-Mechanical Devices, Systems, and Data Storage Optics, *edited by Gerald F. Marshall*

*Other Volumes in Preparation*

# LASER BEAM SCANNING

## OPTO-MECHANICAL DEVICES, SYSTEMS, AND DATA STORAGE OPTICS

*Edited by*

**Gerald F. Marshall**
*Energy Conversion Devices, Inc.*
*Troy, Michigan*

MARCEL DEKKER, INC. New York • Basel

Library of Congress Cataloging in Publication Data

Main entry under title:

Laser beam scanning.

(Optical engineering ; v. 8)
Includes index.
1. Optical scanners. 2. Lasers. I. Marshall, Gerald F. II. Series.
TK7882.S3L3 1985 621.36'6 85-21804
ISBN 0-8247-7418-3



MARCEL DEKKER, INC.
270 Madison Avenue, New York, New York 10016
Current printing (last digit):
10 9 8 7 6 5 4 3 2 1

PRINTED IN THE UNITED STATES OF AMERICA

*To the memory of*
*Albert and Ethelena*

# About the Series

Optical science, engineering, and technology have grown rapidly in the last decade so that today optical engineering has emerged as an important discipline in its own right. This series is devoted to discussing topics in optical engineering at a level that will be useful to those working in the field or attempting to design systems that are based on optical techniques or that have significant optical subsystems. The philosophy is not to provide detailed monographs on narrow subject areas but to deal with the material at a level that makes it immediately useful to the practicing scientist and engineer. These are not research monographs, although we expect that workers in optical research will find them extremely valuable.

Volumes in this series cover those topics that have been a part of the rapid expansion of optical engineering. The developments that have led to this expansion include the laser and its many commercial and industrial applications, the new optical materials, gradient index optics, electro- and acousto-optics, fiber optics and communications, optical computing and pattern recognition, optical data reading, recording and storage, biomedical instrumentation, industrial robotics, integrated optics, infrared and ultraviolet systems, etc. Since the optical industry is currently one of the major growth industries this list will surely become even more extensive.

Brian J. Thompson
University of Rochester
Rochester, New York

# Preface

The aim of this volume is to provide engineers, scientists, and students with a guideline to the fundamentals of laser beam scanning. It brings together the knowledge and experience of seven specialists in the field, from England, Germany, Scotland, and the United States.

The book covers the recently developed holographic scanners, the well-established polygonal scanners, and the galvanometric and resonant scanners. It includes complementary chapters on gas bearings for rotating scanning devices, the aerodynamic considerations of polygonal scanners, gaussian laser beam diameters, and the optical design of components and systems relating to data storage on optical disks.

Gerald F. Marshall

# Acknowledgments

I would like to extend my thanks to the following persons: Brian Thompson, Provost of the University of Rochester, for inviting me to produce this volume. My six contributing authors for their splendid material. My manuscript reviewers, Emile Bertrand and Michael Spencer of Rexham Corporation, Speedring Division; Warren H. Stevenson of Purdue University, Richard Rallison of Dikrotek International Corp., and J. David Zook of Honeywell Technical Center, who gave their time and helpful comments to selected chapters. My colleagues, Bob Edgerton, Rosa Young, and Jesus Hernandez, of Energy Conversion Devices, Inc.; Fred Nicodemus of the National Bureau of Standards and Leo Beiser of Beiser Inc.; each of whom critiqued my manuscript and made excellent suggestions. Thomas Hayosh of Burroughs Corporation for his thoroughness and effort in reviewing first and second drafts of the manuscripts for five of the seven chapters. John Campbell, retired professor of physics, California Institute of Technology, for integrating the index. Stan and Iris Ovshinsky, the founders of Energy Conversion Devices, Inc., for their underlying support. Irene, my wife, for her interest and generous help. Lastly, J. Christopher Harris, Harold S. Barlow, and W. F. "Bill" Williams, my former educators, whose scholarly teaching abilities and ongoing friendships have played an important part in stimulating my early interest in writing and optical physics.

I feel proud to be the coordinator of these works and appreciate the privilege of being the one to share this treatise with my colleagues in the field.

# Introduction

Laser beam scanning is the controlled deflection of a light beam and is often loosely referred to as "laser scanning." Precision opto-mechanical laser beam scanning intimately involves three main disciplines: mechanics, electronics, and optics. It requires intricate mechanical engineering and manufacturing, clever electronic and magnetic circuitry, and creative optical design. This book brings together contributions from specialists directly and closely involved in opto-mechanical laser beam scanning devices and related components and their applications.

It includes chapters on holographic and polygonal devices in which the active optical component undergoes a rotational motion; and a chapter on galvanometric and resonant devices in which the active optical component undergoes an oscillatory motion.

Other chapters deal with certain aspects relating to these devices, such as gas bearings and aerodynamics. These topics are covered in a way that I believe has not appeared in print previously. The need for Chapter 6 became apparent after I had received the final manuscripts of six excellent pieces of work from my contributing authors. It is written to give a clearer understanding of the meaning of the diameter of a laser beam and to provide specific definitions. Likewise I restore the preferred radiometric terminology that we should be using when describing the radiant properties of a laser beam and an image. No attempt, however, is made to standardize the terminology among the chapters; to do so would have seriously delayed the publication of this book.

The last chapter contains a wealth of practical design information on optical components, laser beam imaging and their performance as they relate to optical disk data storage and laser beam scanning technology.

# Contents

# Contributors

Paul Kuttner Manager, Research and Development, Precision Optics Division, Optische Werke G. Rodenstock, Munich Federal Republic of Germany

Anthony J. Lawler* Development Engineer, Precision Instruments Group, Navigation Systems Department, Ferranti, P.L.C., Edinburgh, Scotland

Gerald F. Marshall Director of Optical Engineering, Energy Conversion Devices, Inc., Troy, Michigan

Jean Montagu Chairman of the Board, Chief Executive Officer, General Scanning, Inc., Watertown, Massachusetts

Joseph Shepherd Design Group Leader, Precision Instruments Group, Navigation Systems Department, Ferranti, P.L.C., Edinburgh, Scotland

Randy J. Sherman President, Lincoln Laser Company, Phoenix, Arizona

Glenn T. Sincerbox Manager, Inspection Science and Technology Department, IBM Almaden Research Center, San Jose, California

---

**Present affiliation*: Research and Development Manager, The Kemble Instrument Co., P.L.C., Burgess Hill, West Sussex, England.

# LASER BEAM SCANNING

# 1

# Holographic Scanners

## Applications, Performance, and Design

Glenn T. Sincerbox
*IBM, Almaden Research Center, San Jose, California*

### 1. INTRODUCTION

Since the invention of holography by Gabor in 1948 and its practical realization by Leith and Upatnieks in 1962, numerous holographic applications have been proposed covering a broad range of areas from three-dimensional television to optical information storage and processing. Although some of these proposals were beyond the state of the art at the time of their conception, and a fraction are still waiting for breakthroughs, many others have progressed at least as far as laboratory devices and feasibility studies. This has resulted in the development of holographic techniques for internal use by different laboratories and manufacturing establishments and the extension of a select few in the commercial market. These entries can be categorized as either stand-alone elements such as holographically recorded diffraction gratings (Lerner and Laude, 1983), display art forms, and jewelry, or as components and subsystems incorporated into nondestructive testing instruments (Heflinger et al., 1966; Powell and Stetson, 1965) and into laser scanning systems (Dickson et al., 1982). The full potential of holography for complex wavefront recording and reconstruction has yet to be commercially realized. Primary emphasis to date has centered on the use of relatively simple holographic elements in the form of diffraction gratings for spectroscopy or beam steering applications.

A large number of well-developed nonholographic beam steering technologies are currently available and can be divided into two groups, those using mechanical motion of mirrored surfaces, and those that involve an inertialess change in the optical properties of a transparent medium. Some members of this latter group involve diffraction of light from an acoustically generated periodic structure (i.e., a transient grating) but do not qualify as holographic devices because the diffraction is not a wavefront reconstruction process. The holographic counterpart of an acousto-optical deflector would rely on diffraction from an optically induced grating hologram with a means for real-time variation of the spatial frequency of the grating (Sincerbox and Roosen, 1982). These opto-optical devices are relatively new, will require considerable development, and do not fit within the general theme of holographic scanning by mechanical methods. In this context, therefore, the discussion of holographic deflection that follows will be restricted to those devices requiring mechanical motion of a permanently recorded holographic structure.

## 1.1. Holography and Light Deflection

The concept of holographic recording and reconstruction—more important, how it deflects light—can best be understood with reference to Figures 1 and 2. In Figure 1a, two wavefronts of equal intensity created from a laser are directed to overlap in some region of space where the recording is to be made. If the optical path difference from the point of beam separation to the region of overlap is within the coherence length of the source, the resulting interference pattern will be stationary in both space and time and will have high fringe contrast. The intensity distribution in these fringes can be exposed onto, or more properly into, a suitable photosensitive medium such as a photographic emulsion. After processing, the recording contains a variation in optical density, refractive index, or optical thickness—sometimes a combination of all three—and is the hologram. When this recording is repositioned and illuminated by one of the wavefronts, for example, the diverging wavefront in Figure 1b, the structure at each point within the hologram diffracts the illuminating light and creates a new wavefront that is an "identical" duplicate of the original, second wavefront—in this case a converging and deflected wavefront. This is the equivalent of a positive or converging lens in combination with a prism converting and deflecting light from a point object to a point image. The efficiency and quality with which this wavefront conversion occurs is directly related to the recording configuration and selection of recording material. As

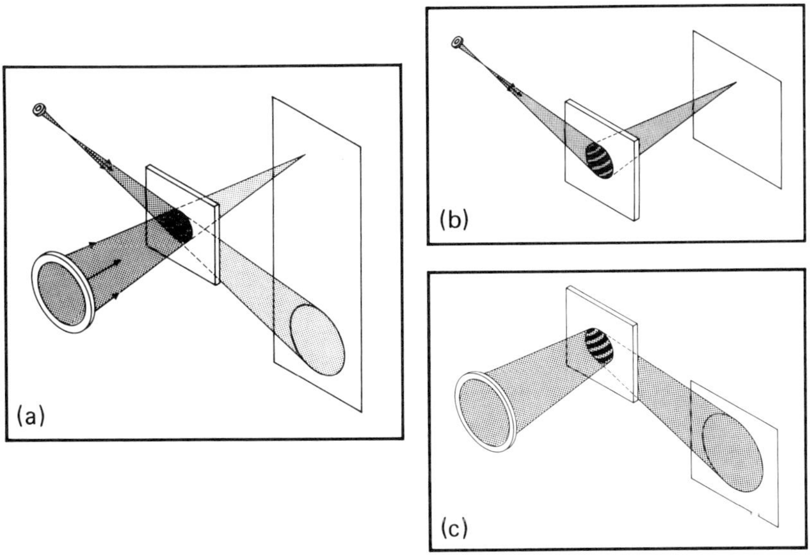


**Figure 1** Simple holography: (a) recording the hologram; (b) reconstruction of the convergent wavefront (positive lens); (c) reconstruction of the diverging wavefront (negative lens).

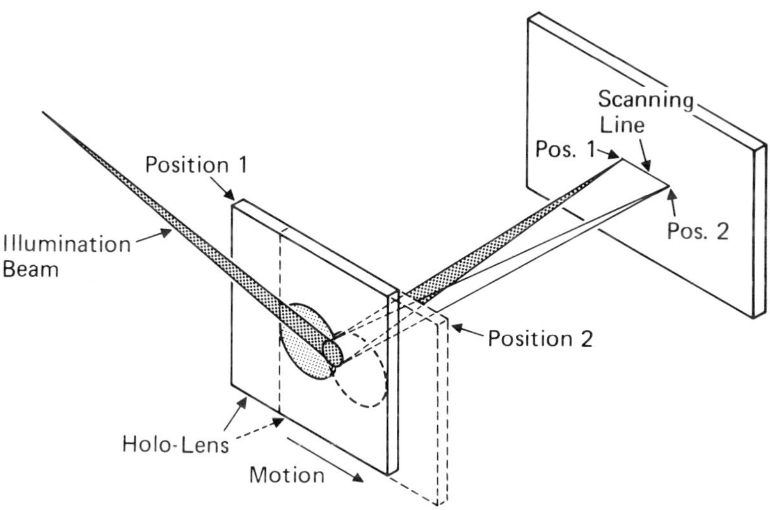


**Figure 2** Principle of holographic deflection.

will be seen later, the ideal material should have a linear response to exposure, be sensitive to a broad range of wavelengths, produce a diffracting phase structure, and require no chemical processing. Returning to Figure 1c, the equivalent of a negative lens is realized by illuminating the hologram with the converging wavefront, thereby reproducing the original diverging wavefront. Holographic recording of complex multidimensional objects can be treated as the recording of a superposition of individual spherical waves from all the points in the object field.

By using a combination of small-area illumination and hologram translation, the reconstruction geometry of Figure 1b can be used to create simultaneous light deflection and focusing. This is shown in Figure 2, where the hologram is initially located at position 1 and has a small subarea, on the right side, illuminated by a diverging wavefront. The light is focused by this hologram area to a point in the image plane corresponding to the location, with respect to the displaced hologram, of the original converging wavefront, labeled "position 1." As the hologram is translated, different subareas of the hologram are passed under the illuminating beam and the reconstructed image point is caused to translate by the same distance, in this case to position 2. This is completely analogous to the deflection and focusing that would occur if a conventional lens were illuminated off axis with a collimated beam and the lens displaced normal to its optic axis. A continuous back-and-forth motion in either case produces the same motion in the focused spot.

This is a rather simplistic view of holography and holographic deflection and is intended only to present the concepts involved. The realization of a functional design is obviously a complex process requiring consideration of numerous interdependent parameters and evaluation of the resulting design trade-offs. The driving factor is, of course, the type and quality of scan line that is required for the particular application under development. These requirements, together with illumination specifications, largely determine the format of the hologram(s), the type of motion to be employed, and the recording material itself. It is this design process and parameter considerations that will form the bulk of the chapter.

### 1.2. Attributes and Limitations

Before discussing details, it is appropriate to establish the motivation for why, and perhaps why not, holograms should be considered in light deflection applications as alternate deflection elements. Of particular importance is an added degree of freedom gained by combining several functions—focusing, in addition to deflection—into a single component. This unique property is fully exploited in a multifaceted design where functional performance is improved by using different rather than identical holograms. Hence each facet can deflect light into a different direction with focal distances of varying lengths and, perhaps, with other than point images. This is accomplished with a device having mechanically smooth and continuous surfaces, devoid of edges and attendant air turbulence, as well as asymmetries introduced when attempting to create displaced deflection lines with variably inclined facets. The end effect is a multidimensional, arbitrary image generator, providing considerable freedom in application design and innovation.

Just as with conventional optical components, holographic elements can operate in two modes: as deflectors generating moving spots of light, and as scanners collecting some portion of the light that is scattered by some object field. An important feature of the retrocollection mode [patented by Pole and Werlich (1978) and demonstrated by Pole et al (1978)] is that the location of the return image is invariant with the position of the moving spot in the scan field. Resolution enhancement beyond the limits set by the numerical aperture of the scanning beam becomes possible by detector plane filtering. The hologram also serves as an ambient light filter, directing only light within the proper spectral bandwidth to the detector plane—a factor of interest in a point-of-sale terminal application.

Mechanically, the size and weight of the hologram are negligible compared to its substrate, and this in turn can be a relatively simple surface. Consequently, the packaging can be easier, a smaller volume can be used, and the drive system needed to create the required motion will generally be loaded with a smaller mass.

A principal concern is the ease by which holographic deflector elements can be manufactured on a mass production scale. Although it is often possible to sacrifice cost for the purpose of adding performance features and functions, or to address a new market, the economics of production will determine ultimate success. The creation of each holographic device as an original recording (master), particularly in the case of a deflector having different facets that would necessitate a dynamic recording system, would be difficult to implement effectively in a manufacturing environment where high throughput and yield are required. The alternative is a replication process whereby a master, or submaster, is employed to generate numerous copies. Depending on the choice of holographic mechanisms, volume phase, or surface relief, either optical copying or mechanical replication can be considered (Vanin, 1978). In optical copying, a holographic master with medium efficiency (i.e., ~40%) is illuminated and the reconstructed wavefront is combined with the transmitted zero-order wavefront. The resulting interference is used to create a second holographic recording. This is done using a very small separation between master and copy and should not be confused with contact printing of the actual diffracting structure, as in semiconductor lithography. The replica is, in all respects, a hologram of a hologram. As will be discussed in detail later, this is one of the very few ways of fabricating an aberration-free hologram for use at a wavelength other than the copy recording wavelength. In mechanical replication, the diffracting structure is in the form of a surface relief structure and is replicated by any number of techniques, including hot pressing, injection molding, and ion milling. Much of this technology has been developed and is used today in the manufacture of holographic diffraction gratings, video and optical storage disks, and for the replication of optical components.

A discussion of the attributes of holographic deflection must include a discussion of the factors that restrict its utility. Significant limitations are presented by the lack of spectral broad-bandness inherent in the recording and reconstruction process. The hologram is a highly tuned structure, expecting the illuminating wavefront to be of particular shape and wavelength, and as a result introduces severe aberrations into the reconstructed wavefront if the appropriate conditions are not satisfied. As will be described later, it is only in the case of plane waves that the aberrations generated by a wavelength change can be compensated for by a change in illumination angle. This problem is further compounded by a lack of high-efficiency holographic materials sensitive to red and infrared wavelengths. One of the best materials available today, dichromated gelatin, provides diffraction efficiencies in excess of 80% but has no recording sensitivity to wavelengths longer than 515 nm. In addition, these holograms must be environmentally encapsulated to protect the diffracting structure within the gelatin from moisture. Finally, deflectors having diffraction-limited focusing properties and highly repeatable deflection characteristics over large fields require that additional consideration, and cost, be directed to the optical and mechanical quality of the substrate(s).

## 2. APPLICATIONS

The concept of using holographic scanning has been around for more than a decade (Cindrich, 1967) and during this time many different applications have been suggested, but few have been demonstrated and an even smaller number have made it into the marketplace. At the time of this writing there are only five announced commercial products: point-of-sale scanners by IBM, Nippon Electric Company, and Fujitsu and electrophotographic printers by Benson and General Optronics. Another printer system under development by Ricoh will be included in the discussion based on technical literature.

## 2.1. Point-of-Sale Scanners

Point-of-sale scanners have been with us for almost a decade and are gaining rapidly in acceptance by both the merchant and the customer. Based on a universal product code (UPC) adopted in 1973, labels bearing the now familiar bar codes can be found on nearly every grocery item. The equivalent codes in Japan are the Japanese article number (JAN) and in Western Europe the European article number (EAN). For reading, the operator brings the product toward the exit port or window of the scanner with the label facing forward and also perhaps downward. An item sensor is activated, usually a light beam, causing a shutter to be opened and the scanning laser beam to exit the window. The scanner itself (hologram disk, multifacet mirror, etc.) is constantly moving, so that there is no startup delay. The dense scan pattern traverses the product as it moves forward and some of the light scattered by the surface is collected by a photodetector that converts the optical modulation to an electrical signal. The resulting digitized signal is examined by a computer and that portion of the signal corresponding to the unique bar code pattern is identified. Only those signals that represent a scan of the entire label are selected and used to determine product identity; usually agreement from multiple scans is required as verification. Data are then fed back to the terminal about product price, and provisions are available to update inventory, gather statistics, and so on.

The primary factors in the maturation of these devices has been a quest for a higher and higher throughput rate with a matching improvement in productivity. This has required the relaxing of previous restrictions on product orientation so that the operator need only direct the label toward the window, in contrast to placing the label against the window. The measure, of course, is not only how fast the product can be moved through the reading space, but also how accurately the label can be read. Reducing the number of misreads and the need for additional scanning is of primary importance. This requirement has, in turn, demanded the design of forward-looking scanners with large depth of fields and, in some cases, the addition of side-scanning systems that combine to produce a wraparound pattern.

As pointed out by Dickson et al. (1982), the requirement to read a bar code with a resolution of 0.2 mm is quite easily satisfied over the 12.5 mm depth of field requirement contained in the current UPC specification, but is incompatible with the larger, 150-mm fields required of forward-looking scanners. This is particularly true for low- to medium-contrast labels. This requirement can be satisfied by creating scans having different focal distances with respect to the window. An individual line will not improve depth of field, but the total of all the lines will, by generating a three-dimensional scan. To accomplish this with classical scanning systems, a long scan line must be broken into segments via mirrors and each segment passed through a different optical system prior to exiting the window or at least delayed in space with respect to each other to produce multiple focal distances. The unique ability of holography to utilize both deflection and focal power in one element and the freedom to make each facet different can be utilized to create a high-density, large-depth-of-field scan pattern.

### 2.1.1. IBM 3687 Supermarket Scanner

The IBM 3687 supermarket scanner was announced in 1980 and shipped to customers in the following year. A systems overview has been described by Moore (1981) and the optics and holography by Dickson and colleagues (1981, 1982). The concept used in the IBM 3687 is shown schematically in Figure 3. The scanner disk is 195 mm in diameter and consists of a sandwich of two identical glass disks encapsulating the holographic recording medium. The hologram material is dichromated gelatin (DCG) and the disks are manufactured using a master/copy process as described in Section 9.2.1. Diffraction efficiencies as high as 80% have been achieved, with the residual 20% loss caused by Fresnel reflections at the air-glass surfaces, absorption and scattering in the gelatin, and by a very small undiffracted (zero-order) transmission. Environmental encapsulation between the glass disks is necessitated by the moisture sensitivity of DCG. Optical and mechanical tests performed in conjunction with temperature/humidity cycling have indicated no degradation in optical performance or diffraction efficiency for this sandwich configuration. There are 21 wedge-shaped facets, 50 mm long, of various angular extents between 10 and 22.5° that produce 20 external scanning beams and a single internal diagnostic beam. The disk is illuminated by a 1.5-mW collimated and

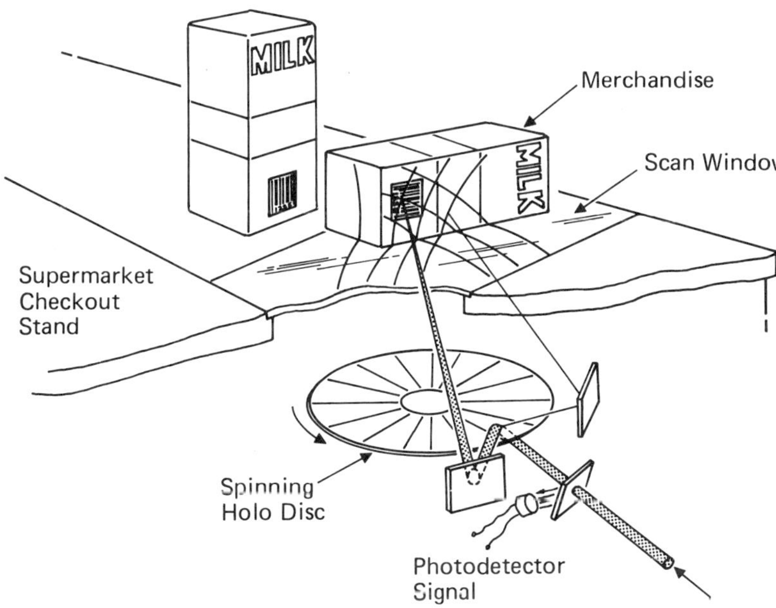


Figure 3 Schematic of a holographic supermarket scanner.

s-polarized HeNe laser beam inclined toward the rotation axis at 22° to the disk normal. Each facet diffracts light into a different direction and with a different focal length. The diffracted and focused beams range in f number from 200 to 300 and are reflected by mirror assemblies to pass out through the scan window in different directions to produce the wraparound effect. The window is glass, 9.5 × 15.25 cm in size, and the pattern in the plane of the window is shown in Figure 4a. Figure 4b shows the pattern in a plane normal to the window. The horizontal lines project forward at about 45° to read vertical bar labels on the front or bottom of the package, and the X lines project inward and forward to read horizontal bar labels on the front, sides, and bottom. Finally, the pairs of outer skew lines project forward and inward at about 45° to read vertical bar labels on the sides and bottom. Light is diffusely scattered by the package and some small fraction passes back through the window and mirror systems and is incident on the facet creating the deflection. The hologram sees this return light as emanating from a point source, conjugate to the object beam, and attempts to form a conjugate reference beam, in this case a collimated beam, back toward the laser. The

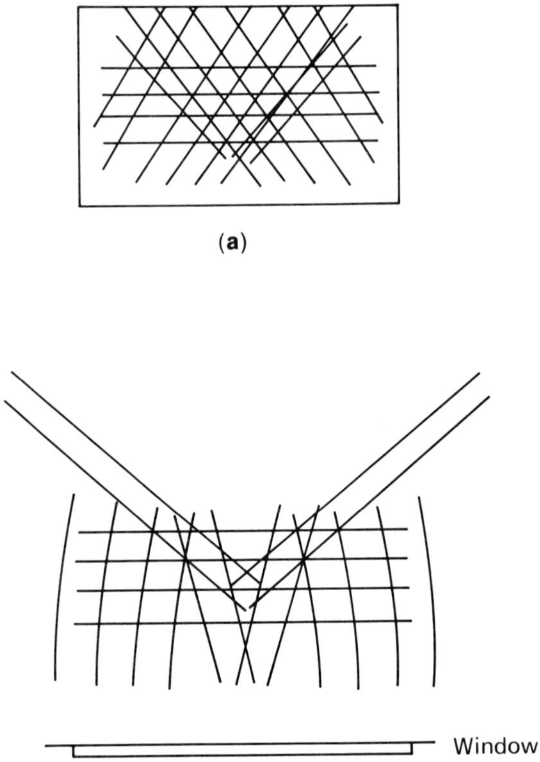


(a)

(b)

Figure 4 Scan pattern of a holographic supermarket scanner: (a) in the plane of the window; (b) in a plane normal to and in front of the window.

diameter of this beam is determined by the entire facet and most of the beam bypasses the small mirror used to introduce the narrow illuminating beam. This light passes on to an aspheric collection lens and is imaged onto the detector. The angular width of each facet is selected to equalize the collection solid angle over the entire scan pattern, which in turn reduces the dynamic range requirement for the detection system.

### 2.1.2. Fujitsu Point-of-Sale Scanner

The Fujitsu holographic point-of-sale scanner, FACOM 3721A, also uses a disk configuration to provide deflection, focusing, retrocollection scanning, and ambient light rejection. Used for scanning the Japanese article number (JAN), development models of these systems were installed in Japanese supermarkets in 1980. As described by Ikeda et al. (1976, 1978) and later by Ikeda et al. (1979b), the holographic disk used in this scanner, termed an interferometric zone plate (IZP) is fabricated using a collimated reference with a converging object wavefront. Reconstruction is performed with diverging spherical wave illumination that causes the scanning plane to move away from the original focal plane by the convex lens behavior of the IZP. This introduces a 4× scan length magnification and reduces the required disk size to less than 30 cm in diameter. Under these conditions, however, the image plane distance varies greatly with deflection angle and reduces the useful operating range to less than ±10° for image distance errors of less than 10%. In a technique described by Ikeda et al. (1979a), an oblique recording and reconstruction geometry is proposed and demonstrated to extend the deflection range to 30° with minimum aberration. In the scanner system, a rectangular lattice of 24 × 8 scan lines is generated over a scan field 67.5 mm × 150 mm in size with a resolution of 0.26 mm. Signal detection of the scattered HeNe light is performed using lenses embossed into the substrate, adjacent to the hologram ring, that direct the light to a detector/spatial filter assembly. The holographic material and manufacturing process are unreported.

### 2.1.3. NEC Point-of-Sale Scanner

The Nippon Electric Company point-of-sale scanner, announced in 1980 as the OBR-80-1, is described by Ono et al. (1982) and Ono and Nishida (1982) as an embossed surface relief holographic disk containing seven facets on a 176-mm-diameter substrate. Transmission surface relief holograms are manufactured by recording a master on Kodak 120-01 plates, and duplication of the amplitude hologram is accomplished by contact printing onto photoresist by the method of Nakano and Nishida (1979). The developed resist is converted to a nickel master by electroforming and multiple copies produced by hot pressing into transparent thermoplastic sheets. The resulting diffraction efficiency is approximately 35%. The seven parallel scan lines generated by facet rotation are converted to an 18-line delta pattern by an auxiliary mirror arrangement to produce all direction symbol reading. In other publications (Shirakabe et al., 1981, 1982) the scanner is reported to rotate at 100 rev/sec to generate scan angles varying from ±20.9 to ±30.4° with a 70-mm depth of focus and 0.2-mm resolution over a 180 mm × 120 mm area. Hologram illumination with a 2.0-mW HeNe laser provides a backscattered signal with label modulation that is collected and directed to the detector by retrocollection through the scanning facet.

## 2.2. Laser Printers

The use of holographic scanners to replace rotating and oscillating mirror systems has not been restricted to point-of-sale scanner applications. Indeed, a logical extension is to consider their application to laser electrophotographic printer systems as a direct replacement for the polygonal mirror. The performance requirements in this case are far more demanding and the multifunction advantage of holography is not as relevant. In particular, the principal requirements for a laser printer application are high repeatability and uniformity (i.e., a true emulation of a polygon). Scan quality factors, such as scan line straightness and spot quality, are far more critical. Consequently, the freedom to tailor each facet as a different optical "system," diffracting light into different directions with variable focal lengths and corrected light collection apertures, is lost and with it some of the motivation to adapt holography to this application. Other advantages still exist: cost, system size, drive requirements, scan speed, and certain mechanical tolerances and, as such, represent economic and implementational rather than functional advantages.

Two printer systems have been announced recently, but unfortunately, no significant amount of technical data has been released except the fact that they both use holographic deflection. These systems are the Benson Laserwriter 100, announced in June 1982, and the General Optronics Holoscan I, announced in September 1982. The Benson product is a tabletop plotter and CRT hardcopy device that uses light from a laser diode that is deflected by a holographic grating to write image points on a photoreceptor drum. Printing is by charge transfer to dielectric paper and by toning with monocomponent toner. Vertical and horizontal resolution is 208 pels/in. with a maximum write-line length of 10.5 in. The General Optronics system, specified as a 28-page/min printer, uses a laser diode operating at 780 nm with 5-10 mW of output. The output is raster scanned by holographic deflection onto a photoreceptor and the image is transferred to plain paper by conventional toning with a vertical and horizontal image resolution of 300 pels/in.

A third system for laser printing, under development by Ricoh, was recently reported by Funato (1983). This system uses a HeNe laser and is in a disk format containing 10 facets, each a simple linear grating. Each grating is a surface relief hologram fabricated directly into positive photoresist with diffraction efficiencies of 84%. Bow correction of the scan line is improved by a factor of 40 by using anamorphic optics to image the point of diffraction (on the hologram) to the scan plane. This is in comparison to the nearly straight line scan that can be achieved with particular recording geometries, as described in a later section. These additional optical components also reduce the effect of disk wobble by providing less than unity magnification in the direction perpendicular to the scanning direction. Print resolution is reported to be 300 pels/in., and with the addition of a single-element f-θ lens, scan distortion is less than 1%.

### 2.3. Other Applications

Over the years numerous proposals have been made for the application of holographic scanners and in many cases laboratory prototypes have been constructed to test feasibility. Applications under consideration have included: wide-band reconnaissance recording (Beiser et al., 1973), opaque document scanners (Pole et al., 1978), reprographics and general data handling (Beiser, 1981), two-dimensional scanning (McMahon et al., 1969; Ih, 1978), and complex pattern generation (Bryngdahl and Lee, 1976; Case and Gerbig, 1981). To what extent any of these applications is still being pursued on a development basis is not known.

## 3. SCANNER CONFIGURATIONS

A key factor in the design of a scanner is the selection of compatible optical and mechanical configuration that will meet the scan quality requirements of the intended application without introducing undue complexity and high manufacturing costs. One of the first issues to be resolved is the choice between reflective and transmissive operating modes. This decision involves consideration of substrate type and quality, drive and bearing specification, and to some degree selection of the holographic process itself.

### 3.1. Optical Considerations

The choice between using a transmission or reflection hologram is dependent on the required scan line uniformity and repeatability. Of particular concern are cross-scan displacements introduced by a nonuniform substrate and/or by mechanical instabilities such as bearing wobble. In a transmission hologram, image quality and scan line repeatability are strongly influenced by the quality and uniformity of the substrate through which both the illuminating and/or the reconstructed wavefront must pass. Although flatness is an important factor, a more stringent demand must be placed on the parallelism of the two (or more) surfaces comprising the substrate and possible protective covering. Any residual wedge angle, $\alpha$, will introduce a deviation, $\Delta\theta$, into a normally incident beam given by

$$\Delta\theta = (n - 1)\alpha \tag{1}$$

where n is the refractive index of the substrate. For example, a total wedge angle of 1 min causes a lack of repeatability during one full disk revolution of ±73 μm in scan line location for a 500-mm working distance. To achieve a deviation of less than ±2.5 μm, the wedge angle requirement becomes 2 arcsec. Although this can be achieved, it adds significantly to the cost.

In addition to wedge angle concern in transmission devices, both transmission and reflection devices are subject to errors introduced by a lack of orthogonality between the rotation axis and hologram plane. This can be a fixed error occurring during the mounting, or a dynamic effect as a consequence of bearing wobble. In either event, the effect on the diffracted beam is to introduce cross-scan displacement. For both transmission and reflection holograms the deviation in the diffraction direction $\Delta\theta$ for small tilt angles $\Delta\alpha$ of the diffracting surface with respect to the illumination direction is given by

$$\Delta\theta = \left(\frac{\cos\theta}{\cos\phi} \mp 1\right) \Delta\alpha \pm \left(\frac{\sin\phi - \sin\theta}{2\cos\theta}\right) \Delta\alpha^2 \tag{2}$$

where the upper sign is for a transmission grating and the lower sign is for a reflection grating. In the case of equal angles ($\theta = -\phi$ for transmission, $\theta = \phi$ for reflection),

$$\Delta\theta = \Delta\alpha^2 \tan\theta \qquad \text{transmission} \tag{3}$$

$$\Delta\theta = 2\,\Delta\alpha \qquad \text{reflection} \tag{4}$$

Under these conditions the reflection hologram behaves like a mirror, while the beam deviation in the transmission hologram is reduced to a smaller, second-order effect. It is interesting to note that the direction of deflection for the transmission hologram is always toward larger angles, independent of tilt direction (i.e., a minimum deviation angle exists at the Bragg angle).

These considerations present two primary options. The first is to use a transmission hologram so as to exploit its wobble-free characteristics and to minimize residual cross-scan effects by using good-optical quality substrates. The other is to use a reflection hologram that can be fabricated on a variety of opaque and transparent substrates. In this case it is only necessary to specify the quality of the surface supporting the hologram. Angular stability must now be provided by the use of precision bearings or by the inclusion of additional optical elements similar to those used for correction of polygonal mirror tilt and wobble (Fleischer et al., 1977). Although these are most often used to introduce focal power in a plane-wave scanner, an anamorphic design has been shown to be effective in providing correction for substrate-induced beam tilt and wobble (Funato, 1983).

## 3.2. Mechanical Considerations

The mechanical design of a scanner must take into account the compatibility of mechanical motion and substrate shape with the shapes and locations of the wavefronts used during recording and playback. This is caused by the highly tuned structure of the hologram and its high sensitivity to any differences between the recording and playback wavefronts. Movement of a hologram subarea into the path of an illuminating wavefront, in the absence of any record/playback wavelength shift, must be done without introducing any angular misalignment between the original recording direction (for that subarea) and the subsequent illumination direction. An angular error will result in a loss of diffraction efficiency, a misdirection of the diffracted beam, and/or the introduction of wavefront aberrations.

The magnitude of this effect can be estimated from coupled wave theory for volume-phase holograms (Kogelnik, 1969). According to this theory, a 10% decrease in diffraction efficiency will occur for an internal angular mismatch $\Delta\phi$ of

$$\Delta\phi \simeq \frac{d}{2\pi T} = \frac{\lambda}{4\pi n T \sin\phi} \tag{5}$$

where d is the recorded fringe spacing, T the grating thickness, $\lambda$ the wavelength, n the refractive index, and $\phi$ the internal half-angle between the recording waves. As an example, using external recording angles of 45° at 633 nm, the resulting fringe spacing is d = 447 nm. Consequently, a 5-μm-thick hologram can be internally detuned by only 0.8° before the efficiency drops to 90% of the perfectly aligned efficiency. This corresponds to an external angular mismatch of 1.5°. As the angular misalignment is further increased, the efficiency will continue to decrease and will become zero at 9°.

It is possible, however, to maintain the required angular alignment by selecting a configuration for which the optimum location of the reference and playback source is invariant with hologram motion. An example of such a motion is rotation. If the center of curvature of the reference source, either as a real source for a diverging wave or a virtual source for a converging wave, is positioned on the axis of rotation, a similarly directed illumination beam (or its conjugate) will have the proper incidence angle for every holographic subarea during its rotation. If this source location is further constrained to be coincident with the center of curvature of the hologram surface, the reference wavefront is reflected directly back upon itself and is its own conjugate. This is advantageous because it eliminates the need to invert the illumination wavefront, a process that may require large numerical aperture optics.

The use of spherical reference, object, and playback wavefronts can also be used to introduce scan angle magnification or demagnification (i.e., a deflection angle that is different from the rotation angle). Although this can occur to a greater or lesser extent for all configurations, it is most easily described for a rotationally symmetric system with the reference and playback sources on the axis of rotation. In such a system, the maximum angle between the reference and object rays, and hence the maximum diffraction angle, occurs with a diverging reference and converging object wavefront, or vice versa. If, for example, the respective source points are equally spaced on opposite sides of the hologram, the deflection will be equal to twice the rotation angle. In general, the deflection angle is equal to the sum of the converging and diverging angles of the recording wavefronts. Demagnification occurs when both wavefronts are either converging or diverging. Additional angular magnification effects occur when the playback source is displaced from the recording reference source location (Beiser, 1983).

Figure 5 is a summary of some of the typical rotary configurations that have been designed and fabricated. Figure 5a shows the original Cindrich (1967) design using a flat substrate and collimated illumination. The radius of curvature of the hologram surface is infinite; hence a normally incident collimated beam is located on the rotation axis and at the same time is coincident with the center of curvature. The conditions for aberration-free

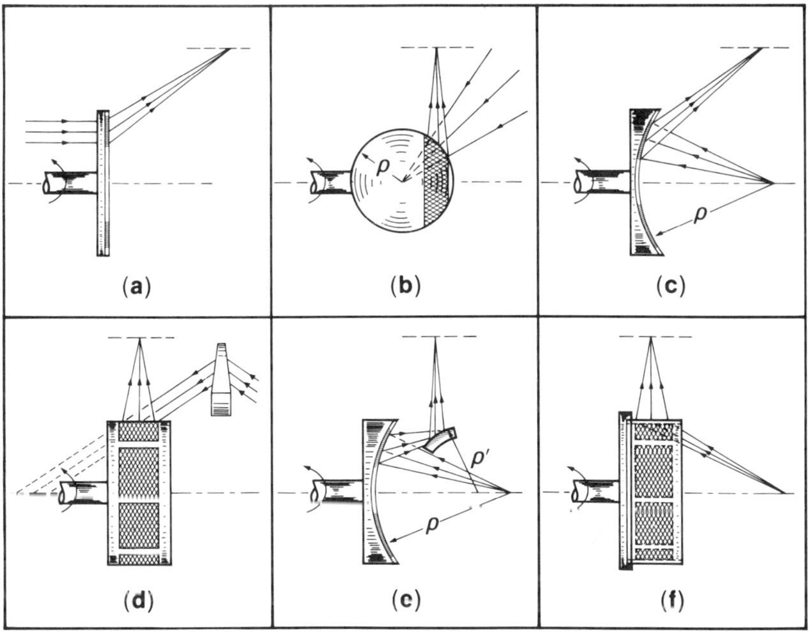


Figure 5 Holographic scanner configurations using rotational motion.

reconstruction are satisfied. The convex and concave configurations of Beiser et al. (1973) are shown in Figure 5b and c. These are reflective systems with the reference wave directed toward or emanating from the center of curvature so as to be normally incident on the surface and self-conjugating. The convex shape, in the form of a spherical ball, has the added advantage of a lower moment of inertia, for a given numerical aperture, and is capable of much higher rotation (deflection) speeds. The concave shape of Figure 5c is particularly interesting because it allows stigmatic operation at a playback wavelength different from the recording wavelength (Flamand and Pieuchard, 1973). The solid reflective cylinder in Figure 5d is an attempt to maximize the hologram aperture for the purpose of multiplexing several channels (scan lines) into a single device (Beiser, 1975). It is an extension of using the outermost surface on the convex shape (Fig. 5b) in a manner that retains low inertia, allows easier beam handling, and provides space for inserting field flattening optics. This approach is somewhat complicated by the need to use conical optics, such as an axicon, to create a line image on the rotation axis. In

Figure 5e, another concave scanner is shown that incorporates an auxiliary reflector (Ih, 1977), to convert the conical scan locus to a planar locus, thus eliminating the oblique incidence angles as in Figure 5a and c. This permits straight-line scanning on a plane perpendicular to the scan locus and is discussed in much greater detail in later sections.

The final configuration (Fig. 5f) is a transmissive drum or cylinder as demonstrated by Pole and Wollenman (1975). Early holographic configurations, such as those shown in Figure 5b, d, and f, are among the few that scan straight lines without requiring additional optics in the scanned image space. Fabrication of this type of deflector is shown in Figure 6a. The object wavefront is directed into the center of the cylinder and reflects from a mirror to converge through the cylinder wall. The point of convergence is in a plane orthogonal to the rotation axis and bisecting the vertical dimension of the cylinder. The reference source is contained on the rotation axis and provides a diverging wavefront. A multiplicity of holograms are recorded by step-and-repeat exposure of the recording material coated on the inside surface of the cylinder. Reconstruction and generation of the scanning spot is from a narrower beam emanating from the same axial point (Fig. 6b) and illuminating a smaller area of each facet.

Fabrication and use of a transmissive disk as in Figure 5a is shown in Figure 7. In this case, a diverging reference wavefront, from an axial source point, is used rather than a colliminated beam. By using a symmetric recording geometry with equal and opposite angles to the surface normal, the diffracting fringe structure is oriented normal to the disk surface. This minimizes the effect of any change in the thickness of the recording material that would distort an inclined structure. Thickness changes are quite common in photographic emulsions, due to removal of unexposed silver halide and to a lesser extent in other materials such as dichromated gelatin. Diffraction-limited operation at reasonable f numbers requires that attention be directed to the effects of shrinkage and process-induced distortions. Figure 7b shows reconstruction using a wavefront conjugate to the original reference wavefront and the curved scan line resulting from disk rotation. For this geometry, the scan is along the arc of a circle concentric with the rotation axis.

Scan line curvature, lack of straightness, is an issue for many applications and is of particular concern in laser printers and document scanners. Solutions using cylindrical configurations and modifications to the convex geometry (Beiser, 1975) address these needs but do so with substrate shapes that are not easy to mass produce by replication and are potentially expensive. The cost of an optical quality transmissive cylinder is an obvious

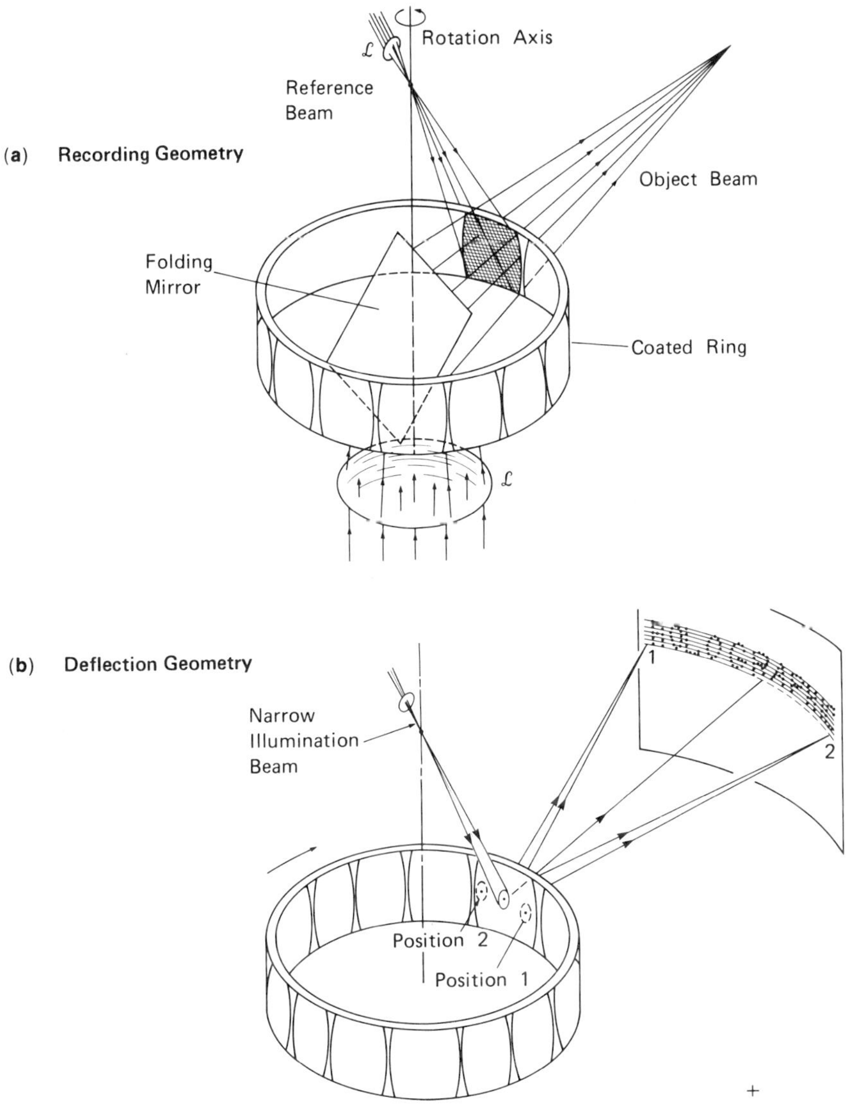


Figure 6 Cylindrical scanner configuration: (a) recording; (b) reconstruction/deflection.

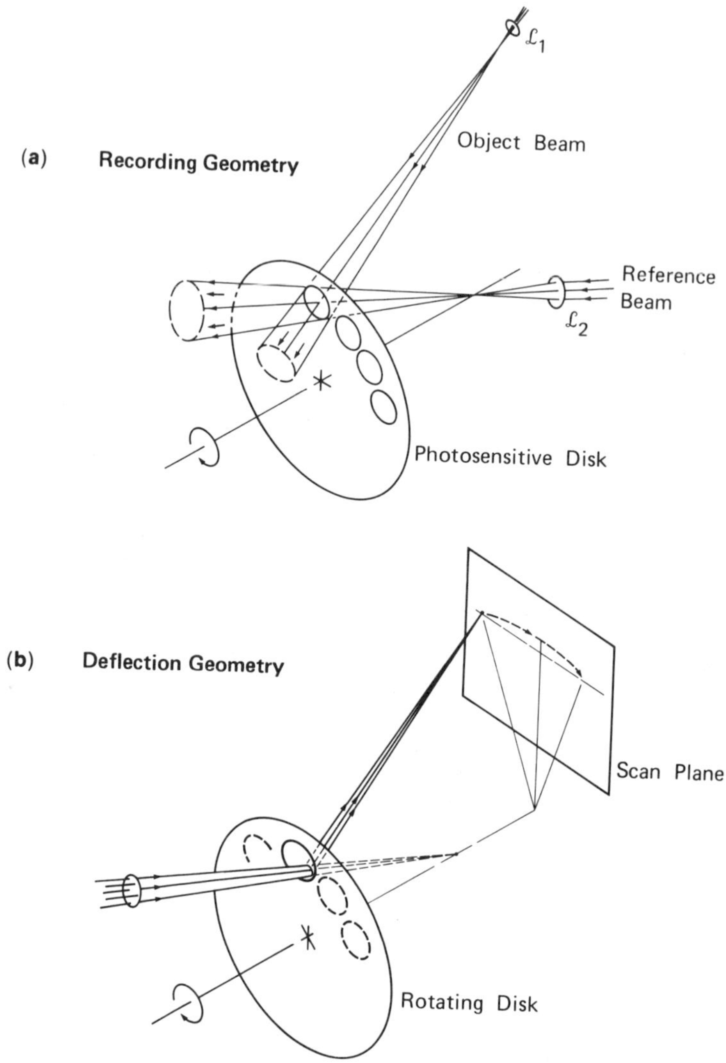


Figure 7 Typical plane disk scanner configuration: (a) recording; (b) reconstruction/deflection using a wave conjugate to original reference wave.

example. Considerable attention has been directed toward finding other methods for creating straight scan lines from a flat disk scanner. One method, by Wyant (1975), accomplishes this by using two identical counter-rotating holographic disks. Used in transmission with an imaging lens and a proper choice of parameters, this configuration also produces a linear scan rate. The use of an auxiliary reflector as in Figure 5e can also be extended to disk geometries to produce scan straightness. A more recent analysis by Antipin and Kiselev (1979) and Kramer (1981) has shown that a scan of usable straightness can be achieved from a single disk by using recording and playback angles near 45°. Although this is the most direct method, it does restrict the design and requires an additional lens for focusing. The last two methods discussed here are treated in greater detail in a later section.

## 3.3. Holographic Media

The selection of deflector type and configuration must take into account the compatibility of available holographic materials and processes. In addition, the application itself may dictate the holographic material and configuration. This is particularly true for those applications where the hologram is used bidirectionally, as in UPC scanners, and hence require higher diffraction efficiency.

In addition to being classified as reflective or transmissive, there are basically three types of diffraction mechanisms that are encountered in the hologram: amplitude, phase, and surface relief modulation. These are usually considered separately; however, there is a good chance that two or more may exist simultaneously and in different degrees. Diffraction from amplitude modulation occurs in materials such as fine-grained photographic emulsions and photochromics. One of the earliest holographic processes, AM diffraction, selectively removes energy from the the illuminating beam absorption and is limited to relatively low diffraction efficiencies, on the order of 6% for thin holograms and 4% for thick holograms. The distinction between thin and thick holograms is generally made on the basis of both the thickness of the hologram and the spacing of the diffracting structure. The important parameter is the number of "planes" that the illuminating beam traverses in passing through the medium. In thicker materials more planes are encountered, such that light diffracted by the planes near the point of entry is subsequently rediffracted into the original beam direction and perhaps even back again as more structure is traversed. Transfer of energy back and forth between the zero-order and first-order directions occurs and must be analyzed by coupled wave theory (Kogelnik, 1969). For example, a hologram recorded with a 1000-line/mm structure (1 μm spacing) would be considered thin for material thicknesses in the range 1-2 μm and thick beyond 5 μm.

The realization of high diffraction efficiency requires the formation of a nonabsorptive phase structure that introduces a periodic variation in the optical path traversed by the illuminating wavefront. This variation can occur either by a localized difference in the refractive index for a hologram of constant thickness, termed volume phase, or by a variation in the physical thickness with material of constant refractive index. The latter results in a surface relief structure and is analogous to classical diffraction gratings.

Early workers attempted to create a volume-phase structure from the absorptive structure of a photographic emulsion hologram by the chemical conversion of the metallic silver to a material of refractive index different from the surrounding gelatin matrix (Rogers, 1965; Upatnieks and Leonard, 1969; Pennington and Harper, 1970; Graube, 1974). Often referred to as bleaching, these processes generally involved the rehalogenation of the silver, for example by exposure to bromine vapor, or by extremely complex processing that either replaces the silver or removes it entirely, leaving voids behind as the diffracting structure (Graver et al., 1980). In many cases the diffraction efficiency of an amplitude hologram can be significantly increased, to the range 30-50%. Attempts to push beyond this range are usually accompanied by an increase in scattering noise and deterioration of the diffracted beam quality. One particular process, developed by Phillips and Porter (1976, 1980), has reported exceptionally high efficiencies with low noise characteristics using a complex process that has been tuned to one particular photographic emulsion. Many of the bleaching processes developed over the past two decades tend to leave behind in the emulsion reaction products that are photosensitive and subject to printout effects. The resulting darkening is particularly active when illuminated with high-power shorter wavelengths, such as those from argon or excimer lasers. Nevertheless, moderately efficient holograms can be realized, and the advantages of photographic emulsion, such as extended spectral sensitivity and high recording speed, may be key factors in evaluating the trade-offs.

The direct formation of a phase diffracting structure has been the subject of intensive research over the years but unfortunately, has resulted in only a few practical materials. Two of the more promising materials of the volume-phase type are photopolymer systems such as those developed by Booth (1977) and Chandross et al. (1978), and dichromated gelatin (Shankoff,

1968; Lin, 1969; Chang, 1980). Photopolymers produce phase structures by selective cross-linking and/or monomer diffusion and are particularly attractive because the hologram formation occurs during exposure in situ without additional processing. Consequently, the recording can be monitored as it occurs and terminated at the desired point by using probe illumination in a wavelength region where the material is not sensitive. The refractive index change introduced into photopolymer systems is very small, $\Delta n \simeq 0.005$ and requires thick coatings on the order of 50 μm to achieve high efficiency. These coatings are not always easy to create in good optical quality. The cylindrical configuration of Figure 6 is particularly suited for making thick, uniform coatings. The cylinder, with end caps, is aligned vertically and rotated about its axis. The material to be coated is introduced into the spinning cavity, and centrifugal force spreads it evenly over the inner surface of the cylinder. Uniform drying is promoted by outward motion of the polymer and the resulting inward displacement of the solvent. Internal trapping of solvent bubbles by the formation of an outer skin, as experienced by casting techniques, is greatly reduced by controlling the rate of evaporation. Dichromated gelatin (DCG), on the other hand, has a much higher index modulation, $\Delta n \simeq 0.05$, and permits high efficiency with thinner, 5-μm, coating thicknesses. A water-based system, dichromated gelatin is much easier to coat by standard techniques, including spinning, dipping, doctor blading, and casting. In some cases, standard photographic emulsions already coated onto glass substrates are used by removing the silver halide by fixation and resensitization with ammonium dichromate (Chang and Leonard, 1979). This eliminates the need for developing coating techniques but does restrict the freedom to control gelatin hardness, an important parameter in achieving high diffraction efficiency and reproducible results. Since dichromated gelatin requires a wet process consisting of multiple cycles of swelling and rapid dehydration, it cannot be monitored in real time to control diffraction efficiency. It can, however, be subjected to additional processing to fine tune the efficiency upward or downward (Chang, 1976). The mechanism of image formation in dichromated gelatin has been the subject of much research over the years and the reader is directed to papers by Meyerhofer (1972), Samoilovich et al. (1980), Sjolinder (1981), and Grant and Jones (1982).

While both the photopolymer and DCG materials are capable of 100% diffraction efficiency, they have the distinct disadvantage of a recording sensitivity to only short-wavelength radiation, $\lambda \leqslant 520$ nm. Attempts have been made by Graube (1973), 1978) and others (Kubota and Ose, 1979; Akagi, 1974; Kubota et al.,

1976) to sensitize DCG to longer wavelengths, with only moderate success. A primary problem is removal of the residual dye to give a complete phase structure and high diffraction efficiency. Without sensitization an optical correction technique is required if the resulting hologram is to be used at longer wavelengths, such as 633 nm from a HeNe laser or in the near infrared as from a laser diode (Latta and Pole, 1979; Winick, 1982; Malin and Morrow, 1981). One particular technique, using a multistep record/copy process developed by Lin and Doherty (1971), is discussed in greater detail in a later section.

The final diffracting mechanism is that of surface relief and is quite analogous to the conventional diffraction gratings used in spectroscopy. In fact, many of the gratings manufactured today are mechanical replications from submasters created from optically recorded, holographic, gratings. The most common material in use is positive photoresist of the type developed for the microelectronics industry (Bartolini, 1977b). A variety of materials and development processes are available that when used with optical techniques and ion milling methods, can be employed to make highly efficient, blazed reflection gratings (Lerner et al., 1982). It has been recently shown by Moharam and Gaylord (1982) and demonstrated by Werlich et al. (1983, 1984) that extremely high diffraction efficiency (>90%) can be obtained for nonblazed transmission holograms in photoresist. Spectral response is again restricted to ultraviolet wavelengths and is still a concern. Nevertheless, surface relief holograms are perhaps the most manufacturable and potentially the lowest cost for high-volume applications.

There is one other surface relief process that uses thermoplastics in combination with a photoconductor as a reversible holographic recording medium (Urbach, 1977). Used primarily for holographic optical storage applications (Lee et al., 1977) and holographic nondestructive testing techniques (Bellamy et al., 1971), these media provide extended spectral response and high recording speed. The principal disadvantages are that this process responds to a limited spatial frequency range centered near 800-1000 lines/mm and because of its shallow modulation does not provide efficiencies much in excess of 30%.

In summary, the most promising materials for holographic deflectors are dichromated gelatin, surface relief in photoresist, and bleached photographic emulsion. The ultimate choice will depend on the required efficiency, cost, and manufacturability. For additional information on these and other more exotic materials, the reader is directed to the review by Bartolini (1977a) and the book edited by Smith (1977).

## 4. SYSTEM DESIGN

The design of a scanner subsystem requires consideration of numerous interrelated parameters that determine both performance and implementation. The interrelationships are generally very strong, presenting only a limited number of solutions to a given application, and in many cases require compromise between contradictory requirements.

### 4.1. Parameter Definition

Design parameters include such items as spot size, scan length, scan straightness, scan linearity, latency, field curvature, and scan speed. The first two, representing resolution, are usually the primary driving factors in the design and permit the least flexibility. Scan straightness is very application dependent—for example, highly important for a laser printer with a cylindrical photoconductor and relatively unimportant in a UPC label scanner—and has a significant influence on selection of scanner configuration and implementation. Scan linearity is defined as a one-to-one correspondence between a unit increment of time during the scan and the incremental displacement of the beam in the scan plane. Lack of linearity will cause nonuniform recording when using a fixed-frequency data clock and hence will require either real-time clock correction techniques or the use of self-clocking patterns. Scan latency refers to that fraction of time during which the scanning beam is unusable. This is caused by the finite size of the beam illuminating the hologram and its overlap onto adjacent holograms at their common boundary. This parameter, often referred to as "dead time," is similar to flyback time in cathode ray tube (CRT) technology and greatly influences data rate and throughput. Excessive latency makes inefficient use of the scanner and may require the use of high burst data rates to accomplish the task. Field curvature has an impact on image spot quality and scan line linearity. Spot quality, as characterized by spot size uniformity across the scan line, results from deviation of the focal waist location from a true plane. Scan nonlinearity, of course, arises from the projection of equally spaced points on a curve onto a plane and becomes more exaggerated with curvature. The final performance parameter, scan speed, is one of the more independent parameters, having a reasonable degree of freedom. Although it is determined by both the number of facets and rotational speed, the latter is constrained only by mechanics and strength of materials. This is not intended to minimize these considerations but rather to indicate their lesser impact on the other parameters. The implementation parameters (a distinction that should be made with some reservation) include factors such as working distance, number of facets, individual angular widths, radial distance of hologram from rotation axis, orientation and shape of the recording and playback wavefronts, and rotation speed. Working distance is intimately related to scan length and spot size as well as being the determining factor in fixing the size of the overall scanner assembly. It is perhaps the bridge between performance and implementation and may, in fact, be the pivotal factor in the design. The elements of both parameter sets are present in every application, but the importance or weight assigned to each may vary significantly. For example, scan straightness, spot uniformity, and linearity are vital in a laser printer application and considerably less important than facet aperture and working distance in a UPC label scanner application. Hence there is no single solution.

### 4.2. Parameter Trade-Offs

A study of the parameter trade-offs is so application dependent that an exhaustive treatment is beyond the scope of this chapter. Some general comments are in order and will be discussed in regard to a laser printer application (Latta and Sincerbox, 1981, 1982). The starting point in determining these design trade-offs is to consider the desired scan length and spot size required over this scan. The spot size is governed by the f number of the diffracted wavefront, focal length divided by illumination aperture, while the scan length is determined by the focal length and deflection angle. The small size of both the total optical package and the holographic disk requires a short focal length f and a large deflection angle $\theta$ to maintain scan length, and a relatively small number of facets on the disk to minimize the disk radius. As $\theta$ increases, however, the risk of misalignment also increases and off-axis aberrations can be introduced that will degrade performance. Thus there is some compromise value required for f and $\theta$ to achieve a given scan length with acceptable spot quality. In addition, since the latency is determined by the ratio of the illuminating beam diameter to the facet width, this consideration argues for a larger disk radius to maintain a given spot size.

As an example, a typical interrelationship between these parameters is developed for a straight-line scanner using the spherical auxiliary reflector configuration (Ih, 1977) of Figure 8 to achieve a 350-mm scan length with a 200-$\mu$m $1/e^2$ spot diameter.

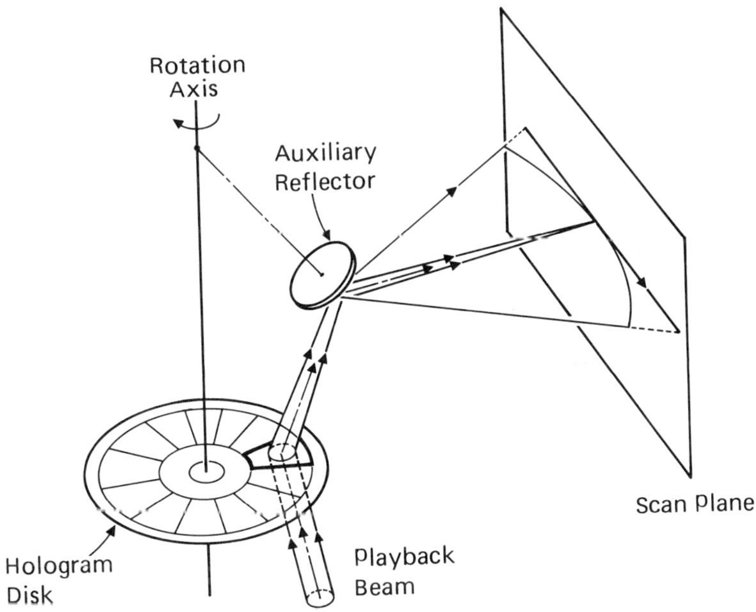


**Figure 8** Straight-line scanning using a spherical auxiliary reflector.

In Figure 9 is plotted the optimum focal plane location (i.e., working distance) as a function of the number of facets for several different disk radii. The optimum focal plane is selected so as to equalize the spot size deviation between the center of scan and the end of scan. For a given disk radius, as the number of facets increases, the distance to the best focal plane also increases. This is, of course, due to the fact that the scan angle is smaller for a disk with more facets and thus requires a greater path length to achieve the same length of scan. The family of curves also shows that a smaller disk radius requires a greater overall path length, again by virtue of the smaller scan angle. In this case the reduced scan angle is caused by the smaller useful aperture of the facet with respect to the illumination size needed for the 200-μm spot size. For example, a 50-mm-radius disk of 14 facets requires a 975-mm path length, whereas a 75-mm-radius disk with the same number of facets requires only a 800-mm path length. Figures 10 and 11 show the variation in scan linearity for these same families of curves. Scan latency varies

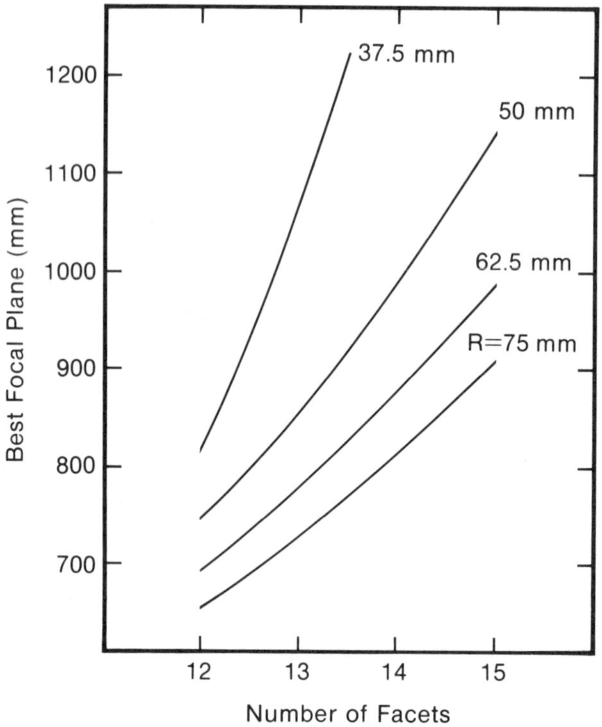


**Figure 9** Focal plane location for a 260-mm scan length and 200-μm $1/e^2$ spot size as a function of the number of facets and hologram disk radius.

in the same way as distance to the best focal plane. That is, as the number of facets increases or the disk radius decreases, the latency value increases. Thus for the 14-facet case, a 75-mm-radius disk exhibits 12% latency (dead time), significantly better than the 24% latency for a 50-mm-radius disk. Deviation from linearity shows the opposite behavior due to the increasing field angles associated with fewer facets and/or larger radius. Note that the various curves for different disk radii have become inverted in position, with the largest disk radius on top. Thus for the 14-facet case, the 50-mm-radius disk produces a better linearity value than does the 75-mm disk. An example showing a detailed plot of linearity is shown in Figure 12 for a 62.5-mm-radius disk with 14 facets. This plot shows the scan deviation of each spot from its ideal position for half of the scan (zero is the optical axis located at midfield). If the focal plane is chosen so as to provide the correct spot position at full field, ±175 mm, the maximum scan deviation is 0.825 mm ≈ 0.8% of the field position. If a midfield correction is used to equalize the deviations at the center and ends of scan, the scan deviation is +0.5 mm (0.5%).

A typical printer design goal could be to minimize the overall path length while holding the latency to a maximum of 0.20 and the deviation from linearity to a maximum of 0.750 mm over a full-field scan length of 350 mm. Thus a design point of 14 facets with an operating radius of 60 mm can be chosen. By studying Figures 9, 10, and 11, one can see the effects of changing this

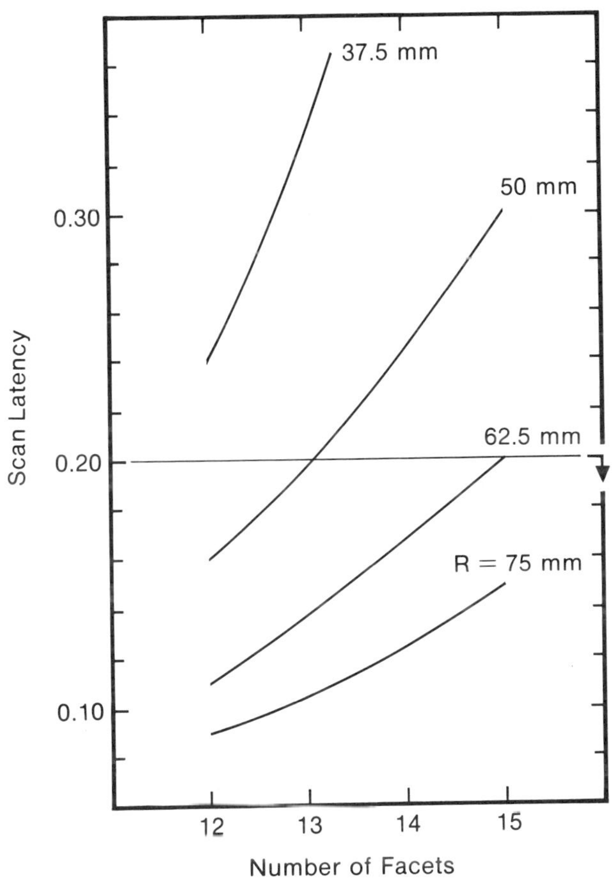


Figure 10 Scan latency as a function of the number of facets and disk radius.

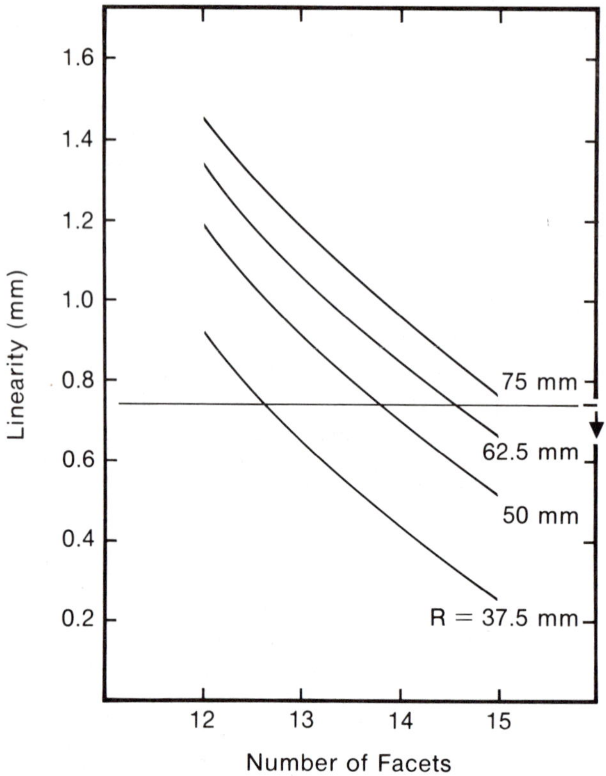


**Figure 11** Scan linearity as a function of the number of facets and disk radius.

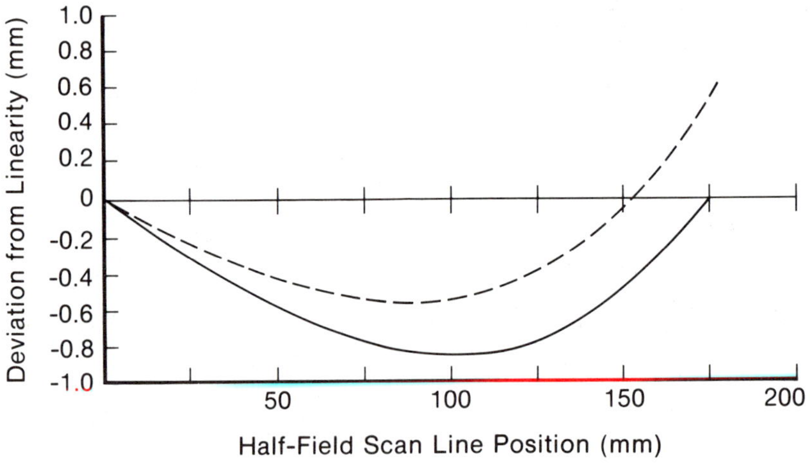


**Figure 12** Deviation from scan linearity as a function of scan field position.

design operating point. For example, with 14 facets and a disk radius of 60 mm, the linearity is 0.8 mm, latency is 0.18, and the best focal plane lies at a distance of 900 mm. Changing the radius to 50 mm reduces the linearity to 0.7 mm but increases the latency to 0.24 and the best focal plane distance to 975 mm. If the radius is increased to 75 mm, latency falls to 0.12 and the best focal plane lies at 800 mm, but the linearity increases to 0.975 mm. As indicated earlier, such an analysis is application dependent and will vary according to the deflector configuration chosen. The design task is further compounded by considering optical and mechanical tolerances, material behavior, and ultimately, the manufacturability of the unit.

## 5. DESIGN SIMULATION

Many of the optical and mechanical configurations do not lend themselves to closed-form solutions and the creation of trade-off curves as in the preceding section. In addition, variations from ideal conditions caused by loose mechanical tolerances, wavelength shifts, and off Bragg illumination lead to deviations from expected performance and create scan aberrations. It is useful therefore to model a particular design, examine the influence of parameter changes to optimize performance, and establish assembly and operational tolerances.

### 5.1. Diffraction Model

A principal element in this simulation is the ability to determine the direction of diffraction from a holographic element subject to the recording and playback conditions, particularly if an intervening hologram motion or wavelength change has occurred. Over the years, numerous ray tracing techniques have been developed to describe holographic diffraction (Spencer and Murty, 1962; Ludwig, 1973; Close and Moss, 1973; Close, 1975; Margerum, 1975). The discussion that follows uses the technique of Margerum (1975) to create a simple but general closed-form solution for accurate ray tracing. By assuming that the hologram is recorded with continuous wavefronts that are either converging or diverging, each point on the hologram can be treated independently, as if it were formed by the interference of two plane waves. The resultant microhologram has a sinusoidal grating structure independent of the recording layer thickness. Since we are interested only in the direction of the diffracted rays, and the expected departure from the Bragg condition is small, the reconstruction efficiency need not be considered. As a result, this volume structure can be treated as a surface grating, with the periodicity and orientation determined by the intersection of that structure with the surface. By combining the three-dimensional grating equation with the appropriate boundary conditions, an equation can be written that relates the direction of the diffracted light to the directions of the recording and playback wavefronts:

$$\left(n_d \vec{S}_d - n_i \hat{S}_i\right) - m \frac{\lambda_i}{\lambda_r}\left(n_0 \hat{S}_0 - n_r \hat{S}_r\right) - \Gamma \hat{n} = 0 \qquad (6)$$

where

$\hat{S}_r$ = reference beam unit vector

$\hat{S}_0$ = object beam unit vector

$\hat{S}_i$ = readout beam unit vector

$\vec{S}_d$ = diffracted beam vector

$\hat{n}$ = unit normal to hologram surface

$n_r, n_0, n_i, n_d$ = refractive indices

$\Gamma$ = measure of departure from Bragg condition

$\lambda_r$ = recording wavelength

$\lambda_i$ = playback wavelength

m = diffraction order

Note that the equation can accommodate differences in the refractive index in the spaces traversed by the individual wavefronts and that the effect of any wavelength shift between the record and playback steps is included. This model of the holographic diffraction process can be conveniently interpreted in terms of vectors. With reference to Figure 13a, assume that the hologram point in question is located at the center of an x-y-z coordinate system and that the two wavefronts being recorded have directions denoted by the unit vectors ($\hat{S}_r$ and $\hat{S}_0$ at this point. Illumination of the developed hologram with a wavefront propagating in a direction according to a third unit vector $\hat{S}_i$ gives rise to a diffracted beam that must satisfy the equation

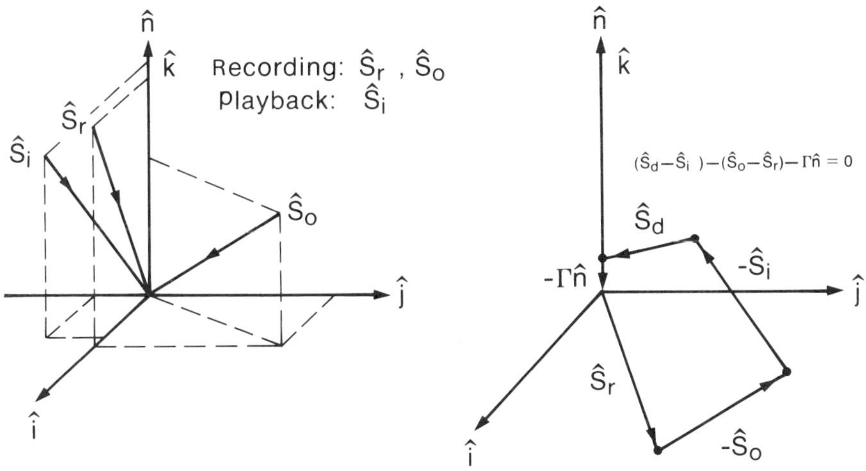


Figure 13 Vector model of holographic diffraction: (a) unit vectors representing recording and playback beam directions; (b) vector closure condition giving direction of diffracted wave.

$$\hat{S}_d - \hat{S}_i - \hat{S}_0 - \hat{S}_r - \Gamma\hat{n} = 0 \tag{7}$$

In this example it is assumed that there is no wavelength shift and that all the refractive indices are equal. The solution is shown in Figure 13b and is the unit vector, $\hat{S}_d$, that must be added to the other unit vectors with appropriate signs to produce a resultant vector $\Gamma\hat{n}$ aligned along the normal to the hologram at the point in question. This amounts to a vector closure of the sum with the constraint that the closing vector be along the hologram normal. The value of $\Gamma$ can of course be zero and, if so, represents the exact matching of record and playback conditions. A wavelength shift is accounted for by adjusting the lengths of the recording vectors by the ratio of the playback/record wavelengths, $\lambda_i/\lambda_r$, and applying the same conditions while still using unit vectors for the illuminating and diffracted vectors.

The closure equation (6) can be solved exactly be rewriting it slightly and imposing the boundary condition that the diffracted vector be a unit vector:

$$n_d \vec{S}_d = \vec{V} + \Gamma\hat{n} \tag{8}$$

where

$$\vec{V} = n_i\vec{S}_i + m\,\frac{\lambda_i}{\lambda_r}\left(n_0\hat{S}_0 - n_r\hat{S}_r\right) \tag{9}$$

but

$$|\,S_d\,| = 1 \tag{10}$$

Physically, this means that the wavelength of the playback light does not change during the diffraction process. The resultant quadratic equation has two solutions;

$$\Gamma = -(\vec{V}\cdot\hat{n}) \pm \sqrt{(\vec{V}\cdot\hat{n})^2 - \vec{V}^2 + n_d^2} \tag{11}$$

where the positive sign applies to a transmission hologram, and the negative sign to a reflection hologram. Imaginary solutions represent illumination angles or wavelength shifts for which no solutions exist.

## 5.2. Geometrical Spot Diagrams

The shape and size of the diffracted spot can be determined by expanding the analysis to include a bundle of rays representing illumination by a finite wavefront. Considering each ray in the bundle separately, they can be traced through the recording and diffraction stages and examined in arbitrarily defined image planes as conventional geometrical spot diagrams. When this is done for a multiplicity of hologram positions representing hologram translation with respect to the illuminating beam, the scan pattern can be examined in detail at any field position and in any image plane. This is further expanded by adding pre- or post-scan optical elements, including lenses, plane mirrors, and as in much of the treatment here, mirrored spherical surfaces. Ray tracing is continued through these elements by operating on the diffracted vector set and successively applying the laws of reflection or refraction at each surface encountered along the ray's path. Although this is straightforward, the presence of optics in the recording geometry must be treated with more caution, particularly when simulating a displacement of the hologram during reconstruction. Since it is necessary to know the description of the recording rays that were incident on the particular microhologram area that has moved into the path of the illuminating ray, the corresponding rays from each recording source must be determined by iteration. That is, the location or intersection of the recording rays in the hologram plane is known, but the direction that these rays make to the plane normal is not known. Except for very simple geometries, a brute-force method is employed. Rays are propagated backward through the recording optics from the hologram point in question and the one that passes within a predetermined tolerance of the source point is selected as the correct recording ray. It should be emphasized that the overall design process described here is one of iteration rather than design by closed-loop convergence to satisfy a set of figures of merit. Incorporation of this more conventional lens system approach is warranted only if numerous systems are to be designed.

In the case where the scanning element also collects light, the ray tracing process may be inverted to determine the effective solid angle and hence, the collection efficiency of individual facets. This is important in systems where the facets are different from one another, each producing wavefronts of different curvature and direction. The methodology is to start with a narrow cone of rays, symmetric with the chief ray, and systematically to increase the half-angle of each ray while tracing its propagation back through the optical systems. The increase in half-angle is terminated for any ray that fails to make it back through both the optical system and hologram aperture. The result is a convoluted cone representing the maximum extent of the bundle that can be collected. In the case of a UPC label scanner this allows adjustment of the mirror apertures and angular width of the holograms in order to equalize the collection efficiency between facets. This in turn reduces the dynamic range over which the photodetection system must be designed to operate.

## 6. DESIGN EXAMPLES

For the purpose of example, the design of a straight-line scanner will be discussed using two different disk configurations. The first case uses the spherical auxiliary reflector configuration (Ih, 1977) introduced earlier and shown in Figure 8. The other employs plane waves and a particular angular geometry in conjunction with an imaging lens (Kramer, 1981). The use of a spherical reflector is particularly interesting because a unique radius of curvature and center of curvature location can be determined that results in a mathematically perfect scan line

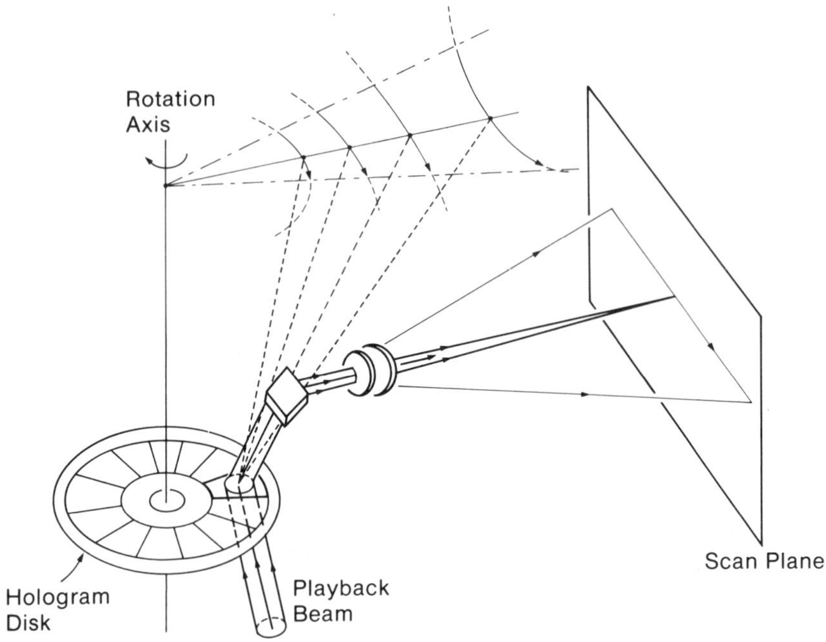


**Figure 14** Straight-line scanner using the plane waves and a focusing lens.

straightness. The use of an extra element in the system to create straightness can be avoided, however, by using a collimated beam geometry at a unique set of angles near 45°. As shown schematically in Figure 14, there is one particular set of angles at which the normally conical scan locus straightens. The straightness is no longer mathematically perfect, but it is more than adequate for most applications. The advantages of using collimated beams are numerous, as will be seen shortly. The need to employ an additional lens for focusing is somewhat disadvantageous and constitutes a departure from the multifunction "advantage" so unique to holography.

In the sections to follow, the design considerations for each of these systems are treated in detail and comparisons made where appropriate. The term SAR is used to represent configurations using a spherical auxiliary reflector.

## 6.1 Spherical Auxiliary Reflector Design

While any reflective surface having axial symmetry, be it spherical, conical, elliptical, or other, can be used to convert a conical scan locus to a straight line (planar locus), the concave or convex segment of a spherical surface is the easiest to implement. In most instances, a metalized simple lens of appropriate curvature can be used. The side view of a SAR configuration is shown in Figure 15. The radius of curvature ρ and the location of the

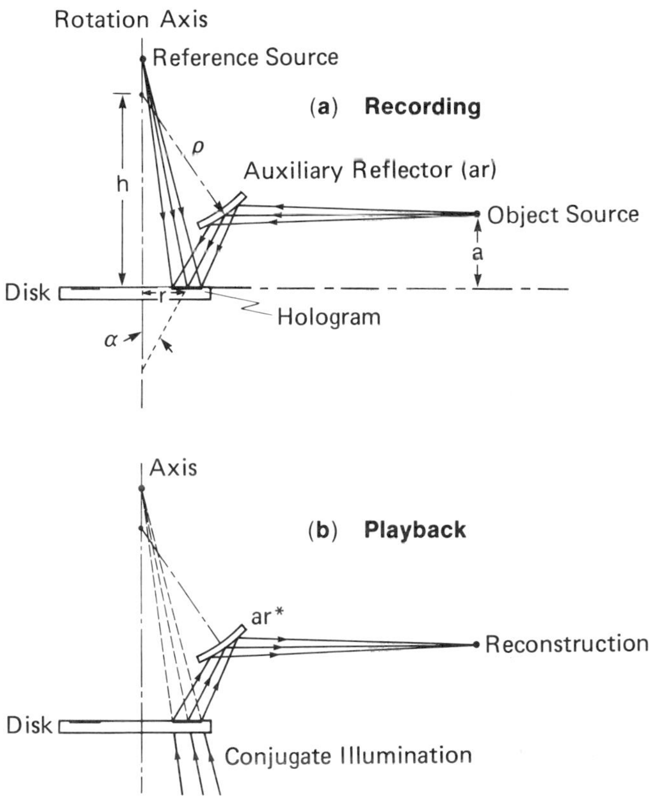


Figure 15 Side view of spherical auxiliary reflector configuration; (a) recording; (b) playback with conjugate illumination.

center of curvature h (h is assumed to be on the axis of rotation), which are necessary to deflect the incident beam into a direction parallel to the hologram plane, are given by

$$\rho = \frac{r + a \tan \alpha}{\sin (45 - \alpha/2)} \tag{12}$$

and

$$h = a + \frac{r + a \tan \alpha}{\tan (45 - \alpha/2)} \tag{13}$$

where $\alpha$ is the diffraction angle from the hologram, a the desired height above the hologram plane of the scan line, and r the radial location of the hologram from the rotation axis.

### 6.1.1. SAR Spot and Scan Quality

To maintain scan line straightness, both the spherical surface and the reference wavefront must have their centers of curvature located on the axis of rotation. During reconstruction, shown in Figure 15b, the illumination wavefront must be directed and shaped so that it is a conjugate of the original reference wavefront (i.e., it must be directed to the same point on the rotation axis). Using the techniques for design simulation described in the preceding section, departures from this ideal geometry can be examined using geometrical spot diagrams to determine their effect on scan shape and spot quality (Latta and Sincerbox, 1982). In general, a misalignment of the recording wavefront with respect to the rotation axis (and a corresponding playback misalignment) produces scan line curvature and astigmatism. Scan line curvature has been found to be linearly related to the angular misalignment of the reference source point from the rotation axis (Fig. 16). The relationship is independent of wavefront curvature; all wavefront curvatures give the same scan curvature as long as their source points are on the same tilt angle line. From the figure it is evident that to obtain scan line curvatures of less than 50 μm, the tilt angle misalignment of the wavefront must be less than 0.1°. The astigmatism introduced into the reconstructed spot is very small at these angles. Even at a 1° misalignment the spot elongation is less than 25 μm.

This, of course, assumes that the playback wavefront is perfectly aligned with the recording wavefront; however, if we

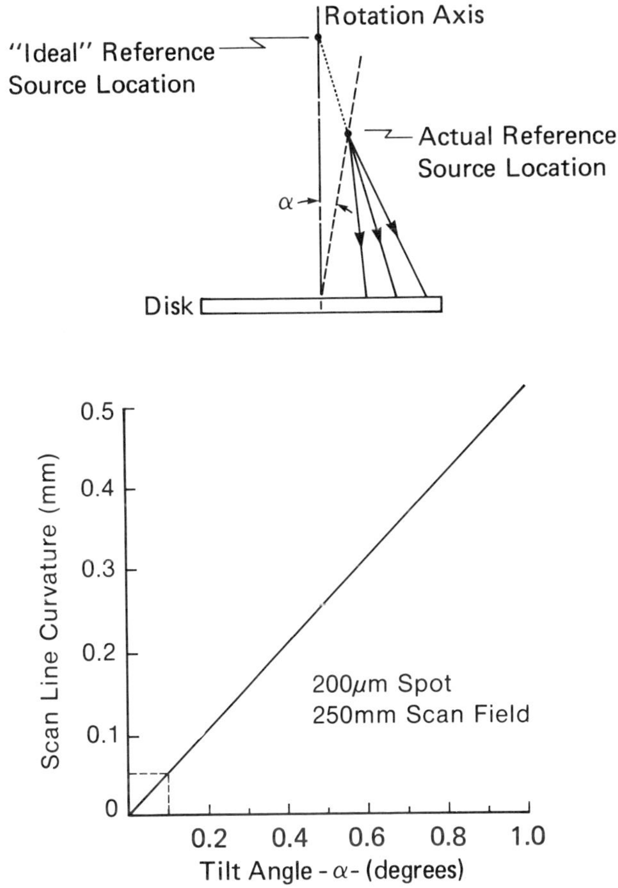


**Figure 16** Scan line curvature as a function of angular misalignment of the recording source from the rotation axis.

have an alignment error in one wavefront, we can expect to have an error in the other. The effect of this additional misalignment on the spot diagrams is to introduce significant spot growth as well as curvature. In the example above, simulations have shown that the relative angular mismatch is the key parameter. A conjugation alignment error of ±0.03°, for example, will cause 50 μm of spot growth. The effect is linear with increasing angles. Scan curvature and spot growth represent two different boundary conditions—a scan curvature tolerance determines the accuracy of the recording wavefront location and a spot growth tolerance establishes the relative accuracy of the playback wavefront location.

Many other parameters and tolerances affect scan line quality and are important in the design of a practical system. This is particularly true in the auxiliary reflector case because the recording object wavefront undergoes a highly off-axis reflection prior to interfering with the reference wave. Such a reflection introduces spherical aberration which is subsequently recorded. Playback under ideal conditions inverts this process and the aberrations are canceled, thus maintaining good spot quality. Differences may exist, however, between the reflector in the master recording system and those in the various playback units. One such critical tolerance is the radius of curvature. The general effect of a radius mismatch is to introduce spot elongation in the most off-axis rays (i.e., those most sensitive to aberration). Simulations have shown that this elongation, together with spot growth and scan line curvature, remain at acceptable levels if the differences in radius of curvature are held to less than 0.2%.

The most critical tolerance has been determined to be the variability in the location of the hologram with respect to the axis of rotation. Any differences introduced during fabrication and mounting or during use as a result of decentration and/or bearing wobble will cause a severe lack of repeatability in the scan pattern. This deviation of scan location may become unacceptable in a laser printer application, where uniform scan-to-scan spacing must be maintained. If this condition is not satisfied, the summation of intensities from two adjacent scans will not produce the proper overlap and subsequent density variations will be observed. The cause of this effect is shown in Figure 17. The rays in the space between the reflector and hologram are not parallel; hence a radial displacement of a hologram element from point a or b to the illumination point, o, will produce reconstruction rays that strike a different region of the reflector. The angular deviation is magnified upon reflection, causing beam displacement in the scan plane. Typical SAR configurations have magnification effects on the order of 10x. Hence, for a nominal scan spacing uniformity of ±10%, a 240-pel/in. printing resolution would require a repeatability of ±10 μm for an allowable radial runout of ±1 μm. This is difficult to achieve even under the best of conditions and virtually impossible in a manufacturing process where the hologram is made on one system and used on another. One way to reduce this effect is to use a reference source location that is symmetric to the virtual image of the object source created by reflection from the spherical surface (Sincerbox and Ih, 1981). To what extent these corrections are required will depend greatly on the application and the corresponding tolerances that are required.

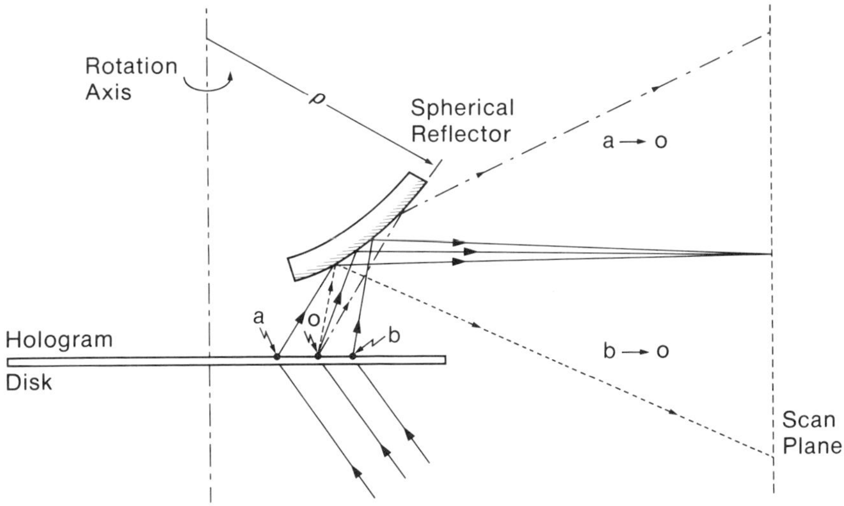


Figure 17 Scan line displacement in a spherical auxiliary reflector configuration caused by changes in the radial location of the hologram.

### 6.2. Plane Wave Scanner Design

As indicated earlier, sufficient scan line straightness can be obtained from a disk scanner without the addition of an auxiliary reflector, by using plane gratings in conjunction with imaging optics. Such a system was schematically presented in Figure 14 and is repeated in side view in Figure 18. The angles $\alpha$ and $\beta$ are the plane wave reference and object beam incidence angles used during recording, and $\gamma$ is the illumination beam angle.

#### 6.2.1. Plane Wave Scan and Spot Quality

Analysis of this configuration shows that the shape of the scan line is extremely sensitive to the recording and playback angles. A typical result presented in Figure 19 shows scan line straightness as a function of field position for various recording and playback angles. In this case, the recording angles are equal and opposite and conjugate playback illumination is used. For small angles, the diffraction follows a conical path concentric with the rotation axis. As the angles are increased, the curvature of the scan line inverts and moves in the opposite direction. As is evident in the figure, the optimum design point is in the

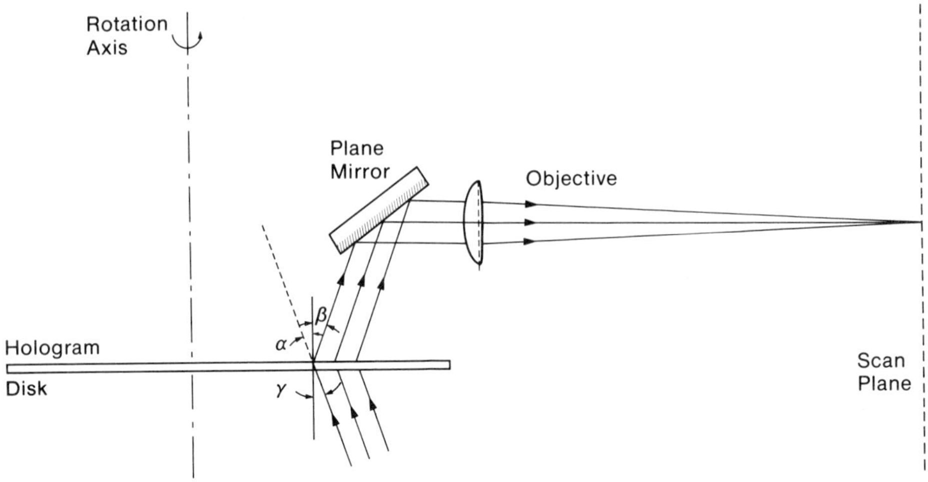


Figure 18 Side view of a plane wave scanner using focusing optics.

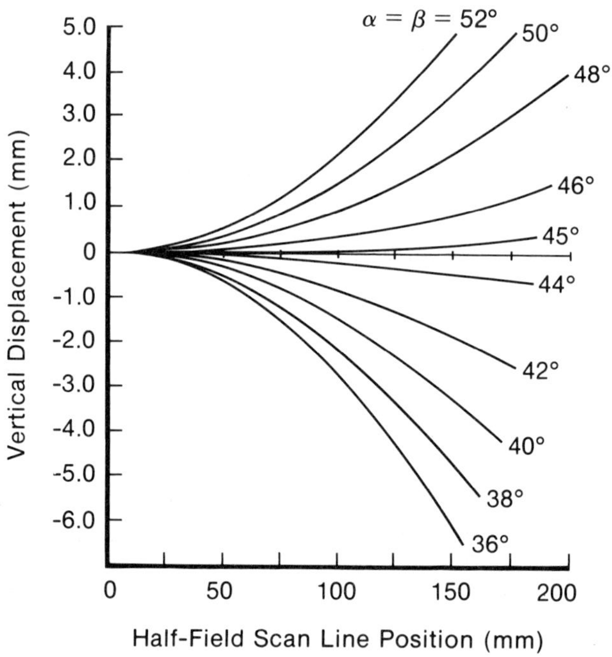


Figure 19 Scan line straightness as a function of field position for various equal and opposite recording angles and conjugate illumination.

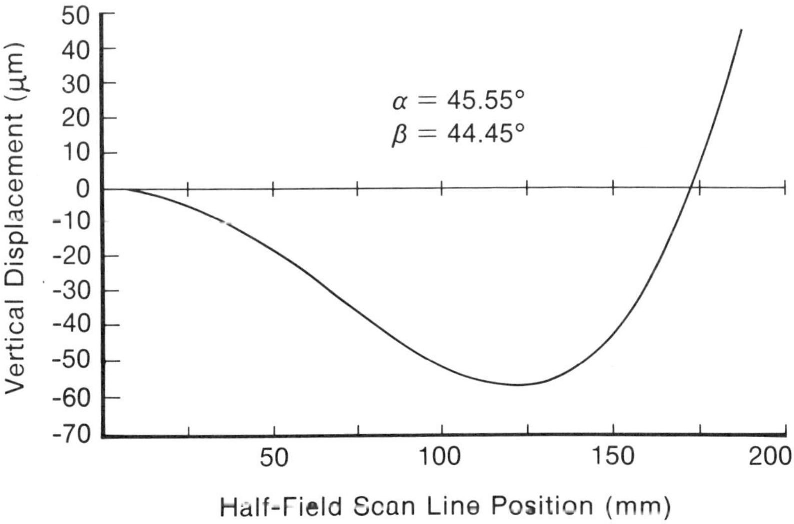


Figure 20 Scan line straightness optimized for a ±175-mm scan length.

neighborhood of 45°. This region has been examined in finer detail for different sets of recording angles to determine the minimum deviation from straightness for a half-field scan distance of 175 mm. As before, the illumination is selected to be conjugate to the reference direction. The result, shown in Figure 20, is a plot of the optimum scane line for recording with $\alpha = 45.55°$ and $\beta = 44.45°$, $\alpha + \beta = \pi/2$. Under these conditions, the deviation from straightness is less than ±30 μm. An alignment error of 0.1° will increase the deviation from straightness to ±40 μm. For other scan lengths, the optimum operating point shifts to other angular combinations. Compared to the use of an auxiliary reflector for achieving straightness, this technique is more direct and perhaps easier to implement. The price that must be paid, however, is a loss of freedom in selecting the recording and playback angles. The design is now restrained to be near 45°, which means that the holographic fringes must be recorded with a relatively high spatial frequency $\nu$:

$$\nu = \frac{2}{\lambda} \sin\theta = \frac{\sqrt{2}}{\lambda} \tag{14}$$

At 633 nm this is 2234 lines/mm and as such may restrict the choice in recording materials.

Another side effect of operating near 45° is the polarization sensitivity of the diffraction process. The reconstruction efficiency of a volume-phase hologram is, according to coupled wave theory (Kogelnick, 1969):

$$\eta = \sin^2 \left[ (\hat{p} \cdot \hat{q}) \frac{\pi \, \Delta n \, t}{\lambda \cos \theta_0} \right] \tag{15}$$

where $\hat{p}$ and $\hat{q}$ are the polarization directions of the incident and diffraction rays. For s-polarized light,

$$(\hat{p} \cdot \hat{q}) |_s = 1 \tag{16}$$

for all angles; for p-polarized light,

$$(\hat{p} \cdot \hat{q}) |_p = \cos\left(2\theta_0\right) \tag{17}$$

At $\theta = 45°$, the s- and p-polarization efficiencies become

$$\eta_s = \sin^2\left(\frac{\sqrt{2} \, \pi \, \Delta n \, t}{\lambda}\right) \tag{18}$$

$$\eta_p = 0 \tag{19}$$

Light of p polarization is not diffracted, nor can any p component in the scattered, depolarized light be collected. Hence applications requiring p polarization and/or collection of unpolarized light must be considered with caution.

A significant feature of the plane wave configuration is the absence of any radial decentration effects. These are accentuated by the SAR design or, for that matter, any design containing optical power. In contrast to the effect shown in Figure 17, a radial displacement of a fixed-frequency plane wave hologram does nothing to the direction of the diffracted rays, and there is no scan line displacement to reduce repeatability.

Spot quality of the focused light in the scan line is now determined by the post-deflection lens. The use of this type of lens is quite common in polygonal and oscillating mirror systems, and their designs can be carried directly over into holographic applications. A primary advantage of adding focal power separately from deflection is that the lens can be designed to convert equal increments of angular space into equal increments of linear positioning. This design, termed f-θ, provides linearity for the scan such that a fixed-frequency data clock can be used with the laser modulator. Without an f-θ lens, a portion of the deflected beam can be used to scan a fixed pattern, such as a grating, in order to generate and/or update the data clock.

## 7. WAVELENGTH CONSIDERATIONS

A distinct limitation of holographic optical elements is created by their highly tuned structure and resulting lack of spectral broadbandness. In general, the faithful reproduction of an object wavefront requires that the shape and wavelength of both the recording reference and the playback wavefronts be identical. If this condition is not satisfied, image distortion as well as loss in efficiency will occur. This problem is further compounded because there is a lack of high-diffraction-efficiency holographic materials with a recording sensitivity at the wavelengths usually desired for commercial applications (i.e., red and infrared radiation from HeNe and GaAs lasers). There are some exceptions, of course, such as bleached photographic emulsions requiring complex processing (Phillips and Porter, 1976), two photon recording methods that require very thick layers to produce even moderate efficiency (Gerbig et al., 1983), or specially sensitized silver halide emulsions not usually available in this country (Kalashnikov et al., 1980). The most widely used holographic materials are dichromated gelatin and photoresist, both sensitive only to wavelengths shorter than 500 nm.

In the case of plane waves, a simple angular adjustment can be used to compensate for wavelength changes. The fringe spacing of a plane wave recorded structure is given by

$$d = \frac{\lambda_0}{2n \sin \frac{1}{2} (\theta_r' + \theta_0')} \tag{20}$$

where $\theta_r'$ and $\theta_0'$ are the internal recording angles after refraction. This simplifies to

$$d = \frac{\lambda_0}{2 \sin \theta_0} \tag{21}$$

when the external incidence angles, $\theta_0$ are equal and on opposite sides of the normal. At any new illumination wavelength $\lambda$, maximum efficiency will occur if the angle of incidence $\theta$ is altered so as to match the angle that would have been required to record the same structure. In particular,

$$\frac{\lambda}{\sin \theta} = \frac{\lambda_0}{\sin \theta_0} \tag{22}$$

This condition can be satisfied simultaneously at every point in a plane wave hologram by a single angular change. This is, of course, a corresponding angular change in the diffraction direction.

A simple modification of one of the recording wavefronts from a plane wave to a spherical wave results in a recorded structure that varies continuously in spacing and inclination over the region of overlap. This condition requires a different angular realignment at each point. If plane wave illumination is required, then, at best, the condition can be satisfied only for the central rays. The degree of mismatch will depend on the size and numerical aperture of the hologram and generally results in diffracted spots with aberrations in one dimension that resemble coma. The result is to require a complex variation in the original reference wave that, although calculable, is not particularly easy to achieve by conventional methods. Even the almost trivial case of a simple holographic lens would require the design and fabrication of a corrector lens (system). It would, of course, be self-defeating to use a corrector lens during reconstruction, so the usual technique is to distort one (or both) of the recording wavefronts so that an illumination wavefront of reasonable shape can be employed (Latta and Pole, 1979). There has been considerable activity recently in the design and use of computer-generated holograms during recording as corrector elements (Winick, 1982).

An alternative approach for spherical wavefronts, and directly applicable to laser scanners, has been developed by Flamand and Pieuchard (1973) using holograms recorded on both concave and convex substrates. In this work, the authors used Fermat's principle and the Rowland circle property of spherical gratings to show that diffraction-limited operation can be obtained for a predetermined wavelength shift by appropriate positioning of the recording and playback sources. This method is most easily implemented for concave devices, as demonstrated by Beiser et al. (1973).

Fortunately, however, there is an alternative solution that is extremely elegant and provides an exact solution when used with thin holograms (Lin and Doherty, 1971). In this method a master hologram is recorded at the wavelength of intended usage and with the desired wavefront shapes: for example, at 633 nm with a collimated reference and convergent object wave on photographic emulsion. The master is processed such that when it is illuminated at a shorter wavelength, the intensities in the transmitted zero-order and diffracted first-order wavefronts are essentially identical. For photographic, absorptive holograms, this is accomplished by overexposure. The reconstructed first-order wavefront is aberrated by virtue of the red-to-blue wavelength shift. This wavefront does, however, overlap and interfere with the zero-order wavefront. Together they are used as sources for a second holographic recording on the blue-sensitive material positioned in their overlap region which is then processed to give a high-efficiency end-use hologram. When illuminated at the original red wavelength, aberration-free diffraction occurs. The "roundtrip" recording process has produced, as an intermediate, a wavefront containing the exact aberration required to compensate for the wavelength shift in the final record/playback process. An example as to how well this procedure works is shown in Figure 21. These curves are the intensity profiles in the focal plane of a SAR deflector obtained by using a moving slit scanner (Latta and Sincerbox, 1982). The solid curve is the spot profile for reconstruction at 633 nm from a photographic master recorded at 633 nm on Kodak 120-01 plates and processed in D-19 to give equal zero- and first-order wavefronts. The dashed curves are from two dichromated gelatin copies made from this master at 488 nm and subsequently illuminated at 633 nm in the same SAR fixture. The peak intensities of the copies have been normalized to match the peak intensity of the master, ~1%. The agreement in spot size is better than 10% and no discernible aberrations are present. The major requirement of this entire process is to have a recording material sensitive at the end-use wavelength that can be processed to give nearly identical zero- and first-order diffraction intensities. Photographic emulsion is well suited for use at 633 nm, and hopefully, materials will be forthcoming for use in the near infrared. In addition, it is important that the master hologram be thin and relatively insensitive to angular mismatch so that the reconstruction efficiency of the aberrated wavefront is uniform.

Reconstruction of aberration-free wavefronts can be expected only when the final wavelength is well known and stable. This cannot always be assured when using laser diodes. Currently available devices exhibit a variation in center frequency of 5-10 nm that is usually due to doping and diffusion nonuniformity. The result will be a scan line shift, focal plane shift, and some

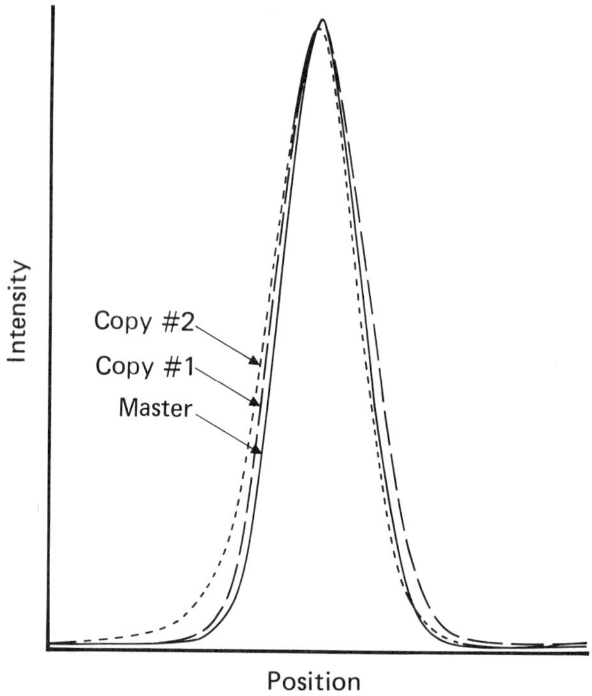


Figure 21 Comparison of intensity profiles for reconstruction at 633 nm from a master hologram (solid wave) recorded at 633 nm and two copy holograms (dashed curves) recorded at 488 nm.

off-axis aberrations. Most of these variations can be adjusted out during initial assembly. There are, however, significant dynamic wavelength variations due to changes in operating temperature and device aging that are beyond correction. Shifts greater than 2 nm can be expected to cause problems even at f/200. These effects are, of course, nonexistent for plane wave scanners, where the focal power is provided by an additional objective lens. Here, dynamic wavelength shifts will cause some spot displacement but generally no aberrations or focal plane shifts. Of greater concern with laser diodes is their finite spectral bandwidth and multimode operation. Each wavelength component will interact with the hologram and produce its own slightly shifted wavefront. The sum of all these contributions results in a blurred spot that may vary with time due to mode hopping. Except for sophisticated closed-loop laser control, very little can be done to correct this situation. Caution must therefore be exercised in designing high-resolution laser diode scanners (Doggett et al. 1981).

## 8. RETROCOLLECTION

The use of a hologram in a bidirectional mode, both as a deflector of light and a collector of scattered light, was first demonstrated by Pole et al. (1978) and was subsequently incorporated into the IBM 3687 supermarket scanner (Dickson et al., 1982). This technique employs a fixed detector and permits resolution enhancement by spatial filtering in the detector plane. Conceptually, it is very simple to understand. In Figure 22a, a hologram is shown deflecting and imaging a narrow cone of light from a source to a point in the scan plane. This could represent a page of a document, UPC label, and so on. Light is scattered backward over a large angle and some of it is collected by the hologram aperture. The hologram sees the incident light as a diverging wave coming from a source on the scan plane and diffracts it backward, focusing it onto the original illumination source. The solid angle of the converging beam is determined by the entire hologram aperture and is much larger than the illuminating cone. Hence this light can be separated from the illumination cone and directed to the detector by using either a large mirror with a hole, as shown, or a small mirror to introduce the illumination beam and swapping detector and source locations. In Figure 22b, displacement of the hologram past the fixed illumination cone causes deflection of the light to another point, and the scattered light is again collected by the displaced hologram aperture and focused onto the detector. As long as the mechanical motion appears to be about the reference source point, the location of the return light scattered from the image plane will be invariant with hologram position and scan spot. This is true for a multiplicity of holograms of different focal lengths and scan directions as long as they share a common reference source point. It should be obvious that a conventional lens will perform this same function. The difference, however, is that the hologram also provides spectral filtering to keep unwanted ambient illumination from the detector. That is, all wavelengths of ambient light scattered from the scan plane in the vicinity of the spot will be focused onto the detector by a conventional lens. The lens is sufficiently achromatic to ensure this. The hologram, however, will diffract and focus each wavelength component into a slightly different position. An aperture in front of the detector would allow only those components of the ambient that are at or near the same wavelength as the deflected spot to enter.

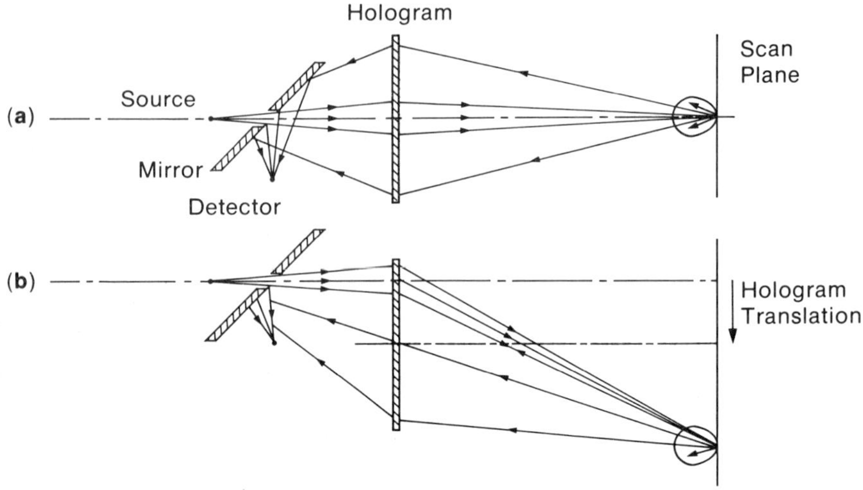


**Figure** 22 Retrocollection of scattered light; (a) on-axis collection; (b) off-axis collection.

Two factors are of importance in using this light collection scheme. The diffraction efficiency of the hologram comes into play twice: once during diffraction and again during collection. Hence a 30% efficient hologram provides at best 9% overall efficiency; 70% holograms are required for 50% operation. To some degree the loss in efficiency can be compensated for by increasing the laser power. This may not only increase system cost, but may present a safety problem, as in the case of supermarket scanners. In addition, an increase in laser power may well lead to a more hazardous classification of the product by the BRH and the necessity of more stringent safety measures. Even with high efficiency, care must be taken because most efficiencies are measured for s-polarized light. This is fine for the deflected beam, but the scattered light may be significantly depolarized such that the collection operates on equal parts of both s and p polarization. As indicated earlier, the diffraction efficiency for s and p polarization may differ greatly depending on the recording angles and fringe modulation. It is very possible that p-polarization efficiency can be nearly zero, as, for example, with 45° angles. The criterion for overall efficiency should therefore be the product of the outgoing, usually s-polarized, efficiency with the average of the collection efficiencies of both return polarizations [i.e., $\eta_s(\eta_s + \eta_p)/2$].

## 9. MANUFACTURING

The choice of manufacturing techniques is limited to either the fabrication of every scanner as an original master recording or the creation of one master recording and subsequent volume production by the generation of multiple copies. The primary factors in selecting a process are the operating wavelength, reconstruction efficiency, need for design flexibility, manufacturing throughput, and of course, product cost.

### 9.1. Production by Direct Recording

In those situations where the operating wavelength and the recording wavelength are identical, consideration can be given to recording each facet of a scanner directly, as an original hologram. Examples would be the use of moderate-efficiency (30-50%) bleached photographic emulsions at wavelengths $\leqslant$ 633 nm, or high-efficiency (>80%) dichromated gelatin or surface relief photoresist holograms at wavelengths shorter than 500 nm. When the recording and playback wavelengths are different, only plane wave gratings can be considered, as the illumination angles can be adjusted to permit use at another wavelength.

Whatever the manufacturing technique—direct recording or copy generation from a master—a conventional split-beam, long-path holographic recording system must be used. Although they are basically the same system, their utilization and hence degree of ruggedness and versatility will be significantly different. The manufacturing version for direct recording must be designed for a high degree of automation, high throughput, and ease of operation by manufacturing personnel. For copy generation, only a few master deflectors are required and they can be generated on a similar recording system that is more of a laboratory installation. In either case, the systems must be extremely stable to mechanical and acoustic vibrations and thermal gradients. As a general requirement, to achieve high efficiency, the interference fringes must remain stable to within $\lambda/10$ during hologram exposure. These exposures may range from a fraction of a second to several minutes, depending on the available power density and recording material sensitivity. Power density will be determined by laser power, facet size, and the degree of uniformity required (i.e., how much filtering and beam expansion is employed). In the cases of DCG and photoresist, yellow filters can be used to allow operation with reasonable visibility, whereas red-sensitive silver halide materials require total darkness.

A particular disadvantage of the direct recording approach occurs for those applications that require a change to be made in the object wavefront between the recording of each facet. Most holographic systems are quite critical as to alignment; beam uniformities are difficult to maintain and hence do not lend themselves to rapid modification—certainly not to automation. This will tend to reduce throughput and require the presence of a skilled operator. If every facet is identical, a step-and-repeat process can be utilized with sufficient time allowed after each motion to allow vibration to damp out. In general, the direct recording of holographic deflectors is risky and should be done only for the purpose of prototyping or custom fabrication.

## 9.2. Production by Replication

The alternative to direct recording is to fabricate a limited number of master holograms in a laboratory environment and then to use either mechanical surface replication or optical copying techniques for volume production.

### 9.2.1. Optical Copying

In Section 7, a technique for wavelength correction was described based on the recording of one hologram from another. This is also the basis for optical copying in which the holographic master is illuminated and used as the source of two new wavefronts that interfere for the recording of a new hologram (Brumm, 1966; Landry, 1967). It should be emphasized that the structure in the holographic master is not being copied directly. This process is not analogous to lithography, where a pattern is transferred from a mask to a photosensitive receptor by contact or projection printing. Although this method was demonstrated by Harris et al. (1966), it is quite restricted in utility, as it requires the use of a vacuum frame and does not work well for fringe spacings of less than 1.5 μm. More recent work by Nakano and Nishida (1979) suggests that high-quality contact printing can be accomplished by copying onto photoresist in intimate contact with the hologram emulsion. This was demonstrated by copying a diffuse object hologram recorded with 1000-line/mm carrier, 1-μm fringe spacing onto photoresist using light from a high-pressure mercury arc lamp. At the contact spacings used, $<10$ μm, there is sufficient coherence in the illumination to suggest that the copy mechanism is optical interference rather than simple pattern transfer.

For optical copying, a high-quality master is required and can be in the form of a prototype multifaceted disk, a composite assembly, or individual facets. Individual facets can be removed from a carousel, positioned over the copy disk with suitable light aperturing, illuminated to record the copy facet, and returned to the carousel. The substrate can then be incremented and another master facet brought into position. The process is repeated until the entire disk has been recorded. This is the most versatile approach, as it allows each master facet to be independently fabricated, tested, and even changed for a new design at a later time. The throughput of this method is restricted by the pick-and-place operation. In the case of identical facets, a single master facet can be rigidly positioned and the copy substrate rotated between each exposure.

At the other extreme is the use of a complete master disk containing all the facets exposed onto the same substrate. The only difference between this master and any subsequent copy is that it has a lower diffraction efficiency in order to equalize the zero- and first-order intensities. The master is placed in contact with the copy substrate with a suitable index matching fluid, such as decahydronaphthalene (n = 1.52), and the exposures performed either as a step and repeat with masking and rotation or with a single-shot flooding illumination. In the latter method, the deflector design must be one that uses a common reference source for all facets (i.e., a reference point anywhere on the rotation axis, including infinity for collimated normally incident light). Disk rotation may still be used to achieve exposure uniformity. This general method places considerable burden on the generation of a good master with every facet meeting specifications. Depending on scan complexity, the fabrication of such a master can become quite involved and difficult to achieve.

The optimum production technique appears to be a combination of both methods discussed above. Individual facets are recorded separately, tested, and then cut into the correct shape and pieced together on a transparent substrate to form a composite disk. In the case of photographic masters, the facets can be attached to the substrate, emulsion side down, with optical adhesive. Appropriate optical and mechanical tooling may be used for alignment prior to adhesion. A typical manufacturing process using this approach is shown schematically in Figure 23. This particular process uses dichromated gelatin as the copy medium, and includes as a final step the attachment of a protective cover plate to provide environmental sealing.

### 9.2.2. Mechanical Replication

The other primary holographic recording material, photoresist, has a diffracting structure in the form of surface deformation or

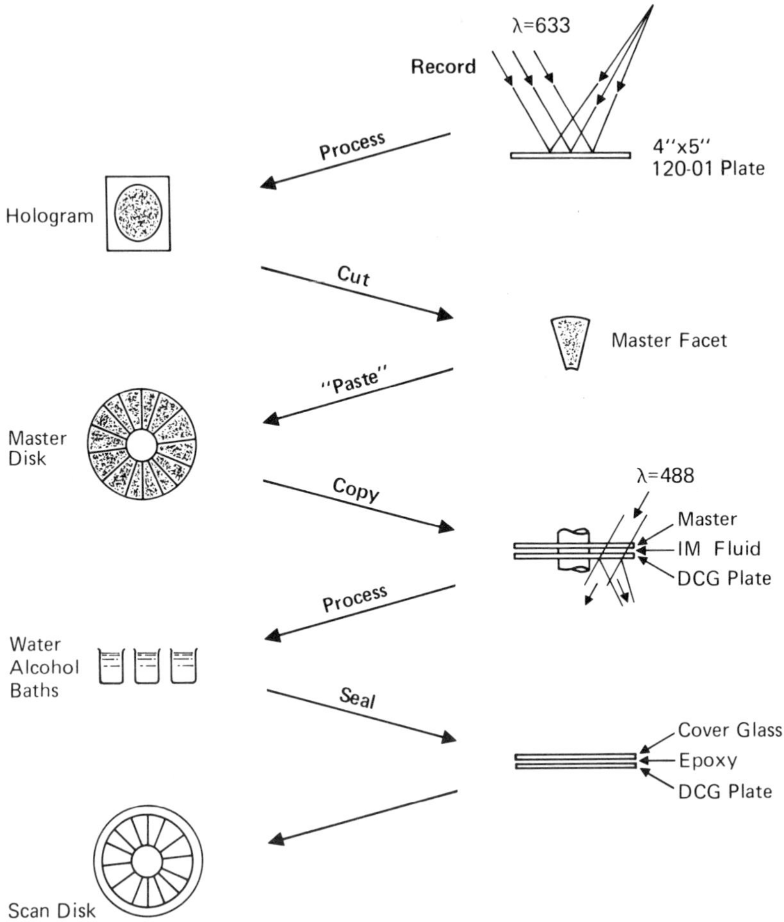


Figure 23 Manufacturing process for a holographic disk containing different facets.

relief, as shown in Figure 24 (Werlich et al., 1983,1984). Consideration can therefore be given to using mechanical replication for mass production. This is not to imply that optical copying techniques cannot be used with surface relief, because they certainly can; either as master or as copy or both. In fact, a photoresist master with 50% efficiency would allow far greater throughput than would a typical silver halide master with 1% efficiency.

Mechanical replication of surface relief is not new, having been developed decades ago for low-cost replication of mechanically ruled gratings. Today, it has become one of the primary manufacturing techniques of the production of high-quality gratings from holographic masters (Lerner et al., 1982). Surface relief replication techniques have also been extended to video disk (Ryan, 1978; Palermo et al., 1977), holographic video tape (Bartolini et al., 1973), and optical storage (Okino et al., 1982) applications. These are very similar in lateral resolution, $\leqslant 1$ μm, but do not have nearly as deep a profile, $\leqslant 0.1$ μm, as that required for high-efficiency diffraction gratings, ~0.7 μm. Replication of much deeper profiles, ~2 μm, at a slightly lower lateral resolution, 1.4 μm, has been reported by Gale et al. (1978) for the recording of zero-order diffraction (ZOD) images for use in optical projectors such as microfilm readers.

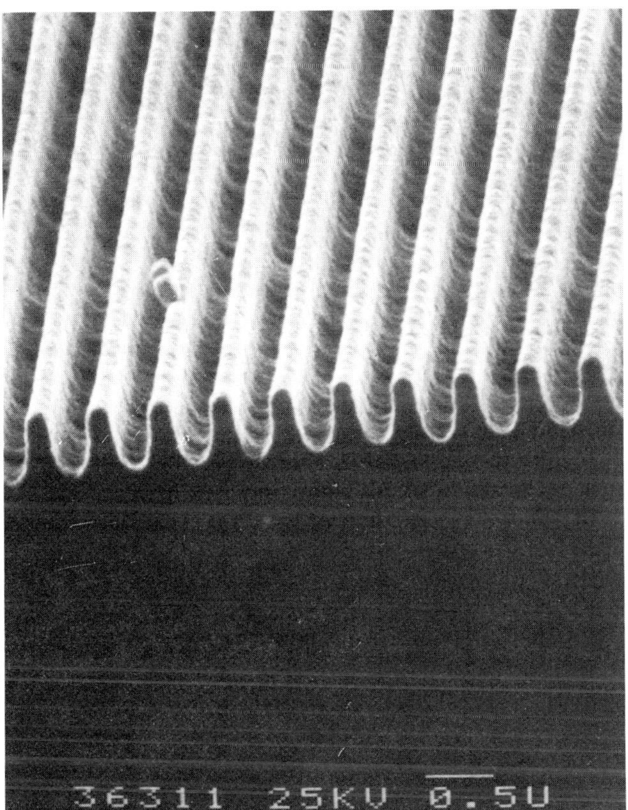


**Figure 24** Surface relief hologram in positive photoresist.

Although replication of a surface relief can be performed directly from the photoresist master, there is some danger that the photoresist may not stand up to repeated mechanical pressures, elevated temperatures, and/or the copy-master release process. Considering the difficulty and potential expense associated with master fabrication, a replication process that permits maximum replication volume is required. This is accomplished by the fabrication of a more durable submaster and usually results in the master itself being sacrificed. A very early technique, borrowed from the audio recording industry, uses a metal "stamper" to emboss or compression mold the relief into a vinyl thermoplastic (Ruda, 1977; Roys, 1978).

In the adaptation of this process (Bartolini et al., 1973; Gale et al., 1978), a very fine grain layer of nickel or gold is deposited by evaporation or sputtering onto the relief to form a conductive conformal coating, typically a few hundred angstroms thick. Nickel formation is then continued by other methods, such as electrochemical deposition, until a thickness of several hundred micrometers is achieved. At this point, the outer nickel surface has no significant relief and it can be attached to a rigid substrate. The sandwich is separated at the nickel/resist interface and the residual photoresist dissolved, leaving behind a rigid metal replication of the relief. This structure is a negative of the original and can be used in either the hot pressing, injection molding, or epoxy replication processes discussed later.

An alternative method of submaster preparation is to transfer the resist relief downward into its own substrate by radiofrequency (RF) sputter etching or reactive ion etching (RIE) techniques (Hanak and Russell, 1971; Lehman and Widmer, 1980). In these methods, the relief surface is removed at a uniform rate by bombardment with accelerated ions or, in the case of RIE, with reactive atoms that react with the substrate molecules to form a volatile gas. The valleys of the resist pattern disappear first and the underlying substrate is exposed and etching occurs. By the time the resist peaks have disappeared, the valley areas are deeply cut into the substrate. Proper choice of photoresist and substrate materials, such as silicon or quartz, and plasma parameters allows the surface relief to be accurately transferred into the substrate and the cross-sectional shape preserved (Matsui et al., 1982). Amplification can be obtained by selecting a substrate that etches faster than the resist—hence grating modulation can be enhanced. A similar technique is applicable to blazed gratings by inclining the ion bombardment direction and using shadow techniques (Aoyagi and Namba, 1976). These processes result in a submaster having a positive replication of the original master, in contrast to the negative shape of the nickel submaster.

Once fabricated, these more durable submasters may be used to generate multiple copies. One such method, thermal mechanical embossing or compression molding, is accomplished by pressing the relief into a heated and softened thermoplastic film such as polymethyl methacrylate (PMMA) or polyvinyl chloride (PVC). Bartolini et al. (1973) rolled the submaster together with a vinyl strip between two heated cylinders. Gale et al. (1973) used a conventional hot stamping press at 150°C and 3 atm to emboss into PVC sheets. A similar pressing technique was used by Iwata and Tsujiuchi (1974), with separation of the copy from the mold performed by sudden cooling and differential contraction.

Replication by pressing tends to introduce considerable strain and other inhomogeneities into the new substrate. It is also quite difficult to maintain parallelism between the surfaces to eliminate wedge effects on scan repeatability. These problems can be overcome by using injection molding techniques that have been developed for high-volume, high-quality fabrication of plastic lenses (Wolpert, 1983). In this case the submaster is one surface and an optically polished stainless steel flat is used as the facing, parallel surface. The appropriate polymer is plasticized to a more fluid state than that used in compression molding and is introduced into the temperature-controlled mold under high pressure (Ryan, 1978). Most of the materials in use are copolymers of polyvinyl chloride (PVC) and polyvinyl acetate (PVA) and acrylic (PMMA) compounds. The acrylic material has the advantages over vinyl of lack of birefringence in the finished substrate, and the stability and ease with which it can be machined and polished.

The final alternative is to use a polymer that can be cross-linked by ultraviolet illumination (Okino et al., 1982). This technique eliminates the need for high-temperature processing and reduces the possibility of induced stresses and dimensional changes upon cooling/curing. An injection mold apparatus can be adapted for these purposes as long as one plate is sufficiently transparent for the ultraviolet illumination. Depending on the use of release agents and relative adhesion, the replica can also be attached directly to a rigid substrate in the same operation.

## 10. CONCLUSION

Holographic light deflectors have been designed and incorporated as subsystems into several laser printer and point-of-sale scanner systems. The point-of-sale scanner application takes full advantage of the unique feature of holography—allowing several functions to be combined into a single optical component. In particular, a multifaceted assembly of individual holograms, each with a different focal length and deflection direction, is used to create numerous scan lines so as to produce a dense scan pattern within a large three-dimensional volume. The application is further enhanced by using the hologram facets to collect the light scattering by the product label, spectrally separate it from the ambient background, and focus it on a photodetector.

In contrast, the requirements for laser printer applications are far more demanding and the multifunctional advantage of holography is no longer as relevant. Current laser printer designs emphasize such factors as line-to-line repeatability, scan line straightness, spot quality, and uniformity. Because these requirements are satisfied using polygonal mirror light deflectors, the choice of an alternative near-term technology must be based on considerations of cost, size, rotation speed, and packaging rather than added function. Ultimately, the design of new printer architectures will be required to better exploit the attributes of holography, such as two-dimensional deflection.

It is clear that future applications—scanners, printers, displays, facsimiles and so on—require improved manufacturing techniques for high quality and economical mass production. Associated with this is an equally important need for the development of high-diffraction-efficiency red- and infrared-sensitive recording materials and processes. With an established manufacturing process and aberration-free performance at a greater selection of laser wavelengths, more applications should be forthcoming.

## ACKNOWLEDGMENTS

I would like to acknowledge the contributions and assistance of many of my colleagues at IBM, in particular Milt Latta for the analysis of parameter trade-offs in the auxiliary reflector system, LeRoy Dickson for making the IBM 3687 supermarket scanner a reality, and special thanks to Harald Werlich and the late Robert Pole for their pioneering efforts in the development of holographic scanners and for serving as a continuing source of encouragement and inspiration.

## REFERENCES

Akagi, M. (1974). *Photogr. Sci. Eng.*, *18*, 248.

Antipin, M., and Kiselev, N. (1979). *Tech. Kino Telev.*, *6*, 43.

Aoyagi, Y., and Namba, S. (1976). *Opt. Acta*, *23*, 701.

Bartolini, R. A. (1977a). *Proc. SPIE*, *123*, 2.

Bartolini, R. A. (1977b). In *Holographic Recording Materials* (H. Smith, ed.), Springer-Verlag, Berlin, p. 209.

Bartolini, R., Feldstein, N., and Ryan, R. J. (1973). *J. Electrochem. Soc.*, *120*, 1408.

Beiser, L. (1975). *Proc. 1975 Electro-Opt. Syst. Des. Conf.*, p. 333.

Beiser, L. (1981). *Proc. SPIE*, *299*, 175.

Beiser, L. (1983). *Appl. Opt.*, *22*, 1264.

Beiser, L., Darcey, E., and Kleinschmitt, D. (1973). *Proc. 1973 Electro-Opt. Syst. Des. Conf.*, p. 75.

Bellamy, J. C., Ostrowsky, D. B., Poindron, M., and Spitz, E. (1971). *Appl. Opt.*, *10*, 1458.

Booth, B. (1977). *J. Appl. Photo. Eng.*, *3*, 24.

Brumm, D. B. (1966). *Appl. Opt.*, *5*, 1946.

Bryngdahl, O., and Lee, W.-H. (1976). *Appl. Opt.*, *15*, 183.

Case, S. K., and Gerbig, V. (1981). *Opt. Commun.*, *36*, 94.

Chandross, E., Tomlinson, W., and Aumiller, G. (1978). *Appl. Opt.*, *17*, 566.

Chang, B. J. (1976). *Opt. Commun.*, *17*, 270.

Chang, B. J. (1980). *Opt. Eng.*, *19*, 642.

Chang, B. J., and Leonard, C. D. (1979). *Appl. Opt.*, *18*, 2407.

Cindrich, I. (1967). *Appl. Opt.*, *6*, 1531.

Close, D. H. (1975). *Opt. Eng.*, *14*(5), 408.

Close, D. H., and Moss, G. E. (1973). *J. Opt. Soc. Am.*, *63*, 1324A.

Dickson, L. D., and Sincerbox, G. T. (1981). *Proc. SPIE*, *299*, 163.

Dickson, L. D., Sincerbox, G. T., and Wolfheimer, A. D. (1982). *IBM J. Res. Dev.*, *26*, 228.

Doggett, D., Barasch, S., and Wegener, M. J. (1981). *Proc. SPIE*, *299*, 151.

Flamand, J., and Pieuchard, G. (1973). *Proc. 1973 Electro-Opt. Syst. Des. Conf.*, p. 322.

Fleischer, J., Latta, M., and Rabedeau, M. (1977). *IBM J. Res. Dev.*, *21*, 479.

Funato, H. (1983). *Proc. SPIE*, *390*.

Gabor, D. (1948). *Nature*, *161*, 777.

Gale, M. T., Kane, J., and Knop, K. (1978). *J. Appl. Photo. Eng.*, *4*, 41.

Gerbig, V., Grygier, R. K., Burland, D. M., and Sincerbox, G. T. (1983). *Opt. Lett.*, *8*, 404.

Grant, B., and Jones, C. (1982). *IBM Research Report RJ-3612*.

Graube, A. (1973). *Opt. Commun.*, *8*, 251.

Graube, A. (1974). *Appl. Opt.*, *13*, 2942.

Graube, A. (1978). *Photogr. Sci. Eng.*, *22*, 37.

Graver, W. R., Gladden, J. W., and Eastes, J. W. (1980). *Appl. Opt.*, *19*, 1529.

Hanak, J. J., and Russell, J. P. (1971). *RCA Rev.*, *32*, 319.

Harris, F. S., Sherman, G. C., and Billings, B. H. (1966). *Appl. Opt.*, *5*, 665.

Heflinger, L. O., Wuerker, R. F., and Brooks, R. E. (1966). *J. Appl. Phys.*, *37*, 642.

Ih, C. S. (1977). *Appl. Opt.*, *16*, 2137.

Ih. C. S. (1978). *Appl. Opt.*, *17*, 748.

Ikeda, H., Ando, M., and Inagaki, T. (1976). *Jpn. J. Appl. Phys.*, *15*, 2467.

Ikeda, H., Ando, M., and Inagaki, T. (1978). *Dig. 1978 Conf. Lasers Electro-Opt. Syst.*, p. 6.

Ikeda, H., Ando, M., and Inagaki, T. (1979a). *Appl. Opt.*, *18*, 2166.

Ikeda, H., Matsumoto, S., and Inagaki, T. (1979b). *Fujitsu Sci. Tech. J.*, *15*, 59.

Iwata, F., and Tsujiuchi, J. (1974). *Appl. Opt.*, *13*, 1327.

Kalashnikov, S., Molochev, V., Pilipovich, A., Popov, Y., Semenov, G., and Shmatin, S. (1980). *Sov. J. Quantum Electron.*, *10*, 1054.

Kogelnik, H. (1969). *Bell. Syst. Tech. J.*, *48*, 2909.

Kramer, C. (1981). *Laser Focus*, *17*(6), 70.

Kubota, T., and Ose, T. (1979). *Appl. Opt.*, *18*, 2538.

Kubota, T., Ose, T., Sasaki, M., and Honda, K. (1976). *Appl. Opt.*, *15*, 556.

Landry, M. J. (1967). *Appl. Opt.*, *6*, 1947.

Latta, M. R., and Pole, R. V. (1979). *Appl. Opt.*, *18*, 2418.

Latta, M. R., and Sincerbox, G. T. (1981). *J. Opt. Soc. Am.*, *71*, A1613.

Latta, M. R., and Sincerbox, G. T. (1982). *IBM Research Report RJ-3456.*

Lee, T. C., Lin, J. W., and Tufte, O. N. (1977). *Proc. SPIE*, *123*, 74.

Lehman, H. W., and Widner, R. (1980). *J. Vac. Sci. Technol.*, *17*, 1177.

Leith, E., and Upatnieks, J. (1962). *J. Opt. Soc. Am.*, *52*, 1123.

Lerner, J. M., and Laude, J-P. (1983). *Electro-Opt.*, *15*(5), 30.

Lerner, J., Flamand, J., and Thevenon, A. (1982). *Proc. SPIE*, *353*, 68.

Lin, L. (1969). *Appl. Opt.*, *8*, 903.

Lin, L., and Doherty, E. (1971). *Appl. Opt.*, *10*, 1314.

Ludwig, U. W. (1973). *J. Opt. Soc. Am.*, *63*, 1105.

Malin, M., and Morrow, H. E. (1981). *Opt. Eng.*, *20*, 756.

Margerum, E. (1975). *Research Note ETL-0052*, U.S. Army Engineer Topographic Laboratories, Fort Belvoir, Va., pp. 1-7.

Matsui, S., Moriwaki, K., Aritome, H., Namba, S., Shin, S., and Suga, S. (1982). *Appl. Opt.*, *21*, 2787.

McMahon, D., Franklin, A., and Thaxter, J. (1969). *Appl. Opt.*, *8*, 399.

Meyerhofer, D. (1972). *RCA Rev.*, *33*, 111.

Moharam, M. G., and Gaylord, T. K. (1982). *J. Opt. Soc. Am.*, *72*, 1385.

Moore, E. A. (1981). *Electronics*, *54*(14), 139.

Nakano, M., and Nishida, N. (1979). *Appl. Opt.*, *18*, 3073.

Okino, Y., Sano, K., and Kashinhara, T. (1982). *Proc. SPIE*, *329*, 236.

Ono, Y., and Nishida, N. (1982). *Appl. Opt.*, *21*, 4542.

Ono, Y., Kosuge, K., and Nishida, N. (1982). *IECE Jpn. Tech. Dig.*, 2-217.

Palermo, P., Korpel, A., Dickinson, G., and Watson, W. (1977). *Opt. Laser Technol.*, *9*(4), 169.

Pennigton, K., and Harper, J. (1970). *Appl. Opt.*, *9*, 1643.

Phillips, N., and Porter, D. (1976). *J. Phys. E.*, *9*, 631.

Phillips, N., Cullen, R., Ward, A., and Porter, D. (1980). *Photogr. Sci. Eng.*, *24*, 120.

Pole, R. V., and Werlich, H. (1978). U.S. Patent 4,113,343.

Pole, R. V., and Wollenmann, H. P. (1975). *Appl. Opt.*, *14*, 976.

Pole, R. V., Werlich, H., and Krusche, R. (1978). *Appl. Opt.*, *17*, 3294.

Powell, R. L., and Stetson, K. A. (1965). *J. Opt. Soc. Am.*, *55*, 1694.

Rogers, G. (1965). *J. Opt. Soc. Am.*, *55*, 1185.

Roys, H. E., ed. (1978). *Disc Recording and Reproduction*, Dowden, Hutchinson & Ross, Stroudsburg, Pa.

Ruda, J. C. (1977). *J. Audio Eng. Soc.*, *25*, 702.

Ryan, R. J. (1978). *RCA Rev.*, *39*, 87.

Samoilovich, D., Zeichner, A., and Freisim, A. (1980). *Photogr. Sci. Eng.*, *24*, 161.

Shankoff, T. (1968). *Appl. Opt.*, *7*, 2101.

Shirakabe, K., Kawamata, F., Miyazaki, H., Yamoda, T., Yamaguchi, Y., Shinoki, T., Ono, Y., and Nishida, N. (1981). *NEC Tech. J.*, *34*(4), 64-66.

Shirakabe, K., Yamaguchi, Y., Tanaka, Y., and Miyazaki, H. (1982). *NEC Tech. J.*, *35*(11), 40.

Sincerbox, G. T., and Ih, C. S. (1981). *IBM Tech. Discl. Bull.* *25*, 3702.

Sincerbox, G. T., and Roosen, G. (1982). *Appl. Opt.*, *22*, 690.

Sjolinder, S. (1981). *Photogr. Sci. Eng.*, *25*, 112.

Smith, H. M., ed (1977). *Holographic Recording Materials*, Springer-Verlag, Berlin.

Spencer, G., and Murty, M. V. R. K. (1962). *J. Opt. Soc. Am.*, *52*, 672.

Upatnieks, J., and Leonard, C. (1969). *Appl. Opt.*, *8*, 85.

Urbach, J. C. (1977). In *Holographic Recording Materials* (H. Smith, ed.), Springer-Verlag, Berlin, p. 161.

Vanin, V. A. (1978). *Sov. J. Quantum Electron*, *8*, 809.

Werlich, H., Sincerbox, G., and Yung, B. (1983). *Dig. 1983 Conf. Lasers Electro-Opt.*, p. 224.

Werlich, H., Sincerbox, G., and Young, B. (1984). *J. Imaging Tech.*, *10(3)*, 105.

Winick, K. (1982). *J. Opt. Soc. Am.*, *72*, 143.

Wolpert, H. D. (1983). *Photonics Spectra*, *17*(2-3), 68.

Wyant, J. C. (1975). *Appl. Opt.*, *14*, 1057.

# 2

# Polygonal Scanners

## Applications, Performance, and Design

Randy J. Sherman
*Lincoln Laser Company*
*Phoenix, Arizona*

## 1. INTRODUCTION

Polygonal mirrors are useful devices for scanning light beams within certain boundaries of performance. This section is intended to outline those boundaries as an aid to the system designer in selecting the most practical method of scanning in view of a particular need.

Table 1 lists a number of parameters to be considered when selecting a method of scanning. Opposite the parameters are the values associated with polygonal mirrors when used in conventional optical designs. A note of caution about the use of this table; it should not be assumed that arbitrary values of all the parameters shown (within the ranges given) can be achieved in the same device.

The primary advantages of polygonal mirrors are speed, the availability of wide scan angles, and velocity stability. They are usually rotated continuously in one direction at a fixed speed to provide repetitive unidirectional scans which are superimposed in the scan field, or plane, as the case may be. A common method of producing rotation is to fasten the polygonal mirror directly to an electric motor shaft (see Fig. 1).

The combined inertia of polygon and motor rotor contribute to rotational stability. In some cases an additional flywheel is added to further stabilize rotational rate. The relatively high inertia of polygons and drive motors, on the other hand, render them impractical for applications requiring rapid changes in scan velocity, or start/stop formats.

Table 1 Practical Performance Boundaries for Polygon Mirrors

| | |
|---|---|
| Light beam diameter | 0-70 mm |
| Light beam wavelength | 240-14,000 nm |
| Light beam power (continuous wave) | To several kilowatts |
| Scan angle | 5-360° |
| Scan rate (unidirectional repetition rate) | 30-20,000 Hz |
| Scan efficiency[a] $\left(\frac{\text{active scan time}}{\text{total time}}\right)$ | To 100% |
| Light transmission efficiency | 80-95% |
| Scan resolution (spots/line) | To 20,000+ |
| Scan linearity[b] | 0.02% |

[a]100% scan efficiency can be achieved by filling two polygon facets at a time with the input light beam.
[b]Scan linearity is defined as the percent deviation of angular scan velocity from the mean angular velocity of light reflected by polygon facets.

Regular polygonal mirrors are available from the inventories of standard manufacture from several suppliers. Table 2 lists typical commercial-grade polygon dimensions and characteristics.

Rotational speeds up to 120,000 rpm are practical for alternating-current (AC) motors. For applications where rotational rates exceed the capability of electric motors, gas turbines provide an alternative. Rotational speeds in the range 90,000-1,000,000 rpm are achieved with such devices (see Fig. 2). Turbine-driven rotating mirrors should be considered as an alternative only when short duty cycles (a few minutes) can be tolerated and relatively high costs can be afforded. Frictional heating of the rotor due to air drag at speeds approaching or exceeding Mach 1 at the periphery can determine the maximum practical run time.

**Figure 1** Synchronous motor/polygon assembly. The polygon is fastened directly to the motor shaft, which is supported by ball bearings, one at each extreme end of the rotor. (Courtesy of Lincoln Laser Co.)

Table 2 Range of Dimensions and Characteristics Typical of Commercial-Grade Polygons

| Feature | Range |
|---|---|
| Number of facets | 1-60 |
| Inscribed circle diameter | 2-14 in. |
| Pyramidal error | 5 arcsec-2 arcmin |
| Facet flatness | $\lambda/2$-$\lambda/10$ at 550 nm |
| Facet-to-facet angular error | 5-30 arcsec |

## 2. TYPES OF POLYGONAL SCAN MIRRORS

Four common types of polygonal scan mirror have been developed and put into use over the years:

1. Regular polygons
2. Irregular polygons
3. Inverted polygons
4. Pyramidal polygons

The purpose of this section is to define each of the types and give examples of their use to familiarize the reader with the various advantages and disadvantages they present to potential application.

### 2.1. Regular Polygons

A regular polygonal scan mirror is defined as one having a number of plane mirror surfaces (facets) which are parallel to and facing away from a rotational axis, uniformly distant from the rotational axis, and located with equal polar coordinates (see Fig. 3). The significant technical feature of a regular polygon is that it may be used to produce repetitively superimposed straight scans.

Regular polygons are the most common type of polygon found in scanning systems, since their geometry lends itself to

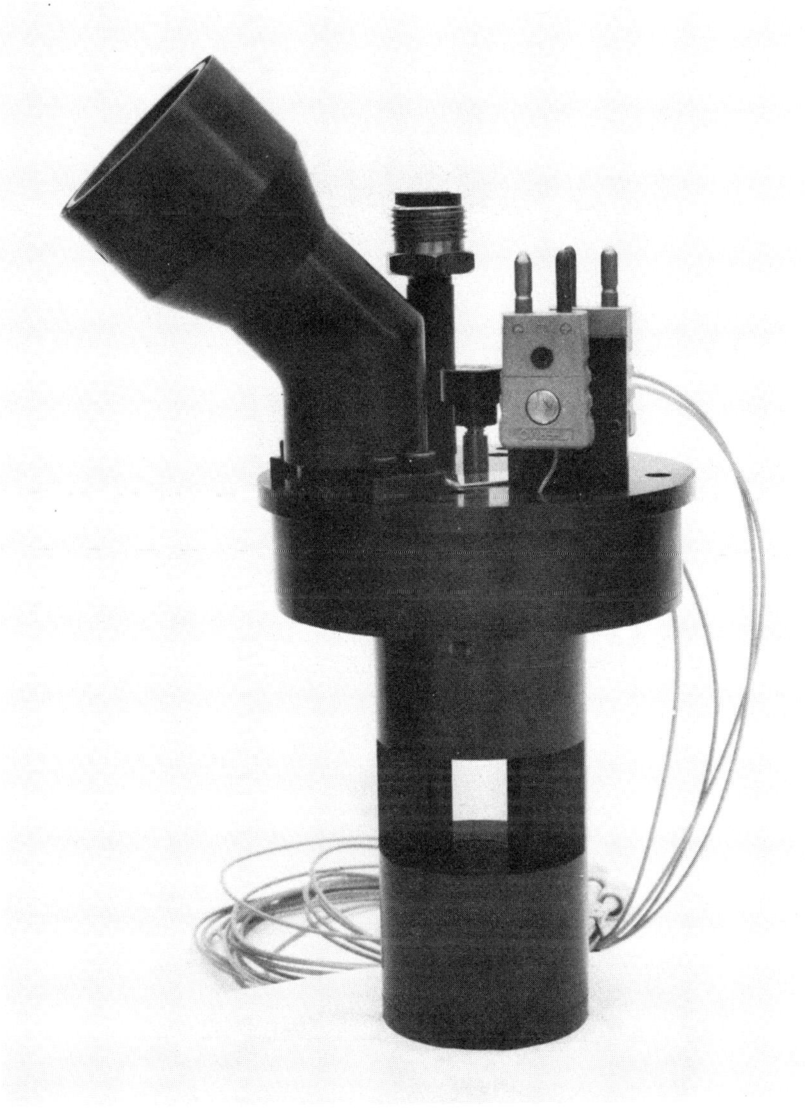

Figure 2 Compressed-air-driven rotating mirror assembly that operates at 498,000 rpm.

Figure 3 Regular 36-faceted polygon used in a production high-speed computer output printer. The facets on this polygon are nickel plated and pitch polished.

mass production and rather low manufacturing cost. The major applications for regular polygons are:

- Nonimpact computer output printers
- Laser inspection systems
- Infrared viewing systems
- Graphics facsimile systems

The symmetry of regular polygons enables the stacking of several polygons on an arbor for processing during manufacture (see Fig. 4). Since the manufacturing cost of major processes involved in the fabrication of the stack is divided over several finished parts, substantial cost reduction is realized.

Figure 4 Stack of polygons on a steel arbor. This group of eight individual polygons has undergone machining, lapping, plating, and polishing as a unit.

## 2.2. Irregular Polygons

An irregular polygon is defined as one having a number of plane mirror surfaces (facets) which are at a variety of angles with respect to, and face away from, the rotational axis. The polar location may also be irregular (a variety of angles) or regular (equiangular) (see Fig. 5).

The significant technical feature of irregular polygons is that they produce nonsuperimposing repetitive scans (i.e., a raster).

Irregular polygons are commonly used in applications where a coarse (few lines) raster is of value, such as:

- Point-of-sale scanners for reading universal product code (UPC) symbols
- Multiple-detector infrared scanners
- Laser heat-treating systems
- Intrusion-alarm scanning systems

Because of their asymmetry, irregular polygons can seldom be stacked or grouped together during processing and therefore tend to have a higher manufacturing cost than that of regular polygons. The most common methods of manufacture are replication and single-point diamond machining.

When considering irregular polygons for use, it is recommended that a design employing diametrical symmetry be used, for the sake of dynamic balance. Figure 6 describes preferred

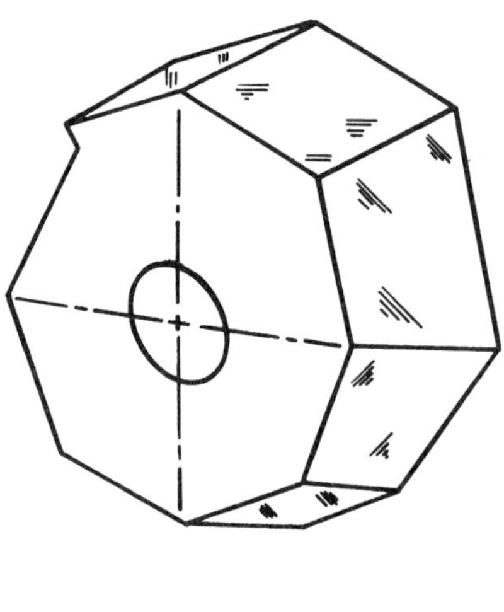

Figure 5 Irregular polygon.

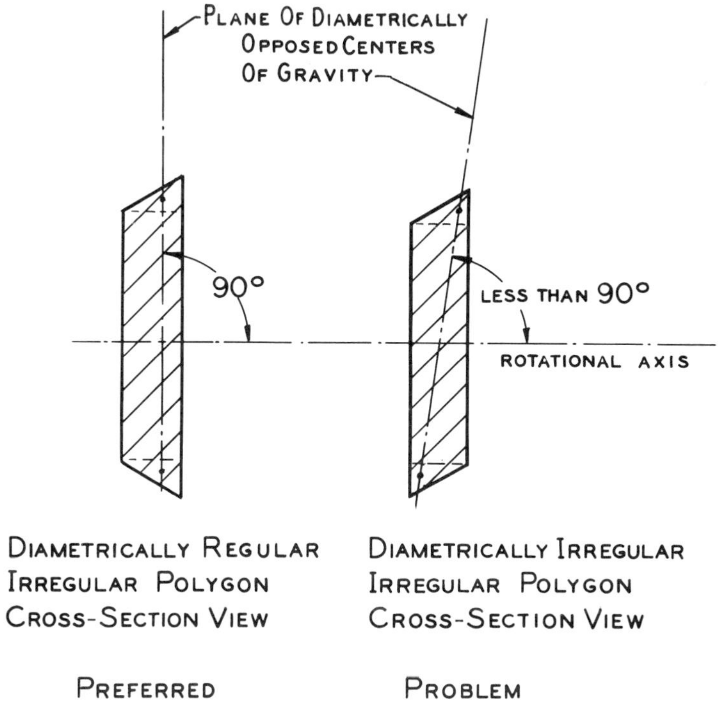


Figure 6 Dynamic balance considerations in irregular polygon construction.

and problem configurations. At high rotation speeds it is necessary to have the centers of gravity of diametrically opposed sections of the polygon in a plane at right angles to the rotational axis.

## 2.3. Inverted Polygons

An inverted polygon is defined as one having a number of plane mirror surfaces which are parallel to and facing toward a rotational axis, usually equidistant from the axis and located with equal polar coordinates (see Fig. 7).

Irregular polygons are also inverted. The significant feature of inverted polygons is that they lend themselves to compact optical designs. Typical applications are scanning systems for missile guidance and reconnaissance.

Inverted polygons are limited to relatively low speed scanning because of their inherent lack of structural integrity and the necessity of using large bearings to support them. The most common fabrication techniques are replication and single-point diamond machining.

## 2.4. Pyramidal Polygons

Regular pyramidal polygons are defined as having a number of plane mirror facets at an angle (rather than parallel) to the rotational axis. In the most common variety the facet angle is 45° to the spin axis (see Fig. 8).

Since the incident rays on a pyramidal mirror may be at other angles than normal (and still produce a straight scan), they are useful devices for producing smaller scan angles with fewer facets than are achievable with polygonal mirrors. For example, a four-sided, 45°, pyramidal scan mirror deflects incident rays (parallel to the rotational axis) through 90° for 90° of pyramid rotation. A four-sided regular polygon deflects incident rays (at right angles to the rotational axis) through 180° for 90° of polygon rotation.

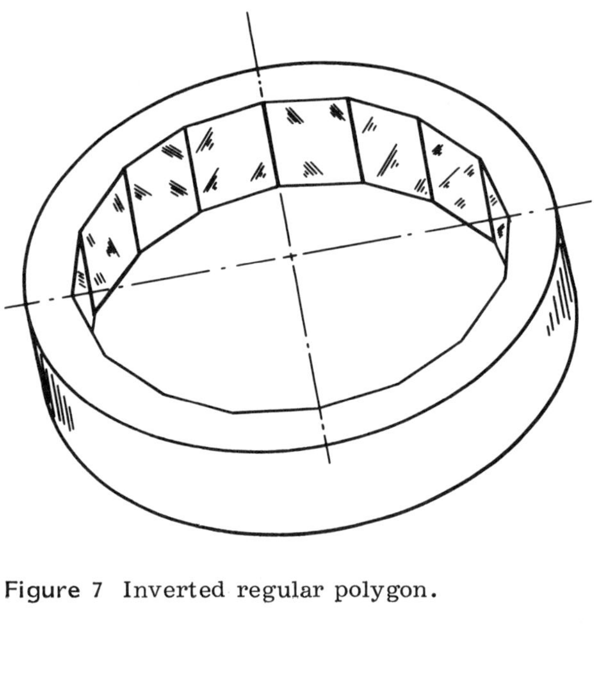

Figure 7 Inverted regular polygon.

Figure 8 Regular 45° pyramidal polygon.

## 3. APPLICATION TO SCANNING

### 3.1. Entrance Scanning

Entrance scanning is a term used to describe the use of a polygon to deflect a ray bundle, which after deflection is finally imaged by a lens or curved mirror (see Fig. 9). This method of scanning places the function of focal plane definition on the lens design rather than on the scanning facet. This is beneficial if the desired focal surface is to be flat.

Several desirable characteristics can be designed into the scan lens when employed in entrance scanning. An example is a lens design commonly referred to as an f-θ configuration. An f-θ lens provides the following features:

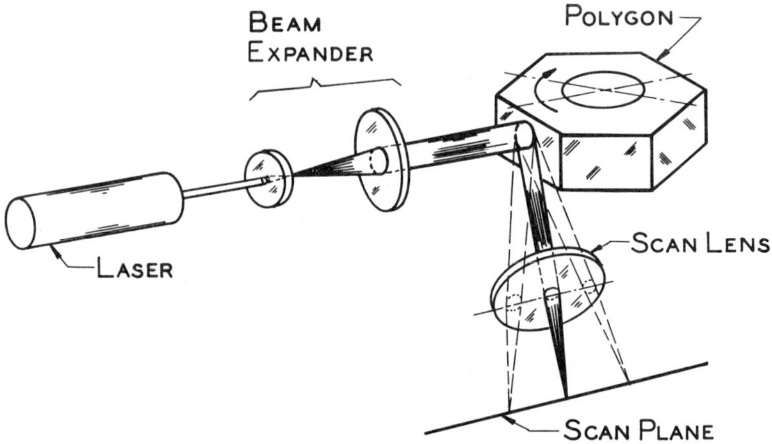


Figure 9 Schematic diagram of a typical entrance scanning application of a polygonal mirror.

A flat focal plane

Uniform spot diameter over the entire scan

Linear spot velocity at the focal plane (assuming constant angular velocity of the entrance beam)

Usually, it is desirable to have the scanning spot move with a highly accurate and constant velocity in the focal plane. Polygonal mirror deflectors provide angular velocity stability in the range 0.005 to 0.05%, depending on the speed and moment of inertia of the scanner. Without the aid of an f-θ lens, however, the spot velocity variation at a flat focal surface will be proportional to the tangent of the scan angle, which for systems involving several degrees of scan means several percent variation.

### 3.2. Exit Scanning

Exit scanning is a term used to describe the use of a polygon to deflect a focusing ray bundle over a focal surface (see Fig. 10). This method places the function of focal plane definition on the polygonal mirror, and the imaging (spot forming) lens is a relatively simple component.

The focal surface of an exit scanning system is curved. The radius of curvature is approximately the distance from the center of the scanning facet to the focal point of the lens. In scanning systems utilizing relatively small scan angles, >20° and large f numbers, f/50 or greater, exit scanners may be used effectively, since the chord depth of the focal surface may be less than the depth of focus of the scanning beam, in which event spot size may be effectively maintained over a flat photoreceptor plane.

## 3.3. Correcting Polygon-Induced Scanning Anomalies

Under circumstances where the combined errors of polygon and drive motor produce undesirable errors in the location of the resulting scans in the direction of scan (timing errors) or at right angles to the scan (tracking errors) some alternatives are available. These become particularly important when the requirements for scanning accuracy become difficult and/or impractical to achieve by placing tolerances of the order of a few arc seconds or less on the polygon/motor assembly. The following correction schemes are those most commonly found in scanning systems requiring high-accuracy scans.

### 3.3.1. Sources of Tracking Errors

There are four major sources of tracking errors that may be contributed to a scanning system by a polygon/motor assembly:

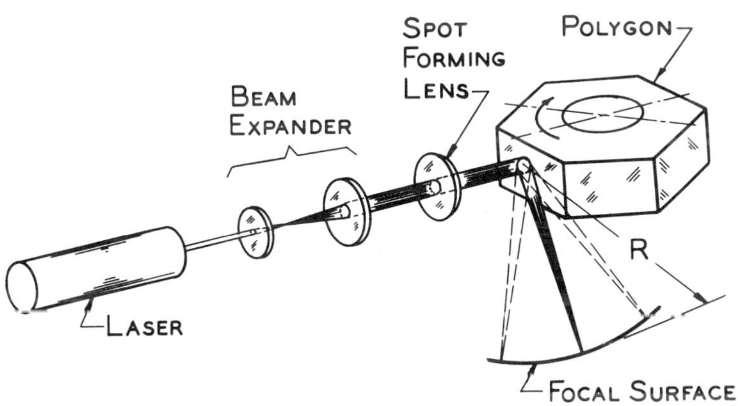


Figure 10 Schematic diagram of a typical exit scanning application of a polygonal mirror.

1. Errors in the angular location of polygon facets with respect to the datum surface of the polygon. These errors are attributable to limitations in the manufacturing process of the polygon.
2. Errors in the fastening of the polygon to the motor shaft.
3. Errors in the rotation of the motor shaft attributable to bearings, lack of rotor stiffness, thermal gradients causing rotor warp, and dynamic unbalance.
4. Errors in facet flatness.

### 3.3.2. Active Methods of Correcting Tracking Errors

Methods for correcting tracking errors fall into two basic categories, active and passive. Active systems involve auxiliary small-angle deflectors located in the scanner optical system to compensate for motor/polygon-induced errors in tracking. If all the errors to be corrected are repeatable, the auxiliary deflector may be programmed to make corrections. If nonrepeatable errors must be corrected, an error detector must be used to interrogate the scanning facet immediately prior to the active time of scan and produce an error signal on which the auxiliary deflector can act. However, to be effective, both of these active methods must rely on the integrity of a scanning facet during its active period of traverse. Errors resulting from items 3 and 4 above may produce undesirable effects which are uncontrollable by programmed correctors or "feedback" correctors. Typically, corrections are made during the "dead time" between scans, rather than during the active period of a scan, hence the dependence on inherent stability of scans between correction periods.

Figure 11 depicts an active correction system for tracking errors that are polygon/motor induced. In this case an acoustic beam deflector (with deflection capability of several arc minutes) is placed in the laser* beam path prior to the beam expander. The small aperture of the deflector dictates a location where the light beam is also small. In this location the acoustic deflector must be able to produce a correction angle which is twice the facet angle error multiplied by the magnification of the beam expanding telescope. The telescope will reduce the deflected beam

---

*Due to the fact that acousto-optic beam deflectors are diffractive devices, they are impractical for light sources other than laser.

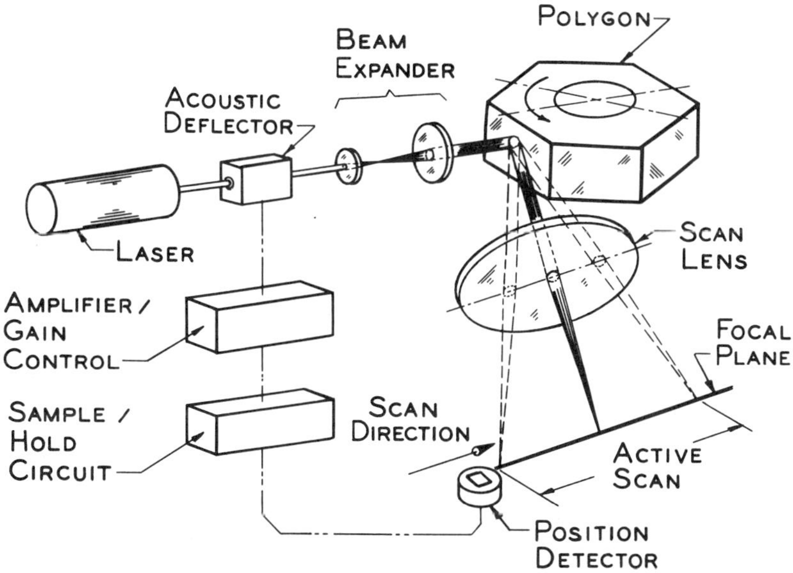


Figure 11 Schematic diagram of a small-angle correction scheme employing a scan position sensor and an acoustic deflector to compensate for polygon/motor-induced tracking errors.

angle in direct proportion to its power, and the scanned beam will have an angular error twice that of the facet angular error.

This scheme is used effectively to correct for facet errors in the range of an arc minute. Polygon/motor assemblies are readily available within this tolerance. A note of caution on this approach, however, is that acoustic deflection devices exhibit small changes in light energy transmitted as a function of scan angle. A difference of several percent (5-7%) can be expected from 0° to 10 arcmin of deflection. It is necessary to employ sample-and-hold electronic circuitry in controlling the deflector, since the correction made must be maintained over the period of the scan of the facet.

### 3.3.3. Passive Methods of Correcting Tracking Errors

Passive correction schemes usually utilize additional optical elements to reduce polygon-induced tracking errors. The most popular method of passive correction involves the use of cylindrical lenses in the scanning optical imaging system (see Fig. 12).

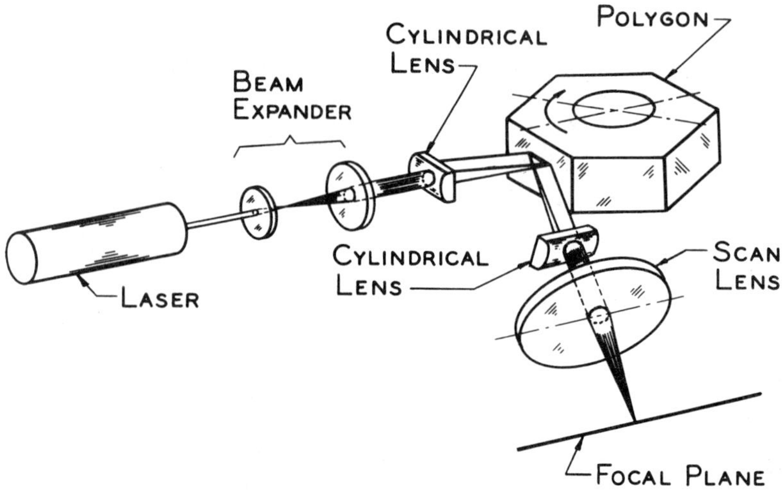


Figure 12 Schematic diagram of an optical system employing cylindrical lenses to correct for polygon/motor-induced tracking errors.

These cylindrical lenses cause focusing in one direction of the parallel rays of light directed toward the polygonal scan mirror. This focusing in one direction creates a line image on the facet with the direction of the line running in the scan direction. The reflected rays from the polygon are recollimated by a second cylindrical lens. The result of this one-dimensional imaging is to permit scanning in one direction, the direction of rotation of the polygon, but it prevents scanning in the direction at right angles to that. Figure 13 describes two views of the light energy emanating from a polygon facet, as shown in Figure 12, showing why a facet error or "wobble" of the rotational axis produces no undesired scanning.

Another novel passive approach involves the use of a retro reflecting prism which accepts the first scan beam from a polygon facet and returns those rays to the same facet. The second reflection will have an error identical to but with the reverse sign of the rays reflected in the first bounce. The result is a cancellation of the errors of the facet (see Fig. 14).

The corrector prism method is simple and effective in correcting for facet dislocation errors. These errors result in a small lateral shift of the beam exiting the prism, but since these rays are collimated, this shift does not result in a motion of the spot in the focal plane.

In the direction of scan the corrector prism behaves as a plane mirror (except for the fixed offset between input and output), so it reflects light back in the direction of the facet. Of course, the input beam to the prism is coming from a moving facet, and so scanning occurs in the prism. The double bounce off the polygon facet causes velocity doubling of the final output beam. As a result, the angular velocity of the output is four times the polygon rotational velocity.

Since the beam returning to the facet is scanning, it migrates across the facet (in the scan direction) (see Fig. 15). The result of this migration in most configurations limits the scan

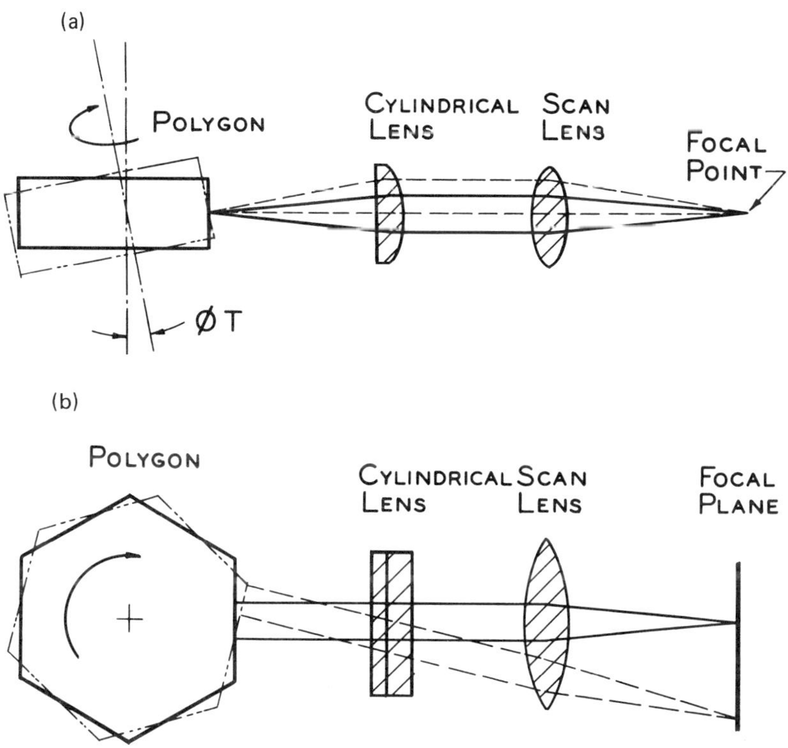


Figure 13 Mechanism of jitter correction in the system as shown in Figure 12. (a) A tilt of the polygon $\phi T$ does not produce a displacement of the focused spot in the direction at right angles to the scan. (b) A rotation of the polygon around its axis produces a scan.

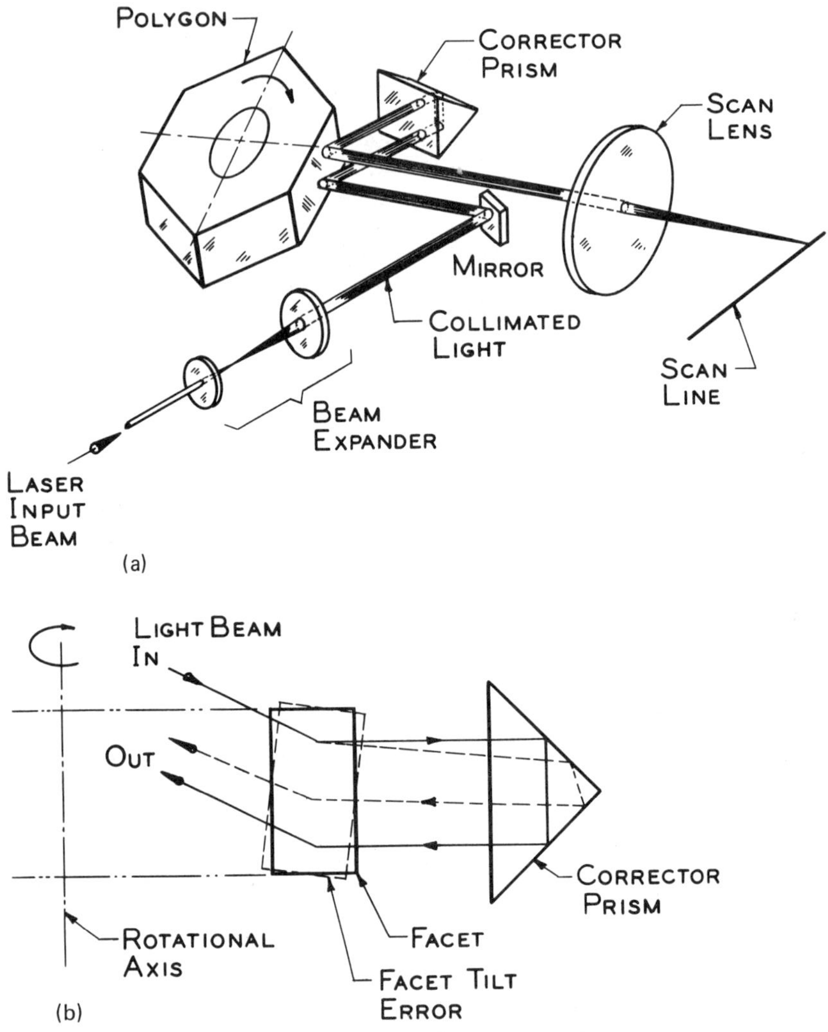


Figure 14 Schematic diagram of the use of a 90° corner prism (PORRO prism) to cancel polygon/motor-induced tracking errors. (a) General arrangement of components; (b) effect of corrector prism as facet tilt error is introduced.

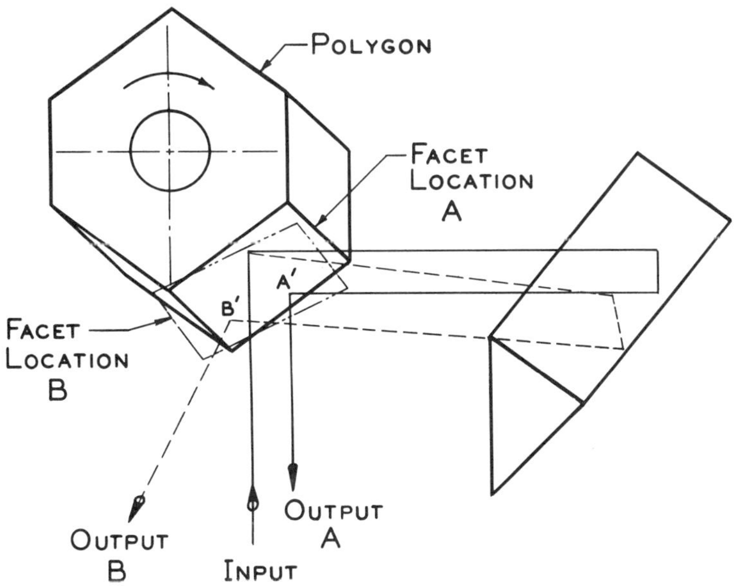


Figure 15 Schematic diagram showing limited-use angle of the polygon when using a prism corrector due to beam migration along the facet. When the facet rotates from position A to B the point of incidence of the output beam moves from A′ to B′.

efficiency (scan time/total time) to 50% or less; however, due to the velocity doubling, there is seldom a sacrifice in scan angle of the output.

### 3.4. Design Trade-Offs

Trading off design possibilities for a scanning system is to some extent an iterative process. A suggested guideline is to define the requirement at the scan focal plane (or focal surface) and work your way backward (up the photon stream) to the light source. Table 3 provides a rudimentary list of parameters that must be developed in order to determine a preferred approach to a scanning system employing a polygonal deflector. It also provides some comments on how the parameters of output relate directly to the polygon.

The key parameters to determining the basic polygon configuration are spot size, wavelength, and scan length. Having specified spot size (most commonly that diameter of spot at which

**Table 3** Design Parameters for a Polygonal Scanner System

| Parameter | Relationship to polygon/motor performance |
|---|---|
| Spot size | Facet size and flatness |
| Scan line length/scan angle | Number of facets |
| Wavelength or wavelengths | Type of reflection enhancement coating |
| Line straightness | Average pyramidal error and flatness of facets |
| Scan rate | Motor speed |
| Scan cross-track accuracy | Polygon/motor tracking accuracy or corrector selection |
| Scan timing accuracy | Motor velocity stability and/or facet polar location accuracy |
| | Facet flatness |
| Scan efficiency | Facet length (scan direction) |
| Scatter losses permissible | Facet surface quality |
| Light transmission uniformity | Facet surface quality |
| | Coating uniformity |
| Mean time between failures | Facet durability (cleanability) |
| | Bearing type and quality |
| | Dynamic balance |
| Signal/noise ratio | Facet roll-off width |

intensity is down to 13% of the peak intensity, i.e., $1/e^2$) and wavelength, we may determine the length-to-diameter ratio of the cone of rays necessary to produce the spot. This length/diameter ratio is commonly called f#, and in the case where the input ray bundle is truncated at the $1/e^2$ diameter, is given by focused spot size/1.8$\lambda$ (see Fig. 16).

For an exit scanner configuration to be used with a planar photoreceptor, it is necessary to calculate the depth of focus of the focusing cone using the formula $D = (f\#)^2\lambda$ as a rule of thumb. This value becomes the permissible chord depth of the scan at the photoreceptor and thereby determines the scan angle and radius.

Knowing the scan radius, we can determine the size of the ray bundle at the mirror facet by

$$\text{Scan radius} \times f\#/\cos\theta$$

where $\theta$ is the tilt angle between the facet and the central ray of the focusing cone.

If the calculated facet length and/or facet angle is impractically long or small, respectively (numbers of facets exceeding 60 are impractical), it is an indication that an entrance-type scanning system is called for.

When determining the polygon characteristics for an entrance-type scanner, it is necessary to examine the required scan

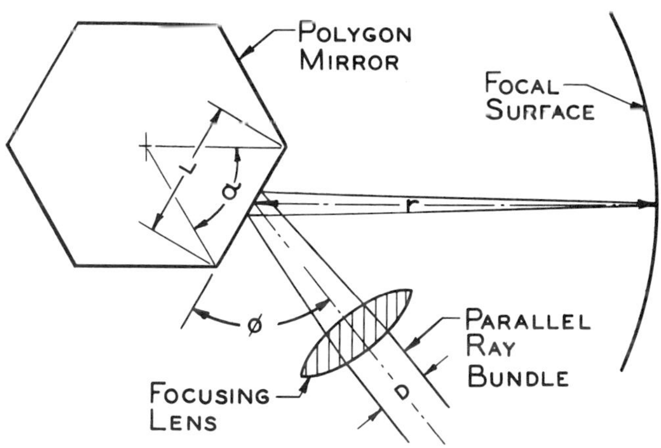


$$\text{Useful scan angle} = 2\alpha \times \frac{D/\cos\phi}{L}$$

$$\text{Useful scan line length} = 2\pi r \times \frac{\text{Useful scan angle}}{360}$$

$$\text{Spot size } (1/\ell^2) = 1.80\lambda \times \frac{F}{D}$$

$$\text{Spots/line} = \frac{\text{Scanline length}}{\text{Spot size}}$$

Figure 16 Schematic diagram of the important criteria for determining polygon facet size for an exit scanning system.

angle and beam diameter at the entrance pupil of the scan lens. Since the scan lens is designed to provide a flat field, the polygon dimensions need not be related to a scan radius sufficient to scan a chord depth within the beam depth of focus.

A particularly useful lens design for scanning is one referred to as an f-θ lens. The design provides a correction for the velocity change that would otherwise result from the tangent of the scan angle on a flat plane. The primary characteristics of an f-θ lens are:

Flat focal plane

Uniformity of spot size over the scan length

Constant scan spot velocity output for a constant angular scan input

Beyond sizing of the polygon there are more subtle concerns of tolerancing and practicality of manufacture. These are dealt with in Section 4.

## 4. CONSIDERATIONS OF QUALITY

### 4.1. Manufacturing Methods

There are numerous methods of fabricating multifaceted polygons. Five are discussed here:

1. Injection molding
2. Assembly construction
3. Single-point diamond machining
4. Replication
5. Conventional polishing

These various methods of manufacture have been refined over the years to rather sophisticated levels. Each manufacturer has proprietary tooling and/or procedures to apply to the task of making polygonal mirrors. These proprietary items, usually referred to as trade secrets, preclude any in-depth disclosure of the precise methods being employed. The intent here is to provide an overview of the most common techniques in use so as to put into perspective their merits in regard to the finished product.

### 4.1.1. Injection Molding

Injection molding has been successfully applied to the making of polygonal mirrors on an experimental basis, and there is some promise that this method will be refined over time to provide very low cost polygons in the medium to low end of the quality spectrum.

Special material formulations are necessary to mold an optical part.

### 4.1.2. Assembly Construction

This manufacturing method involves assembling a polygon with individual planar first surface mirrors fastened to the perimeter of a machined or cast support structure (see Fig. 17). Assembly construction is particularly useful for fabricating irregular polygons.

The individual mirror segments may be made from glass or metal. Methods of fastening include bonding with epoxy adhesives and machine screws. Since the centrifugal stresses become great at high rotational speeds, careful attention must be paid to the potential for distortion of surface flatness and the possibility of fastener failure of assembly construction polygons that are to be used in relatively fast scanning applications.

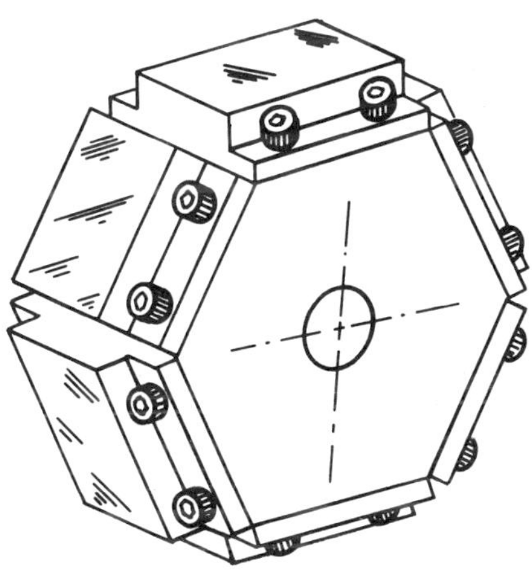

Figure 17 Polygon manufactured by assembly construction.

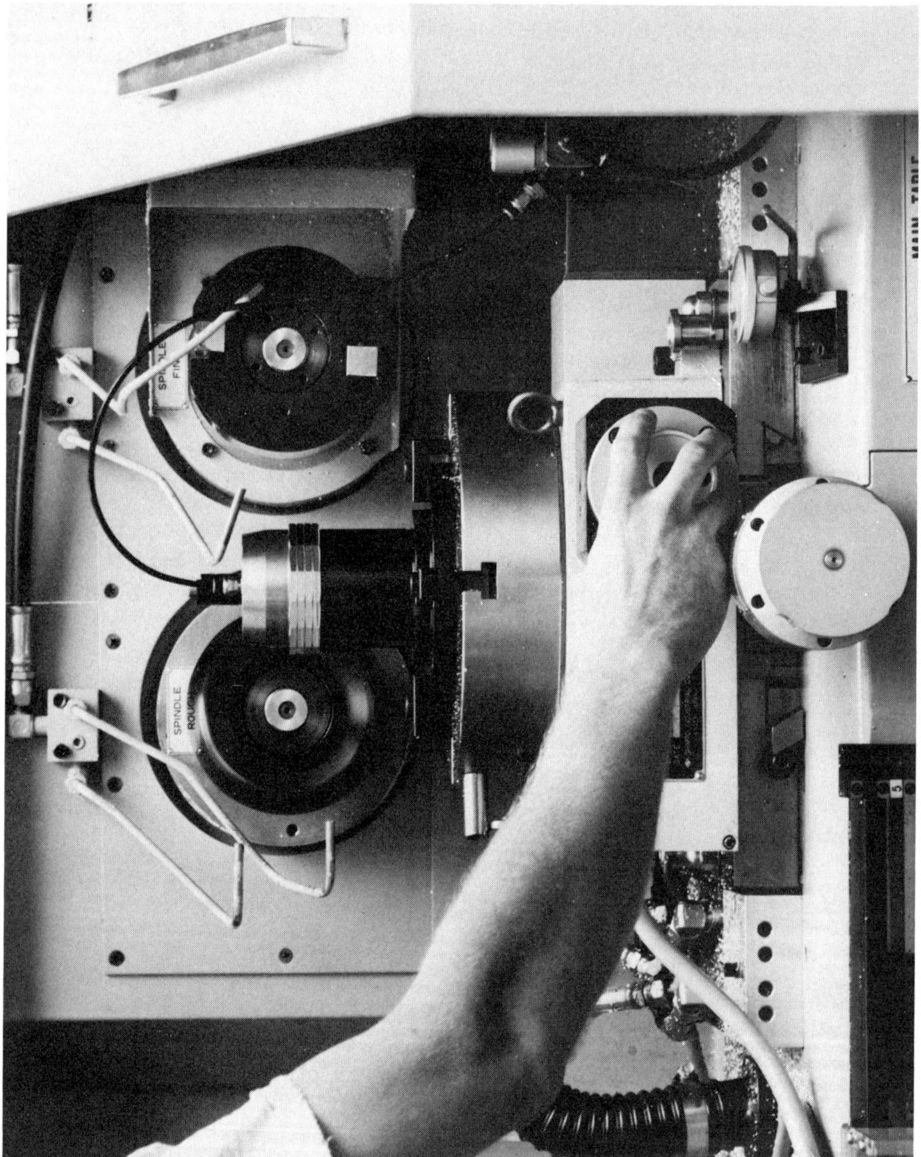
SPINDLE
SPINDLE

### 4.1.3. Single-Point Diamond Machining

Single-point diamond machining is a process of material removal using a finely sharpened single-crystal diamond cutting tool. Diamond-machining machine tools are available in the forms of lathes and millers. The use of ultra-precise air-bearing spindles and table ways enable machining to optical-quality surface figure and finish. Figure 18 shows a diamond-machining system with a polygon in process.

Diamond machining is an efficient process for generating an optical surface, since it can be automated and the technique essentially creates the surface directly from the raw material. The most common substrate material used to make diamond-machined polygons is aluminum, although satisfactory results are achieved on brass, copper, and electroless-plated nickel. Bare aluminum mirror surfaces are fragile and it is recommended that diamond-machined surfaces be overcoated with a harder dielectric layer (SiO) to enhance durability.

To the naked eye, diamond-machined surfaces appear to be perfect mirrors. With optical aids, however, it will be seen that the surfaces resemble a high-frequency grating. As a result, the specular reflectance of the surface degrades at short wavelengths (i.e., ultraviolet). Over the longer wavelengths of the visible spectrum and in the infrared, the grating effects are negligible to immeasurably small. Stacking of polygons can be done to increase yield and reduce cost in the diamond-machining process.

### 4.1.4. Replication

Replication is a process of impressing the pattern of a "master surface" onto the surface of a hardenable fluid. In recent times the fluid is epoxy resin. In the case of fabricating polygons by replication, the polygon substrate is machined to a dimension slightly smaller than the finished article and the space between the substrate and the master is filled up with resin. The master contains a microthin coating of release agent to prevent the resin from adhering to it (see Fig. 19).

Figure 18 Diamond-machining system with a polygon in process of fabrication.

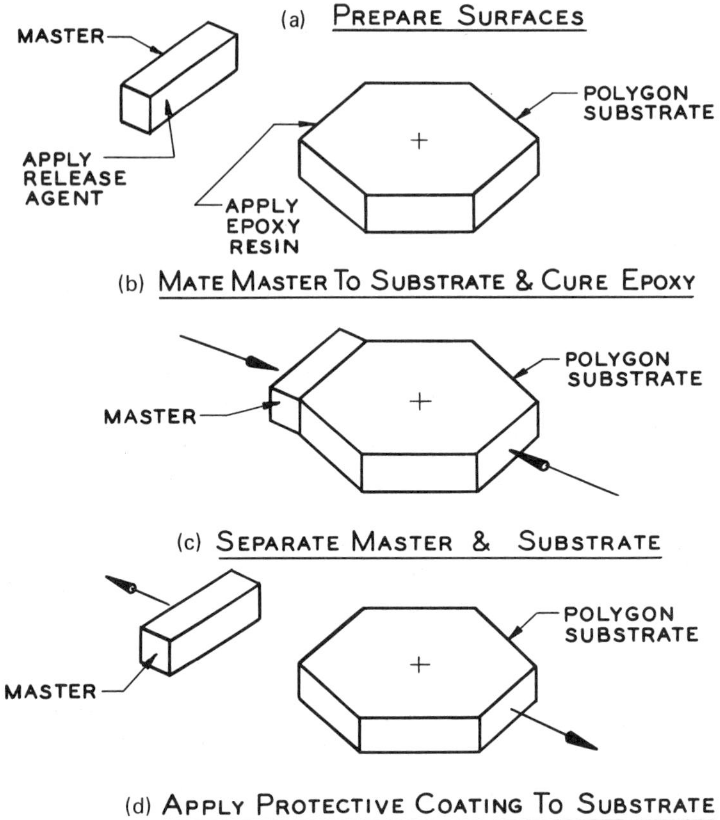


**Figure 19** Schematic diagram of the fundamentals of replication.

The replication process is particularly useful for fabricating pyramidal polygons and types of irregular polygons, wherein the geometry permits the practical fabrication of masters. Due to the shrink characteristics of the best of resins, the replication is usually limited to producing surfaces of $\lambda/4$ or less flatness over relatively small apertures.

### 4.1.5. Conventional Polishing

Conventional polishing in this context is defined as pitch lapping, in much the same manner as glass lenses and prisms are polished. Pitch lapping can be used to produce high-quality surfaces on a number of polygon surfaces, including stainless steel, beryllium, chrome plating, electroless nickel plating, glass, and other glass-like materials (quartz, Pyrex, etc.).

Pitch-resin-faced polishing tools are stroked repetitively over the polygon facet in specially tooled polishing machines that utilize finely divided alumina and water as the polishing compound (see Fig. 20). A large surround plate or "blocking tool" is placed around the facets being polished. This surround is held precisely in the plane of the facets to support the polishing tool and prevent rounding of facet edges, known as "roll-off."

Electroless-nickel-plated aluminum is the most common polygon configuration. This combination provides the low cost and ease of machining of aluminum as the structural material with the superior polishing properties and durability of nickel.

## 4.2. Materials

Material selection for polygonal mirrors involves considerations of performance and cost. A rule of thumb for polygonal or pyramidal mirrors with solid cross sections (except for a central mounting hole) is that if the maximum operating peripheral velocity will not exceed 400 ft/sec, the polygon facets can be expected to stay flat to $\lambda/10$ per inch. If the requirement exceeds that peripheral velocity, analysis will be necessary to determine whether a material with a superior stiffness/weight characteristic is necessary, such as beryllium. This rule of thumb has been derived as the result of a variety of tests conducted over a number of years on various polygon designs, as well as a few finite element analyses which corroborated rather well with testing.

In situations where minor (or major) distortions in facet flatness can be tolerated and peripheral velocities exceeding 400 ft/sec are contemplated, safety must be considered. The speed at which the centrifugal stress will reach the yield strength (causing permanent distortion and dangerously close to the breaking speed) is found using the formula below. For an added margin of safety it is suggested that the circumscribed circle diameter of the polygon be used in the equation (tip diameter)

$$N = 57.3 \sqrt{\frac{32.16S}{(R^2 + Rr + r^2)W}}$$

where

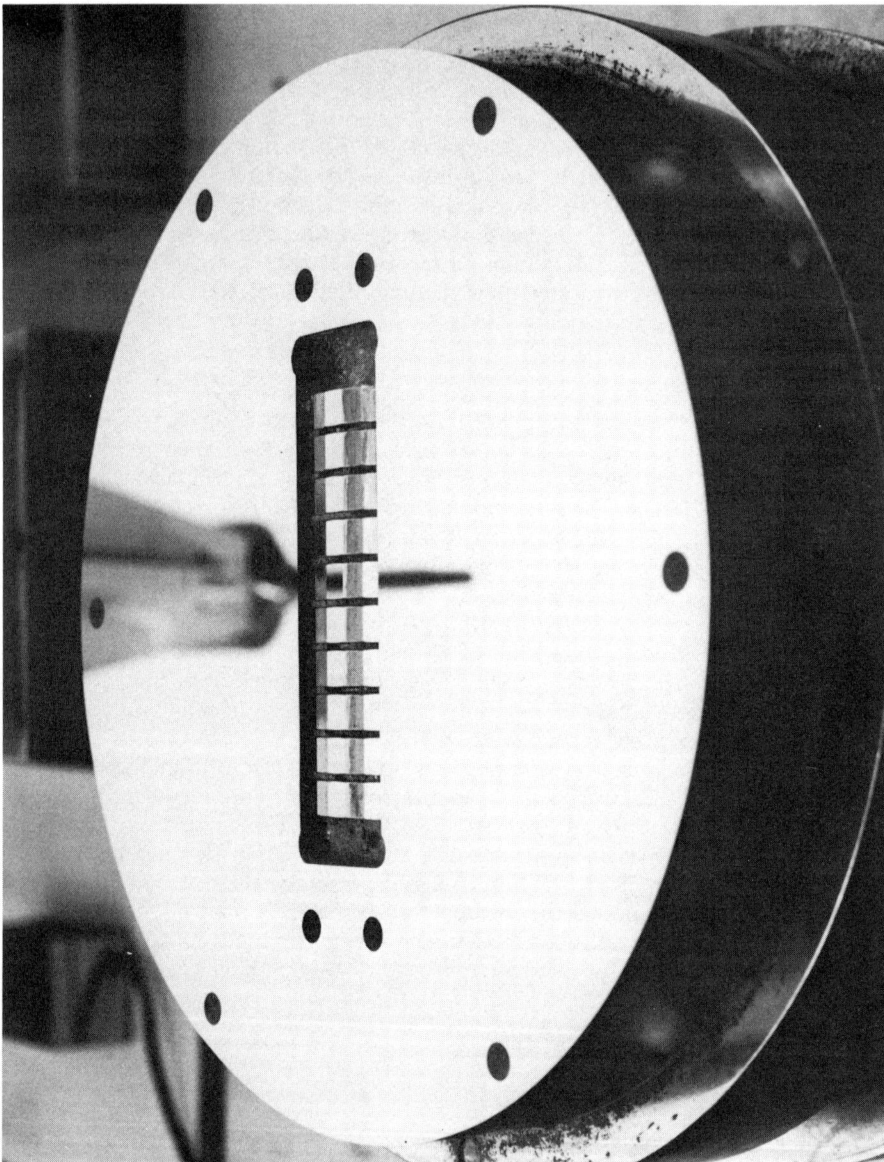

N = limiting speed of the polygon, rpm

S = yield strength of material, lb/in.$^2$

R = radius of polygon circumscribed circle

r = radius of the polygon bore

W = weight of the polygon material, lb/in.$^3$

## 4.3. Angular Tolerances and Measurement

The angular locations of facets on a polygonal mirror are usually specified in relation to two datum surfaces on the polygons. The most common datum surfaces are (1) one side of the polygon at right angles to its rotational axis, and (2) one or more facets, (see Fig. 21).

Facet to datum angles are measured using a measuring autocollimator and a flat surface plate. The polygon is placed datum surface down on the surface plate and rotated to align a facet in the rotational direction in the reticle of the autocollimator. The error is measured and the process repeated for the remaining facets. This technique is employed for angular tolerances of 10 arcsec or less. For tolerances greater than 10 arcsec, a "spin test" fixture is used, wherein the polygon is mounted on a precision spin platform and the facets are used to scan a focused laser beam by a horizontal knife edge with a photodetector behind it. The fixture is calibrated to read tracking error as a function of detector voltage variations produced by varying scan locations.

The polar location of facets is determined by fabrication accuracy and is usually measured one time after fabrication. Measurement of polar location is done using a precision rotary indexing head (angle reference) and a measuring autocollimator. The autocollimator is aligned to one facet to produce a "zero" reading, then the indexing head is rotated to the desired location of the next facet and the deviation of that facet is measured with the autocollimator (see Fig. 22). This method provides true position information.

---

Figure 20 Stack of polygonal mirrors in a polishing support, ready for conventional pitch lap polishing.

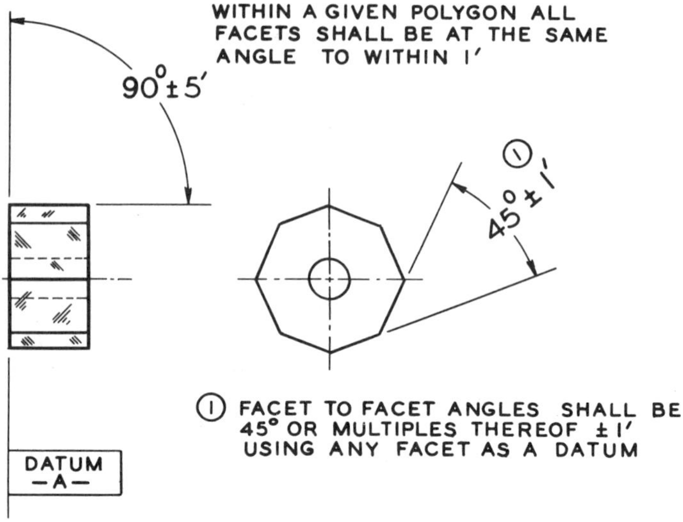


Figure 21 Typical method of specifying facet to datum angles and polar location angles on a polygonal mirror.

## 4.4. Facet Flatness; Static and Dynamic

Facet flatness is important to scanner system resolution. How flat a scan mirror facet must be to produce a given resolving power in a scanner system depends on wavelength, f number, and the range of incidence angles over which the facet will be used. Several factors influence the flatness of polygon facets:

Initial fabrication tolerances

Distortion due to mounting stresses

Distortion due to centrifugal stress when rotating at high speed

Distortions due to stress relief over long periods of time

Figure 22 Measurement of polar location of polygon facets using reference indexing head and measuring autocollimator.

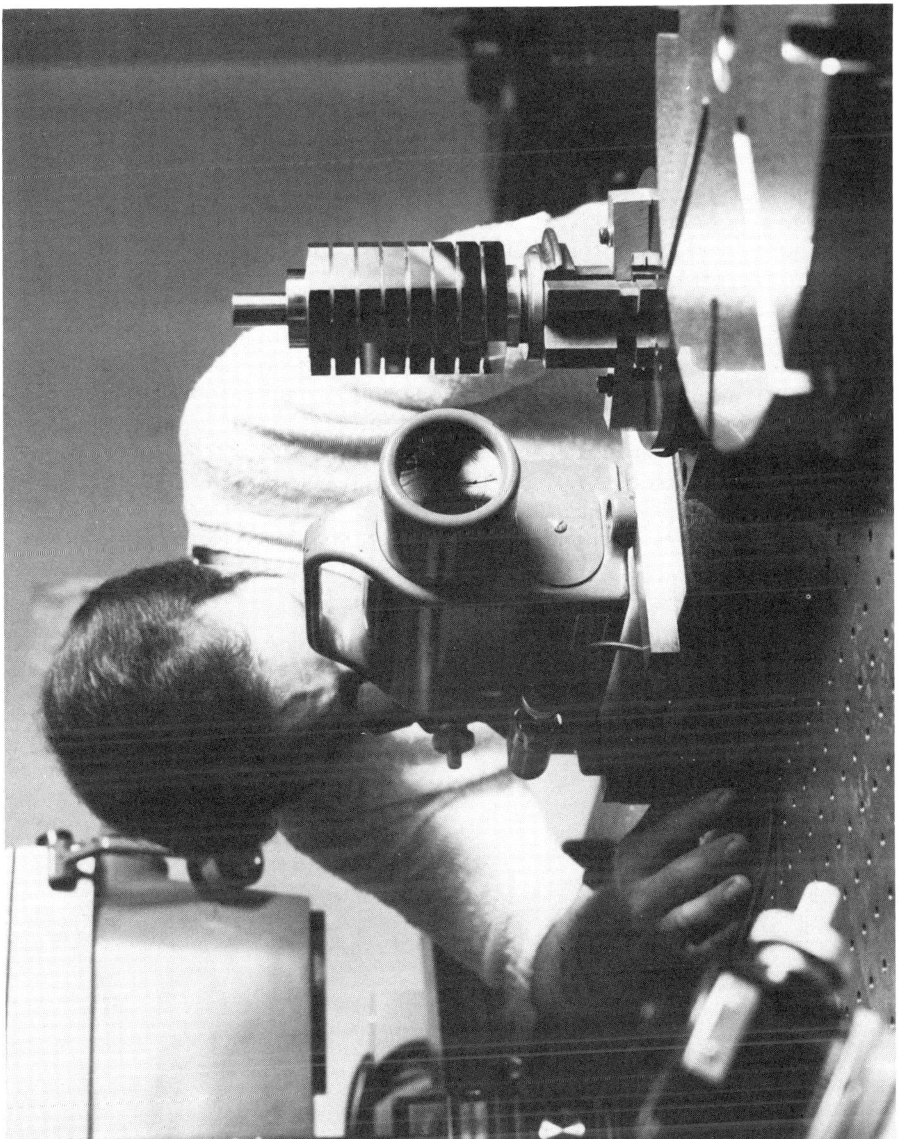

Interferometry is used to measure static facet flatness. Flatness is specified in wavelengths (or fraction thereof) of light. A typical flatness specification will read: flatness – $\lambda/4$ at 632.8 nm. This specification requires that all points on a facet surface lie in a plane to within 632.8/4 nm.

Departure from flatness may have a variety of forms, depending on how the surface has been fabricated. For example, conventionally polished mirror surfaces tend to depart from flat in a regular spherical form, either concave or convex, but more usually convex. Diamond-machined surfaces usually depart from flat in a regular cylindrical form, either concave or convex. Most polygons used in reprographic applications are specified in the range $\lambda/4$-$\lambda/10$ in the wavelength of interest.

To preserve the facet flatness achieved during initial polygon fabrication, it is necessary to fasten the polygon to its drive motor with care, particularly if $\lambda/10$ flatness is required. The polygon-motor interface must be achieved without producing sufficient stresses to cause facet distortion.

A typical mounting scheme is shown in Figure 23. In this case the datum surface of the polygon and the locating annulus of the mounting hub are lapped to optical quality surfaces so that when the two are firmly held together, distortions do not occur.

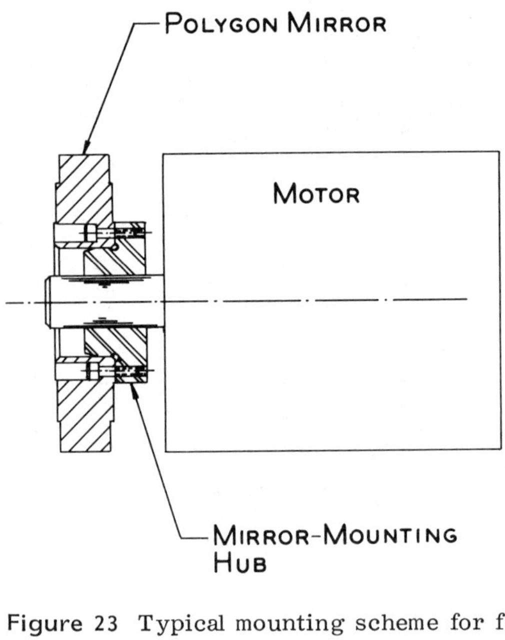


Figure 23 Typical mounting scheme for fastening a polygon to a motor shaft.

Equally important to accurate mating of datum and hub surfaces are cleanliness at assembly and appropriate torquing of the fastening screws. If the fastening screw location is close to the facet, it is desirable to counterbore the screw hole to achieve a minimum land thickness. This will reduce the amount of radiation of stress to the facet.

Polygons operated at particularly high rotational speeds may be subjected to sufficient centrifugal stress to degrade facet flatness. A polygon will stay flat while rotating at high speed if:

It has 12 or more facets.

It is made of aluminum or steel (tempered).

It has a solid radial cross section at least as thick as the facet is wide.

It has a central hole 3/8 the diameter of the polygon-inscribed circle or smaller.

It may then safely operate up to a peripheral velocity (tip velocity) of 400 f/sec with facet distortions of less than $\lambda/10$ at 632 nm. This rule of thumb has iterated from a number of experimental results by testing polygons over the years.

Polygons that fall out of the boundaries of the rule of thumb should be dealt with by dynamic testing. Finite-element analysis may be used to determine close approximation of limiting speeds, but if the analysis shows a borderline case, a test may be in order.

There are a number of methods for measuring facet flatness of a high-speed rotating polygon, and they all involve rather sophisticated optical systems to attain accurate data. One of the simpler and easier-to-use methods is shown in Figure 24.

## 4.5. Surface Quality

In the context of first surface mirrors, a perfect surface will reflect all incident light rays at an angle precisely identical to the incidence angle. Any feature of the surface that causes other behavior, such as transmission through, absorption by, or scattering by the surface is considered a defect in surface quality. Specular is the term used to describe the perfect area of a mirror.

For some time in the past a system of quantifying mirror surface quality (and the quality of refractive elements such as lenses and prisms) known as a "scratch and dig" specification has sufficed. With the advent of machined optical surfaces, new

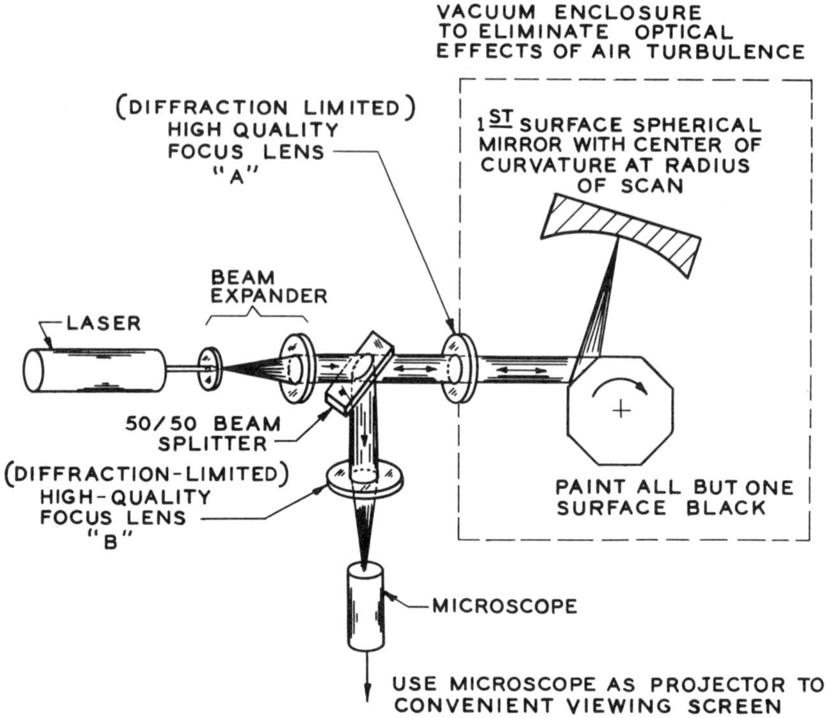


Figure 24 Dynamic facet flatness measuring optical setup. The scanning facet both scans and relays the scanned beam back through the scan lens. Wavefront distortion is measured as a growth or change of shape of the diffraction image formed by lens B. Static and dynamic cases are compared. Due to the high "strobe" frequency, the viewed image appears steady state.

methods have evolved. The U.S. military developed a scratch and dig specification which became documented as MIL 13508 (current revision letter is C) and which has become commonly used in the optics industry. This method of quality determination involves examination of a surface microscopically and identifying by a two-digit, hyphen, two-digit set of numbers, the number of permissible scratches and digs (pits or holes) in a given unit of area. A typical high-quality conventionally polished polygonal mirror will rate 40-20 scratch and dig.

Machined optical surfaces, on the other hand, are made up of a precise regular pattern of scratches (machine toolmarks) which are sufficiently high in frequency and low in peak to valley roughness as to behave aggregately in the manner of an acceptable "plane mirror" at certain wavelengths. The scratch and dig system is inappropriate here. A more useful method is that which specifies root-mean-square (rms) surface finish, since there is an approximate correlation between this value and the value of measured nonspecular reflectance from a machined surface (see Table 4).

## 4.6. Reflectance and Uniformity

The reflectance of a polygon facet and the uniformity of reflectance across that facet and other facets on the same wheel will be determined to the greatest extent by two factors. One is the

Table 4 Relationship Between RMS Roughness and Total Integrated Scatter of a Diamond-Machined Surface

$$\mathrm{TIS} \cong \left[4\pi\delta(\cos\theta)/\lambda\right]^2$$

$\delta$ = rms surface roughness
$\lambda$ = wavelength of interest
$\theta$ = angle of incidence

| Surface Roughness (rms) microinches | Surface Roughness (rms) Angstroms | Wavelength microns | Total Integrated Scatter (%) $\theta = 0°$ | Total Integrated Scatter (%) $\theta = 45°$ |
|---|---|---|---|---|
| 1.0 | 250 | .6328<br>10.6 | 25.3<br>9.0 | 12.6<br>4.5 |
| .8 | 200 | .6328<br>10.6 | 16.2<br>5.8 | 8.1<br>2.9 |
| .6 | 150 | .6328<br>10.6 | 9.1<br>3.2 | 4.5<br>1.6 |
| .4 | 100 | .6328<br>10.6 | 4.0<br>1.4 | 2.0<br>.72 |
| .2 | 50 | .6328<br>10.6 | 1.0<br>.36 | .50<br>.18 |
| .1 | 25 | .6328<br>10.6 | .25<br>.09 | .13<br>.04 |

PNEUMO PRECISION, INC.

specularity or quality of surface and the second is the enhancement coating used to improve the reflectance. Incident light on a mirror surface is either specularly reflected, scattered, absorbed, or transmitted by the surface. Surface reflectance is normally measured by monitoring the specularly reflected component. The scattered component is generally dealt with by specifying surface quality, and it is assumed that the remaining losses are by absorption. In the case where the polygon may be made out of glass, there may be a small component of transmitted light if the enhancement coating is not opaque.

The common measuring tools to determine reflectance are spectrophotometers and laser reflectometers. Most spectrophotometers with a reflectance measuring attachment are able to take only a given sample size, generally circular and approximately a 1/2-1 in. in diameter, to make a measurement. This fact usually precludes measuring the polygon itself but rather, measuring a witness sample which is supposed to be identical to the polygon facet itself. Using a spectrophotometer can be a reliable method of ascertaining the performance of a polygon facet if one has determined that the witness sample is an exact representation of the facet. If this exactness cannot be relied on, it is necessary to use another instrument for making the measurement. In a case where a laser wavelength is to be used in the scanning process with the polygon, the laser itself can be reflected from the surface and the reflected value compared to the incident value (see Fig. 25).

Specific reflectance (i.e., the highest value attainable considering costs) is important from the standpoint of overall system efficiency. The polygon is one of several components in most systems that are optimized to make the most efficient use of the available light. This maximum reflectance value is normally determined by application of an enhancement coating on top of the micromachined or polished surface. Reflectance uniformity, on the other hand, can affect the uniformity of images read and/or recorded by the polygon, so it is necessary to achieve some value of constancy of reflection over the surface of any given facet and from one facet to any other facet on the polygon wheel. In the event that the tolerance on uniformity should be a fraction of 1%, it may be necessary to measure the uniformity of reflectance of a polygon before the enhancement coating is applied. In practice, the reflectance uniformity of rather large polygon facets (e.g., a few inches square) is more difficult to achieve to high tolerances than if the facet is small (e.g., 1/2 in. square). This phenomenon has to do with the length of time of processing of the facet and the degree of difficulty of achieving a number of characteristics

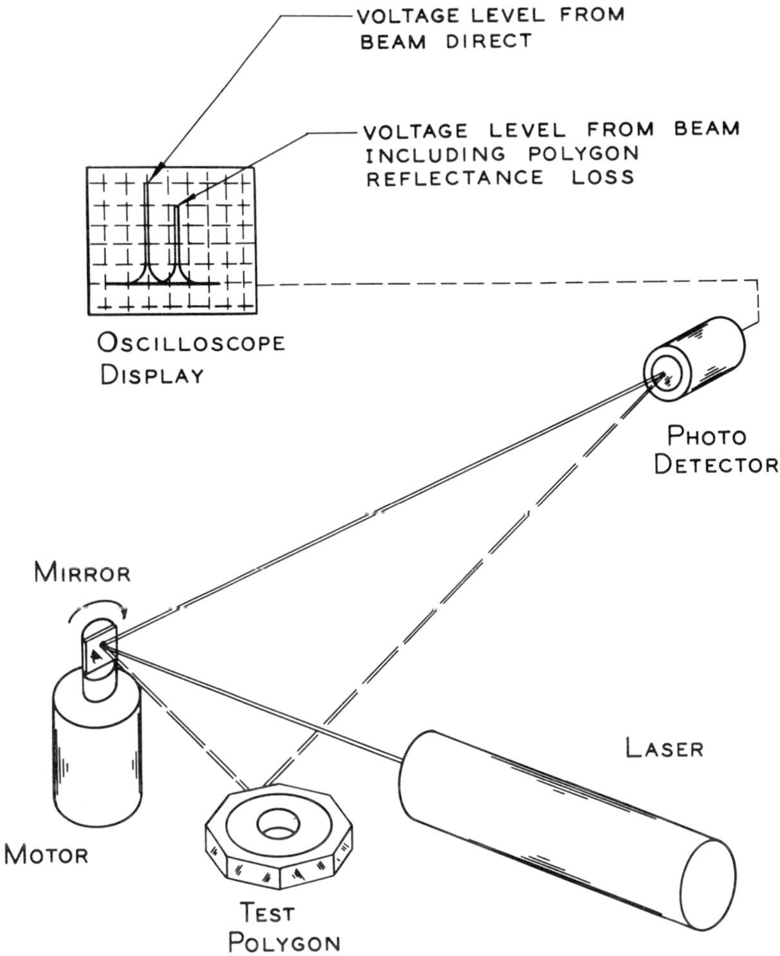


Figure 25 Reflectance test optical system.

simultaneously, such as reflectance uniformity, flatness, facet-to-datum angular tolerance, and facet polar location.

When measuring reflectance uniformity of polygon facets it is good practice to use an optical aperture which reproduces that expected in the system. This will eliminate unwanted contributions by unused areas of the facet. In other words, instead of illuminating the entire facet with incident radiation, the incident radiation should be limited to the effective clear aperture of the facet. As a case in point, polygons used in correction optical systems that produce a line image on the facet have on effective clear aperture that is a very small fraction of the total facet area; this is really the area of interest. In this case a measurement of the total facet may produce unsatisfactory results, whereas measurement of the useful clear aperture may produce satisfactory results.

The measurement of specific reflectance versus the measurement of comparative reflectance (i.e., uniformity) involves two different problems of measurement. In the case of measuring specific reflectance it is necessary to compare the measured reflectance against 100%. It has been the author's experience that sending the same sample to a number of laboratories for a measurement of the specific reflectance in a certain wavelength brings results that vary a few percent one from the other. If reflectance is a critically key element in the successful use of a polygon in a system, it is suggested that supplier and customer routinely and carefully calibrate their reflectance measuring equipment and frequently correlate supplier's and customer's measurements. In a production situation where a good deal of product is being manufactured, sold, and used, this is a good general rule to follow. Reflectance and uniformity are one matter of concern in the process of manufacture and acceptance. They are another matter over the long haul after the polygon has been installed in the system. In use the polygon may be subjected to contaminants from a variety of sources. Typical offenders are cigarette smoke, toner, bearing-oil vapors, paper dust, and fingerprints. Only careful handling and careful attention to the details of the environment in which the polygon operates will maintain high levels of reflectance and uniformity for reasonable periods of time.

It is important to mention that uniformity of reflectance of polygon facets and of mirrors in general is more difficult to achieve at shorter wavelengths than at long wavelengths. At wavelengths in the region 9-14 μm, for example, it is relatively easy to achieve extremely high levels of uniformity. This is due to the fact that the quality of surface necessary to appear specular at rather long wavelengths is easy to achieve. At wavelengths in the range 0.2-0.4 μm, it is a different matter altogether, and very small surface defects disturb the apparent specularity of the surface.

### 4.7. Enhancement Coatings

The matter of optical enhancement coatings is a subject unto its own and we will deal with it here only with regard to the practical application of a few coatings to polygonal mirrors. The characteristics required of coatings to enhance the reflectance of polygonal mirrors are:

High reflectance

Good uniformity

Good adhesion

Good life

There are two major functions of enhancement coatings on polygons: to improve the reflectance of the polished substrate and/or to improve its durability. In the case of diamond-machined polygons, the substrate is usually aluminum (in itself a good reflector over most of the visible and infrared spectrum). A $\lambda/2$-thick coating of silicon monoxide provides a protective "window" on the facet. In the case of replicated surface (epoxy), the enhancement coating adds both reflectance improvement and durability. Nickel-plated aluminum polygons are coated primarily to enhance reflectance, since nickel is a relatively durable surface in its own right.

In most cases the substrate material in a polygon is metal. In a number of cases the base material will be plated to accommodate polishing and to enhance durability. The ground rules on applying vacuum-deposited enhancement coatings to multifaceted metal and/or plated metal polygons are somewhat different from those that may be applied to coating glass or Pyrex or quartz optical elements.

Vacuum-deposited coatings for polygons are usually comprised of two or more layers of material, each a fraction of a wavelength of light thick. They are applied by evaporating or sublimating the materials in a vacuum chamber. The materials disperse in the chamber, adhering to everything that is clean and is in the line of sight of the source.

As a practical matter, coatings must be applied while the polygon rotates in the path of the vapor in order to expose all of it to the line of sight of the source and to ensure uniformity of application. Several polygons may be coated simultaneously on a single rotating shaft.

The first layer deposited is a metal, such as aluminum, silver, or gold. The remaining layer, or layers, is a dielectric material, such as silicon monoxide, magnesium fluoride, or thorium fluoride. The metallic layer is selected to provide the best reflectance at the wavelength of interest. The dielectric layers are added to protect (and in some cases, enhance) the underlying metal.

The index of refraction of some dielectric materials applied can be controlled usefully by controlling (elevating) the temperature of the polygon. A note of caution is in order that the polygon temperature not be elevated to an otherwise destructive temperature during coating. For example, nickel-plated aluminum polygons will distort (go unflat) and visibly "orange peel" if elevated beyond 200°C after polishing. Some aluminum (unplated) alloys will detemper and distort at 310°C.

Silver exhibits very desirable reflectance characteristics over a broad spectrum and is frequently considered as a polygon coating. In practice, however, it is frequently a disappointing choice over the long haul. The slightest pinhole (or minute scratch from cleaning) will expose the silver to reactive contaminants from the atmosphere, which over time (several days or weeks) will diffuse into the silver, producing an ever-increasing size of blemish. Aluminum, which initially exhibits somewhat lower values of reflectance than silver, is far superior in terms of durability.

Crucial to all of the desired characteristics of the coating of a polygon is its cleanliness prior to and during coating. Irrespective of fabrication method, polygons will be handled prior to coating during inspection processes, transport, and installation into the coating chamber. Few coatings can be successfully removed if they are improperly applied or applied on a dirty surface.

The polygon must be cleaned thoroughly to remove foreign material that will degrade surface quality, prevent good coating adhesion, or outgas in the coating system. Typical offenders in achieving this level of cleanliness, and suggested methods of cleaning, are shown in Table 5.

### 4.8. Polygon Durability

In use, a polygon is a rotating part, usually running at several thousand rpm. As such it will have an exceptional opportunity to come into contact with the atmosphere surrounding it. This atmosphere generally contains varying degrees of such bothersome contaminants as cigarette smoke, paper toner, oil mist from the motor bearings, vapors, paper dust, ordinary dust, and others. Good design practice would dictate that the polygon and other delicate optical components in the scanning system be housed in an enclosure to minimize this exposure. Measurements of reflectance of polygon surfaces operated for periods as short

Table 5 Contaminants to Be Alert for on Polygons to Be Coated, Together with Suggested Cleaning Method

| Contamination source | Cleaning method |
|---|---|
| Water vapor on and in the surface (small water droplets resist evaporation for long periods of time at $1 \times 10^{-6}$ torr vacuum) | Glow discharge vacuum chamber with polygon |
| Oil and debris from machining in tapped holes | Retap—ultrasonically clean in Freon TF solvent |
| Mineral residue from dried solvents or water | Clean with cotton[a] and distilled water or solution of 5 drops of liquid detergent in 1 pint of distilled water; then wash with distilled water; rapidly remove residual water with alcohol on a cotton swab |
| Fingerprints (*note*: remove immediately, as etching may occur from some skin oils) | Cotton and acetone or denatured alcohol |
| Stains (*note*: examine microscopically to determine presence of etching, which cannot be removed except by resurfacing) | Cotton and acetone, kerosene, methyl ethyl ketone, or as above for mineral residues |

[a]A good-quality, multiple-layer, soft toilet tissue is an acceptable alternative.

as 50-60 hr at several thousand rpm (unenclosed) in an ordinary office environment (including a modest number of cigarette smokers) show degradation as great as 20-30%.

With proper system design, polygons may be expected to provide 15,000-20,000 hr of operating life over periods of 5 years or longer even at speeds of 25,000-30,000 rpm. Periodic maintenance of the system will include cleaning of the polygon facets. Maintenance and service personnel must be provided with training and procedures for cleaning the polygon without degrading its quality.

It is good practice to design the scanning system so that the polygon motor assembly may readily be removed (and replaced without necessity for optical alignment) for examination and cleaning of the facets. In systems that are less than desirably clean, the polygon will be the greatest contributor to light loss because of the high-velocity airstream at its perimeter.

The foregoing factors should be considered in selecting the type of polygon to use at the outset. There are trade-offs between cost and durability. In this case life-cycle cost should be traded off against initial cost.

## 5. METHODS OF ROTATING POLYGONS

### 5.1. Pneumatic Drives for High Speed

Much of today's scan mirror technology has evolved from the development of ultrahigh-speed polygon/turbine motors for the high-speed photography industry. Compressed air turbines continue to offer an attractive method of rotating a polygonal mirror at speeds beyond the capability of electric motors. The advantages of turbine drives are:

Substantial horsepower can be delivered to the scan mirror to produce rapid acceleration and very high speed (up to 1,000,000 rpm).

They are compact in size and low in weight in proportion to delivered power.

Some versions are equipped with shaft seals so that the scan mirror can be used in a modest vacuum.

The disadvantages of turbine drives are:

They require a compressed air source.

They are asynchronous devices.

They are costly.

They have a relatively short total running life.

Pneumatic drives are only recommended for short duty cycles and where ultrahigh speed is essential.

## 5.2. Motor Drives for Intermediate and Low Speed

Four types of electric motors are commonly utilized for rotating polygonal mirrors:

1. Hysteresis synchronous
2. Induction synchronous
3. Induction slipped
4. Brushless DC

Most applications are in the fractional-horsepower category, so the motors fall into the class of "instrument" as opposed to "utility" motors. Hysteresis and induction synchronous motors synchronize with the AC-induced magnetic flux that drives them. Brushless direct-current (DC) motors run at that speed at which the drive torque (DC dependent) matches the resistance torque of the load they are driving. In the case of rotating polygons, the dominant resistance is air drag.

In all four types of motor the stator is essentially the same, consisting of a multipole laminated iron core wound with a number of coils of insulated copper wire (see Fig. 26). The number of poles (terminated coils) ÷ 2 = the number of times the rotor will go around per cycle of current into the stator.* Therefore, four-pole motors (per phase) run at half the drive current frequency, six poles at one-third, and so on. Using more poles than eight poles is usually impractical. If you want to run the scan motor from the mains, you are limited to the frequency of the mains or fractions thereof of 1/2, 1/3, or 1/4. If you wish to run at other speeds, an inverter must be used to create the desired frequency. An exception to this case is the brushless DC motor, wherein the excitor for commutation (current inversion) is on the rotor so that the stator frequency is always synchronous with (and leading in phase) the rotor frequency.

The rotors of the four motors are substantially different in form and function. The rotor of a hysteresis motor is usually fabricated from a single piece of hardened steel selected out of a

---

*This is also true of brushless DC motors when considering that an electronically commutated inverter is built into these motors to provide AC to the winding.

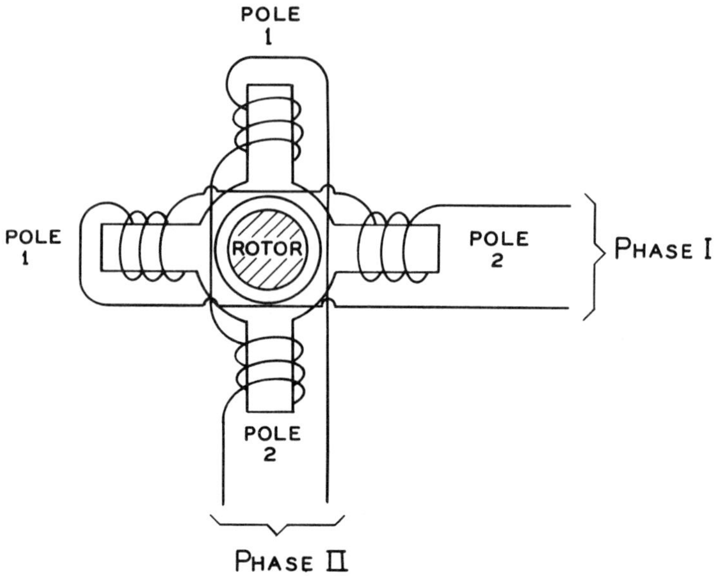


Figure 26 Schematic diagram of AC motor stator.

group (predominantly alloyed with cobalt) that exhibit substantial hysteresis loss. This resistance to the movement of magnetic flux in the material imparts torque to a rotor out of sync with the drive current. This torque is responsible for the motor's ability to start rotation. When the rotor approaches the speed of the stator flux, it becomes permanently magnetized and "locks in" to synchronism with the drive. If the motor is turned off and restarted, the stator flux demagnetizes the rotor and hysteresis takes over again. The synchronous mode of function is more efficient than the hysteresis or startup mode, and in many systems a sync detector is used to reduce drive current after the motor is locked in to save energy and minimize motor heating.

Induction slip motors contain conductor bars in the rotor. Magnetic flux from the stator induces current flow (and therefore magnetic flux) in the conductor bars. The stator flux attracts the rotor. The amount of torque produced is related to the amount of slip or velocity difference between stator field and rotor. Velocity servo systems are used to control the speed of these motors when high accuracy is required.

Induction synchronous motor rotors are more complex in their construction. They contain conductor loops or coils in which current flow is induced by the moving stator flux. This current flow, in turn, produces magnetic flux in the rotor. Torque is developed by magnetic attraction between stator and rotor. Since this torque diminishes as synchronism is approached (due to reduced flux motion in the rotor coil), the rotor must be equipped with permanent-magnet poles to achieve synchronism. Such a motor is a combination of induction slip and permanent-magnet sync.

The rotor of a brushless DC motor is constructed of a shaft, usually made of steel, with a permanent-magnet ring pressed on to it. Both starting and running torque is imparted to the rotor by magnetic attraction.

As an aid to the system designer, the major significance of the design differences among these four motors is given in Table 6. For references dealing with these design parameters in further detail, see the list at the end of the chapter.

Selection of a motor for a polygonal scanner application requires that a detailed performance requirement be generated and a trade-off study be done to weigh the advantages and disadvantages of available designs. The major issues of that trade-off will probably be speed stability and the mechanical rotational accuracy of the output shaft (hence the accuracy with which the polygon will be rotated relative to its datum surface). For systems employing correction optics to correct for polygon tracking errors (see Sec. 3.3), the trade-off may well change to one of motor/controller complexity and cost. For systems requiring accurate tracking (10 arcsec or better), the trade-off will be one of rotor integrity. Complex rotors, those made up of material of various thermal coefficients of expansion, are problematical in this accuracy range. For systems requiring a high degree of velocity stability of the polygon/motor assembly, it may be necessary to close a feedback loop around the motor/controller combination, utilizing a shaft encoder on the motor shaft.

Synchronous motors exhibit a characteristic called phase jitter (hunting). The rotors behave as though they were coupled to the drive waveform by a spring. Within synchronism the rotor springs forward and back in phase at a rate determined by the spring rate (flux density) and the torque/inertia ratio of the system. Typically, the frequency of this phase jitter is in the range 0.5-2 or 3 Hz, at an amplitude of a few degrees (1-6° peak to peak). Under perfect conditions this jitter damps to zero values of amplitude. However, perfection is seldom seen and continual recurrence of jitter may be expected, caused by electrical transients on the input, mechanical shock to the assembly, variable resistance torque of the motor bearings, and so on. For many

Table 6 Design Differences Between Scan Motor Types

| Scan motor type | Prime power | Speed control[a] | Velocity stability | Rotor configuration[b] | Speed capability[c] (rpm) | MTBM[d] | Comparative cost |
|---|---|---|---|---|---|---|---|
| Hysteresis synchronous | AC: one-, two-, or three-phase from mains or separate inverter | Synchronous | Opened loop 0.5-0.01% with feedback externally supplied 0.01-0.002% | One-piece hardened steel | 900-120,000 | 12,000 hr at 24,000 | 1 |
| Induction synchronous | AC: one-, two-, or three-phase from mains or separate inverter | Synchronous | Opened loop 0.5-0.02% with feedback externally supplied 0.01-0.005% | Complex: several materials | 900-24,000 | 12,000 hr at 16,000 | 1.3 |

| | | | | | | | |
|---|---|---|---|---|---|---|---|
| Induction slip | AC: one-, two-, or three-phase from mains or inverter | External feedback loop | Opened loop: load dependent; closed loop: 0.5-0.005% | Laminated rotor with copper or aluminum bars | 1000-36,000 | Bearing dependent | 1.5 |
| Brushless DC | DC | External feedback loop | Opened loop: load dependent; closed loop: 0.01-0.002% | Two-piece shaft/magnet | 0-30,000 | Various (dependent on electronic component reliability of internal inverter) | 1.5 |

[a] See the text with regard to phase stability.
[b] Depends on tensile strength of rotor magnet.
[c] Independent of applied load.
[d] Mean time between maintenance; measured values.

systems the 0.5-0.01% velocity error contribution of phase jitter is acceptably small. For others it is not, and feedback should be employed to reach acceptable values. Figure 27 shows a typical feedback loop configuration for motor speed control.

As mentioned earlier, brushless DC motors, due to rotor-excited inversion of current to the stator (equivalent of brush commutation), have no synchronous relationship with the world outside their housing. Control of these motors is done by varying the input current (voltage). The speed, accuracy, and stability derived by simply applying a fixed DC input current are seldom satisfactory for scanning requirements. It is necessary to monitor the rotor speed and vary the input voltage to regulate speed. This feedback loop resembles the loop used to control AC motors in principle but is different in one major aspect. Input regulation is applied to DC rather than AC.

## 5.3. Bearing Systems for Accuracy and Longevity

The following is a brief discussion on scanner bearings based on the author's experience. For a more in-depth treatment on bearings, refer to Chapter 4.

The current state of the art of accuracy in rotary bearings lies in air-lubricated and oil-lubricated journal bearings. High-precision oil-lubricated journal bearings have developed primarily in the machine tool industry, where minor leakage of lubricating oil under pressure is taken for granted. In systems employing sophisticated optics and electronics, oil is a problem. Air

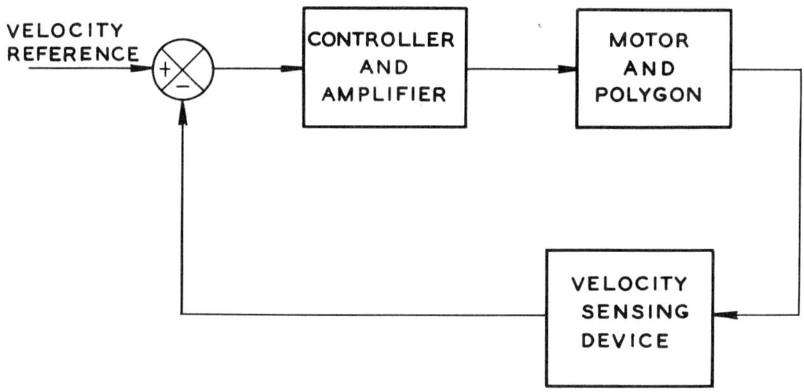


Figure 27 Block diagram of feedback loop for motor velocity control.

bearings, on the other hand, may provide sub-arcsecond tolerances of rotational precision, and they are clean. They are also expensive and generally require substantial support equipment.

Hydrodynamic air bearings require no external pressurization. The rotating shaft is designed to draw in and pressurize ambient air to form a dense lubricating (contact preventing) layer between rotor and stator. This type of bearing has little or no lubrication at low or zero speed. The rotor and stator are made of materials with low friction coefficients to limit wear during stopping and starting. This type of bearing is useful primarily in continuous-duty service.

Hydrostatic air bearings require external pressurization and as long as that pressurization occurs, no contact is made between rotor and stator. During rotation this bearing requires a constant supply of dry, clean, pressurized air. This bearing type is useful for continuous or intermittent duty.

Air bearings of the size and performance range suitable for polygon/motor support are custom designed and manufactured and as such currently fall in the cost category associated with exotic, nonproduction applications.

Precision instrument class ball bearings (ABEC 5 and 7) are currently the most common bearings being utilized in production polygon/motor assemblies, where assembly tracking tolerances of 10 arcsec or greater are common and lifetimes of 10,000 hr and more are specified. The successful use of ball bearings in a polygon/motor assembly (as in any precision instrument requiring optical-quality levels) demands meticulous attention to a number of details, including:

Proper selection of bearing size, type, class, material, and supplier

Proper selection of lubricant and amount

Careful handling during and after installation

Proper installation and fit to shaft

Correct installation in housing

Correct selection and application of thrust preload

Correct attention to rotor dynamic balancing

Careful packaging and handling of completed assemblies

R3 and R4 (3/16 in. ID × 1/2 in. OD and 1/4 in. ID × 5/8 in. OD, respectively) are the most commonly used sizes. This size range is an excellent trade-off of size versus ruggedness for high-quality bearings for the type of duty in a polygon/motor assembly. Full race bearings of 440C stainless steel in class ABEC 5 have a history of exceptional reliability.

It is noteworthy to mention that the only difference between ABEC 5 and ABEC 7 ball bearings is the amount of radial play between inner and outer races. In scan motor use the bearings are thrust loaded to remove all radial play, and in practice ABEC 5 is preferred even though it has a greater radial play. The reason is its ability to "digest" larger particles of contamination without damage. In the occasional situation where preloading is impractical, ABEC 7 is suggested.

Two basic lubrication techniques are applicable to scan motor bearing use. One technique utilizes porous phenolic as the ball separator in the bearing. It is vacuum impregnated with oil, becoming a reservoir of long-term supply. This technique is suggested for polygon/motor assemblies operating at 3600 rpm or less. At higher speeds the oil tends to leak out of the phenolic (due to centrifugal force) and then out of the bearing. The second lubrication technique is to use a grease such as Andock C.* This technique is much less susceptible to leakage out of the bearing and more satisfactory for extended high-speed use.

In very precise scanning applications the amount of radial play in the ball bearings (less thrust preload) may exceed the values desired and thrust load must be used. This is simply done by application of springs (wavy washers or Belleville springs) located axially between the bearing outer race and the assembly housing. This thrust load is parasitic and should be minimized. The mechanism that causes the rotating shaft to exercise radial play is dynamic imbalance. Therefore, the better balanced the rotor is, the less thrust preload must be applied to the bearings and the longer the bearing may be expected to last.

The polygon/motor application of bearings is a relatively gentle use. The operating environment is one of small load (primarily the gravity load of the rotor, i.e., a few ounces or pounds, or in high-speed cases, the centrifugal ball load or imbalance load), usually clean, and hopefully low thermal condition. The major causes of failure, nevertheless, are (in order of importance):

---

*Andock C is a trade name for a bearing grease manufactured by Humble Oil.

1. Abuse (brinelling due to shock from mishandling)
2. Dirt (paper dust, toner, ordinary grit)
3. Heat

A polygon/motor assembly, like a lens, modulator, laser, or other precision optical/mechanical device, will seldom survive a drop to the floor without bearing (or other) damage. Shock isolation of the ball bearings using elastic O-rings between bearing and housing can minimize damage due to abuse, not eliminate it. Shock isolation in shipping containers will limit damage. A 1-in. thickness of foam rubber surrounding a scanner assembly packed snugly with several others in a sturdy cardboard shipping box will prevent shock-induced damage in all but extreme cases of abuse.

Having survived shipment and installation into the scanning system, the polygon/motor assembly should provide several thousand hours of maintenance free service, except for an occasional cleaning of mirror facets. It is said that bearing life is lubricant-life dependent, and that is largely the case at this point. For a bearing to last for 10,000 hr or more of operation (plus many more hours at rest), the lubricant must stay in the bearing for that time. The two main reasons why lubricant disappears from the bearing are leakage and evaporation. In the two lubrication methods mentioned above, leakage normally occurs during the first few hours of operation, when excess lube is slung out. This will ordinarily occur during the test or "burn-in" period at the manufacturer's location. After that, evaporation is the enemy of life. Lubricant evaporation rate is a function of heat and ambient pressure. The vapor pressure of most bearing lubricants at 25°C is in the range 0.01-0.1 torr absolute. The vapor pressure rises exponentially with increase in temperature. As a result, it may be said that lubricant life (and therefore bearing life) is temperature dependent.

A number of things can be done to minimize operating temperature in the polygon/motor assembly, among them:

Minimize the operating drive current to that necessary to achieve performance. All current beyond that will convert to heat in the assembly.

Provide good thermal coupling between the motor housing and an adequate heat sink (the aluminum base plate of the system, for example). Use thermal conductive grease if necessary.

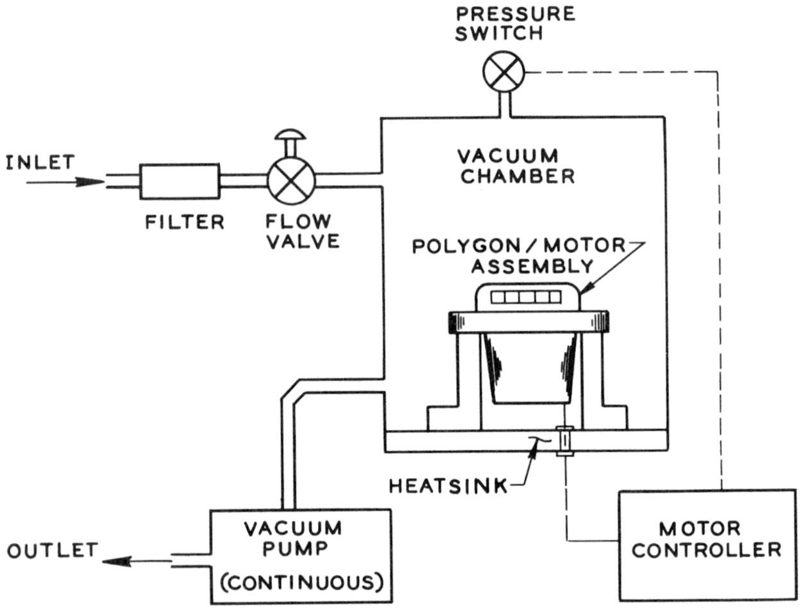


Figure 28 Schematic diagram showing method for providing cooling for vacuum operation of polygon/motor assembly.

In systems where the polygon/motor assembly must be operated in a vacuum to reduce windage and drive power, it is a good idea to configure the system to provide cooling, as shown in Figure 28. A continuously operating vacuum pump maintains a reduced pressure in the vacuum housing. Ambient air is drawn into the chamber through a particle filter (1-µm sieve type, replaceable cartridge), then a flow valve. The flow valve is adjusted to throttle the inlet to control chamber pressure. Ambient air is cooled due to adiabatic expansion into the chamber, which in turn cools the scan motor. A pressure switch is suggested to disable the motor driver in the event that the pressure drops below a set point, normally caused by a clogged filter.

### 5.4. Polygon/Motor Interface

There are two basic mechanical design configurations of polygon/motor assembly. These are known as "cantilevered" and "captured" designs. These terms refer to the location of the polygon on the motor shaft relative to the bearings (see Fig. 29).

The cantilevered approach is commonly used for relatively low speeds (10,000 rpm or less) and/or relaxed tracking tolerances (1 arc minute or greater). The primary advantage of this design is its adaptability to the use of standard commercial motors. Additionally, it is more convenient to those who elect to purchase motors and polygon mirrors from different suppliers, since the motor may be adapted for use without disassembly. The main disadvantages of this design type are:

Limited stiffness. The output shaft must be small enough to go through the bearing.

Difficult bearing replacement. The polygon mounting hub is usually interference-fitted and must be removed to replace the bearing behind it.

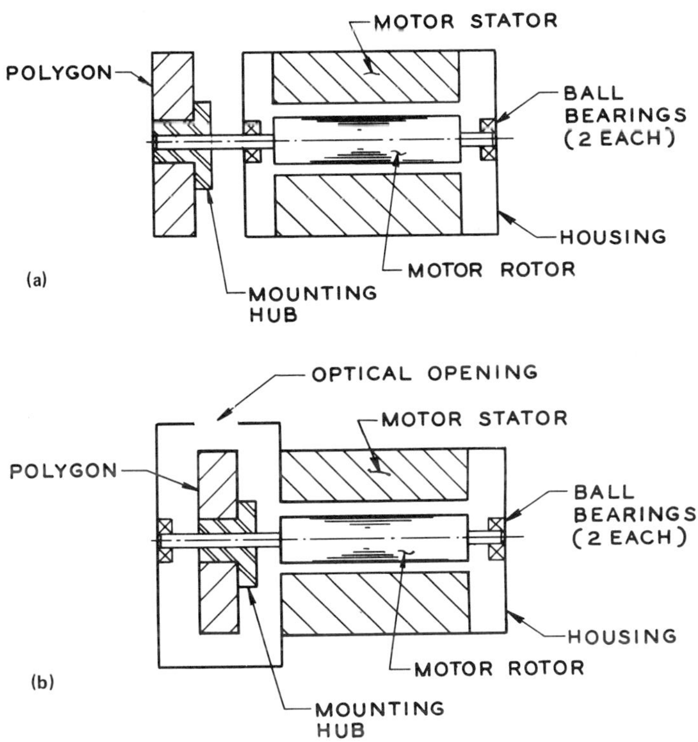


Figure 29 Cross-section drawing showing cantilevered and captured polygon/motor construction. (a) cantilevered construction; (b) captured construction.

Limited dynamic balancing capability. The motor end of the rotor is inaccessible after the polygon is installed.

The captured approach is recommended for applications requiring high speed and/or high tracking accuracy. In this approach the center-to-center distance between bearings is increased, thereby reducing the contribution of bearing runout to shaft angular runout. Also important is the fact that the motor rotor may have a larger diameter near the polygon location (bearing ID not being a limitation), providing greater stiffness and stability.

A major benefit of the captured design is that the entire rotor assembly and bearings may be assembled outside the housing and two-plane dynamically balanced. Additionally, bearings may be replaced without having to remove the polygon mounting hub. The major disadvantages of the captured design are:

Lack of adaptability to standard commercial motors

Rotor must be removed to install or remove polygon

### 5.5. Dynamic Balance

Accurate dynamic balancing of polygon/motor assemblies is important for several reasons:

To enhance bearing life

To enable accurate polygon tracking

To prevent vibrations which may adversely affect other components of the optical system

To minimize acoustic noise

When free to do so, all rotating bodies rotate about their center of mass. If the center of mass does not coincide with mechanically defined rotational axis, an unbalance force is created. This is a force of restraint coupled through the rotor bearings to the bearing holders, in this case the polygon/motor housing. It is known as dynamic imbalance and is expressed in terms weight/radial distance (i.e., oz/in. or g/cm). This force increases with speed. The relationship of the factors influencing imbalance is given by

$$\text{Force exerted against the shaft} = \frac{W\omega^2 r}{g}$$

where

$W$ = weight of the rotating body

$\omega$ = angular velocity, rad/sec

$r$ = distance from the rotational axis to the center of gravity

$g$ = acceleration due to gravity = 32.16 ft/sec$^2$

Polygon/motor assemblies are dynamically balanced on a two-plane balancing machine similar to that shown in Figure 30. This machine supports the opposite ends of the rotor to be balanced on two separate motion transducers whose outputs are amplified and filtered and read out on meters on the machine control. Tho unit is precalibrated for rotor weight and speed.

Signal filtering is employed to eliminate frequencies of vibration occurring at othor than rotor speed. The output of the motion transducer is also used to trigger a strobe light to provide visual cues to the operator as to the phase relationship of the measured unbalance load.

Having measured the amplitude and polar location of the unbalance load at each end of the rotor, weight is added or subtracted from indicated locations to alter the location of the mass axis to more nearly coincide with the rotational axis. Grinding or drilling away small amounts of material from the "heavy side" of the rotor, although a common technique employed in motor manufacture over the years, is not recommended. The opportunity for bearing contamination is too great with this method. A preferred approach is to provide a number of equally spaced, radially drilled and threaded holes at each rotor end. Small machine screws are inserted in appropriate locations and numbers to achieve balance.

The range of dynamic imbalance maximums most commonly specified for polygon/motor rotors is 0.005-0.0005 oz/in. at n rotations per minute, where n is in the range 3600-24,000 rpm. Since the effect of imbalance increases as the square of the rotor speed, the importance of careful balancing increases by the same ratio. Attempts at perfection in balancing at high speed are frustrated by lack of rotor stiffness and thermal inhomogeneities in the rotor materials; further reason for design simplicity and attention to symmetry.

Figure 30 Polygon/motor assembly being dynamically balanced on a two-plane balancing machine.

### 5.6. Electronic Controls for Speed

Figure 31 shows a schematic of a typical two-phase AC synchronous motor controller and a time diagram of its output voltage waveform. Scanner motors are typically fractional-horsepower, low-voltage devices, operating in the range 10-70 V peak to peak and drawing current in the range 0.5-5 A rms per phase. These operating ranges are well within the capability of semiconductor driver amplifiers operating in the square-wave mode.

The primary component of reactance in a synchronous motor is inductance, usually in the range 10-50 mH. Table 7 gives a list of the reactive components of a typical hysteresis scan motor over a range of operating speeds. The significance of the inductive load characteristic is twofold. First, the drive current will lag the voltage delivered to the motor winding by the driver, resulting in a power factor of less than 1. As a result, more current must be delivered to the motor to produce torque than if unity power factor were maintained. For the case where the motor is to be driven at a fixed speed, capacitance may be put in parallel with the driver output to the motor to tune the circuit at operating frequency and improve power factor.

Second, the inductive motor load will distort (round off) the current waveform, particularly at higher frequencies (120 Hz and greater), reducing losses due to stator heating. Similar current and voltage waveform distortions occur in sine-wave drivers in this use, thereby diminishing their desirability as compared to rectangular wave systems.

## 6. TESTING POLYGONAL SCANNERS FOR PERFORMANCE

Component-level measurement and testing are discussed in Sections 4.3-4.6. Assembly-level testing may be approached in one of two ways:

1. A series of individual and combined tests to confirm conformance to a detailed set of specifications, item by item
2. A single "in system" test to verify acceptability on the basis of use

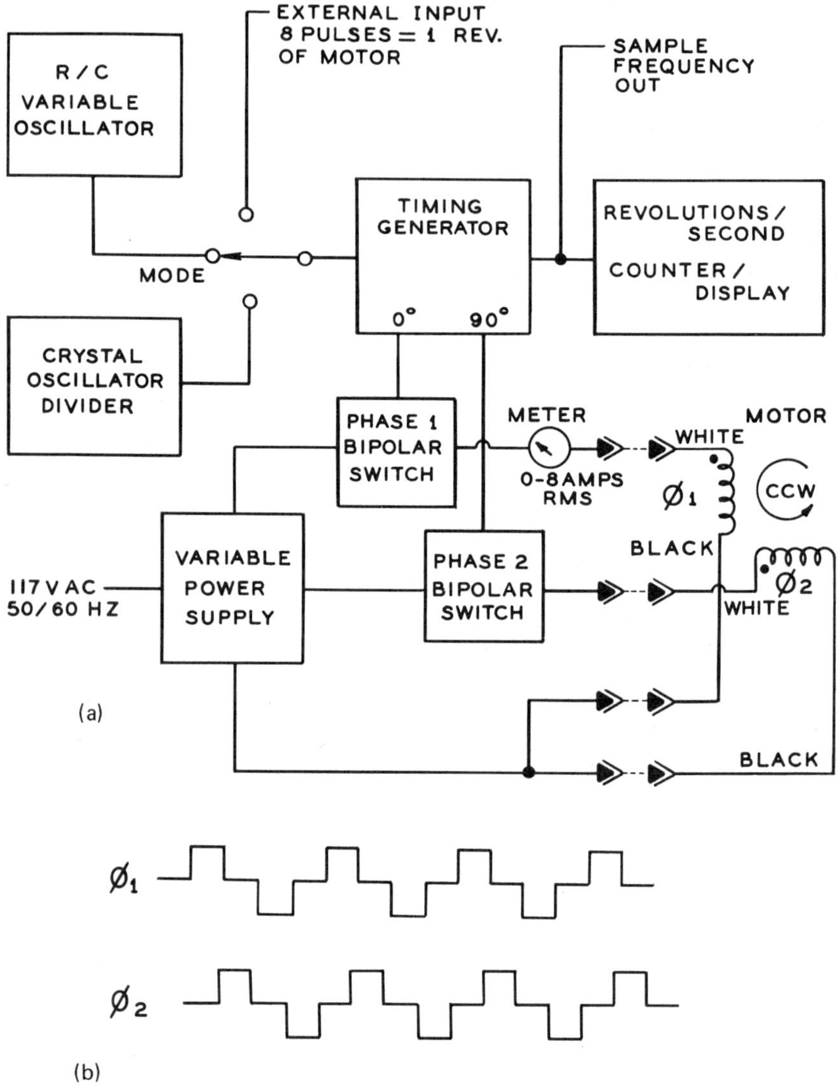


Figure 31 Schematic diagram of a two-phase synchronous motor controller circuit and output diagram. (a) Circuit schematic; (b) output diagram – voltage waveform.

**Table 7** Reactive Components of a Typical Synchronous Scanner Motor

Typical Design Characteristics - Hysteresis Synchronous Motor

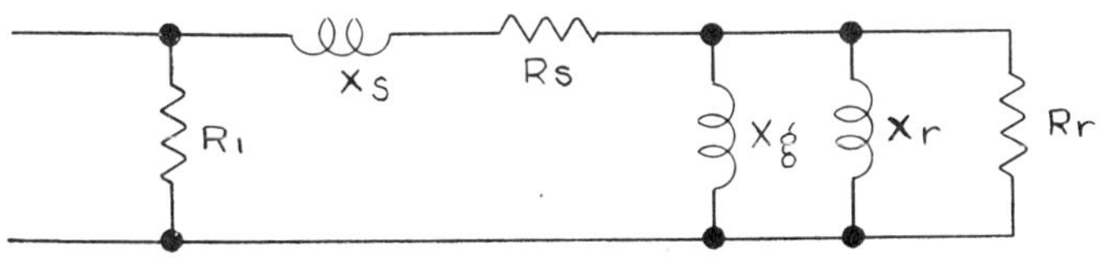


Equivalent Circuit of Hysteresis Synchronous Motor

$R_r$ is the power component of current drawn by the ring. Figuratively this resistor has a shaft sticking out of it, and $I^2R_r$ is horsepower, not heat.

$X_r$ and $X_g$ account for the current drawn to magnetize the ring and the gap, respectively. Values given are ohms reactance, not henries inductance, and therefore vary with frequency.

$X_s$ is reactance from flux that links the stator windings but never gets to the rotor. Small but not negligible. Also ohms-reactive, varying with frequency.

$R_s$ is the resistance of the stator copper. It is the only element that can be measured directly with a meter, and the only one that does not vary with frequency.

$R_i$ represents eddy-current and hysteresis loss in the stator iron. It is not negligible, 8.5 watts at 25,000 RPM, compared with 1.9 watts in the stator copper.

$R_r$ varies with frequency, just as the reactances do. If you hold torque constant but vary speed, you must hold I constant and vary the value of $I^2R_r$. If torque also varies with speed, $R_r$ will vary as $(\text{speed})^2$.

$R_i$ varies as $(\text{speed})^{1.5}$, with current held constant.

All these can be combined into two parallel components, $X_p$ and $R_p$, with values as tabulated below.

| RPM | 25,000 | 15,882 | 10,002 | 6,300 | 3969.6 | 2,500 |
|---|---|---|---|---|---|---|
| Frequency, Hz | 416.7 | 264.7 | 166.7 | 105.00 | 66.16 | 41.67 |
| Parallel Resistance | 12.57 | 7.450 | 4.307 | 2.454 | 1.369 | 0.750 |
| Parallel Reactance | 5.847 | 3.734 | 2.374 | 1.520 | 0.985 | 0.648 |
| Impedance | 5.300 | 3.338 | 2.080 | 1.292 | 0.800 | 0.490 |
| Apparent Inductance (Millihenry) | 2.23 | 2.25 | 2.27 | 2.30 | 2.37 | 2.47 |

The above values are for just before lock-in, when motor is delivering rated torque at nearly rated speed. This is the only place they can be computed, as after lock-in, impedance is a function of load, past history of load, and past history of excitation.

Figure 32 Polygon/motor final test setup.

Each method has merit in certain areas of application. In some instances both types of testing are necessary to establish the acceptability of a polygon/motor assembly in order to find assemblies which, although correct in each item of component detail, may embody tolerance "stack-ups" as an assembly that render it "out of spec."

It is very beneficial for supplier and purchaser to agree on and implement identical procedures and test equipment at each other's facilities to verify compliance with performance specifications. This holds true for polygon/motor assemblies as well as any precision component or subassembly.

When a rigorous, effective, well-documented quality assurance system is employed to verify specification compliance at key locations and times of in-process manufacture of polygon and motor, final inspection should have only to confirm:

1. Motor behavior
    a. Current draw for specified voltage
    b. Achievement of synchronism
    c. Velocity stability
    d. Rotation direction
2. Polygon behavior: tracking accuracy

Figure 32 shows a test setup for measuring the polygon/motor characteristics as given above.

An efficient means of qualifying polygon/motor assemblies frequently is simply to install them in the system and examine the total system output. For example, in a computer output printer, one may rather rapidly "inspect" polygon/motor assemblies by altering a single printing machine to permit the rapid exchange of scanners, so that an actual test pattern may be printed with each one. This is a rapid "acid test" of the scanner. Often defects in the printout may serve as a diagnosis of a problem.

## REFERENCES

Spreadbury, F. G., Fractional HP Electric Motors.

Teare, B. R., Jr. (1940). Theory of hysteresis motor torque. *AIEE Trans.*, *59*, 907.

Veinott, C. G., Fractional and Subfractional Horsepower Electric Motors.

# 3

# Windage of Rotating Polygons

Anthony J. Lawler* and Joseph Shepherd
*Ferranti P.L.C., Edinburgh, Scotland*

## 1. INTRODUCTION

Rotary mirror scanners are often purchased as "standard" motorized spindles to whose shafts a range of mirror-faceted polygons can be fixed. This approach usually offers low unit costs and has proved satisfactory for many optical scanning systems, but for more exacting requirements these units can be wasteful of energy or space or can have inadequate performance.

Many of the advanced systems being contemplated at present require such features as more compact designs, higher scan speeds, complex rotor shapes, mirror polygons placed close to other optics, and reduced power. To incorporate such features successfully, the design of the rotating scanner must be integrated with that of the complete optical scanning system, which can be achieved only if the performance of any shape of rotor, rotating at any likely speed in any surroundings, can be predicted. Such information will allow an optical system designer to consider more fully the effect of possible optical scanning variables and to optimize the design for a particular requirement, reducing the design risk and increasing market potential.

It is the purpose of this chapter to look at the *windage* of rotating polygons, an aspect of performance normally difficult to estimate, and to describe a proven method of predicting the windage for any rotating scanner.

---

**Present affiliation:* The Kemble Instruments Co., P.L.C., Burgess Hill, West Sussex, England.

## 2. WINDAGE

Windage (refer to Fig. 1) is the rotational analog of wind resistance. It is the torque ($T_W$) required to displace the fluid medium surrounding a body when it is rotated. Sometimes it is more useful to consider *windage power* ($P_W$), which is the windage times the rotational speed ($\omega$).

It is a corollary of any object moving or rotating in a fluid medium that it will consume power. The power losses in a rotating mirror scanner are those associated with the drag of the spin axis bearings and the windage of the rotating polygon. In addition, there will be further losses in the drive motor. Of these, windage is not only often the largest component, but it can be altered substantially by seemingly small changes in configuration. Some of the benefits of decreasing the windage power losses are reduced total power input, smaller motor and drive supply, and lower

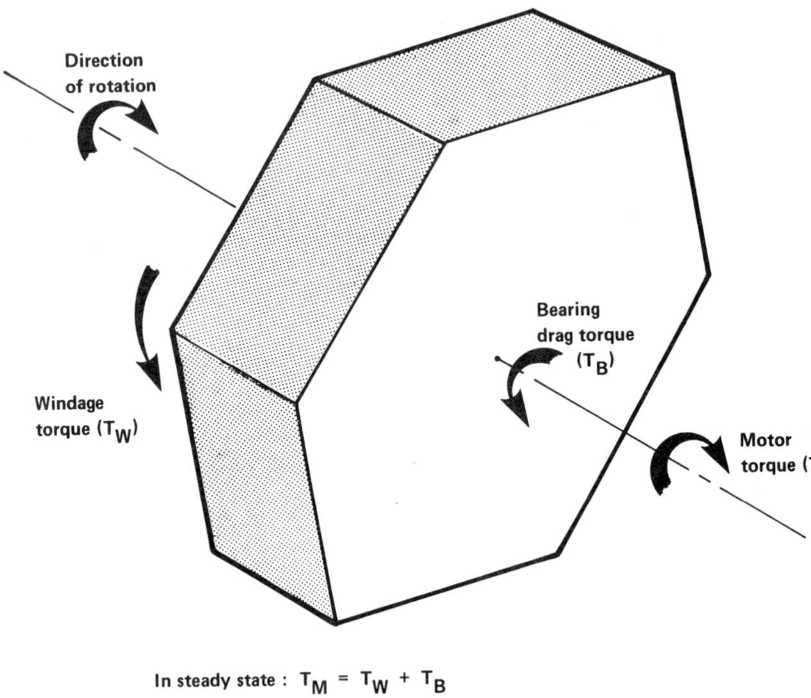


Figure 1 Torques acting on a polygon scanner in steady rotation.

running temperatures or, alternatively, higher possible rotational speeds for the same power input.

Almost no information on the windage of rotating polygons was found in a literature search. What information was available was that published by Daily and Nece in 1960 on rotating disks and that by Dorfman in 1963 on rotating cylinders. It has therefore been necessary to base this chapter mainly on unpublished work carried out at Ferranti. This has been concerned with polygonal rotors, with facets parallel to the spin axis. However, the principle and method described to evaluate windage can also be applied to polygons with facets angled to the spin axis or to angled single-mirror scanners.

Subsequent to the writing of this chapter, a short section on windage, in a paper by Hayosh in 1976, was brought to the author's notice. This is summarized in the appendix to this chapter.

## 2.1. Methods of Predicting Windage

Probably the most common method of predicting the windage of a new rotary scanner design is to base it on measured results from previous scanners, sometimes with the aid of a "windage equation" established from dimensional analysis (see Sec. 3 and the Appendix). This method is useful if the speed, size, and configuration of the proposed design are reasonably close to the previous designs measured, and where only an approximate value of windage is acceptable (e.g., where surplus motor power is available). But what are the limitations on the application of such equations? From experience it is known that small configuration changes can sometimes result in large windage changes. To establish limits on actual scanners would entail many laborious experiments with the variables of speeds, numbers and size of facets, atmospheres, and surroundings. This would be extremely costly.

An alternative approach must be looked for and is to be found in the technique of model testing. This method is strongly advocated as a relatively inexpensive and simple method where it is desired to:

1. Build up a databank of information on a wide range of polygons and surroundings
2. Minimize the windage of a particular design or to take account of complex configurations

## 3. MODEL TESTING

### 3.1. Dimensional Analysis

Model testing, which is one of the most useful techniques of fluid mechanics, is based on the tenets of dimensional analysis. Dimensional analysis states that if a list can be made of all the variables that describe some fluid mechanical situation, be it a ship moving through water or a steel ball falling through glycerine, the complexity of their interrelationships can be reduced by gathering the variables together into "dimensionless groups." These are composite variables organized in such a way that the dimensions (in this case mass, length, and time) of their constituent parts all cancel out.

Considering the case of a polygon rotating in some sort of housing immersed in a fluid, and making the fundamental assumption that the fluid is incompressible, the windage will depend on:

1. Rotor size
2. Rotor shape
3. Speed of rotation
4. Number of facets
5. Housing shape
6. Viscosity of the fluid in which it rotates
7. Density of the fluid in which it rotates

which can be characterized by a relationship of the form

$$T_W = f(\rho, \omega, r, \mu) \tag{1}$$

where

$T_W$ = windage

$\rho$ = fluid density

$r$ = a representative dimension, say, the polygon tip radius

$\mu$ = dynamic viscosity of the fluid

$\omega$ = angular frequency of rotation of the rotor

Dimensional analysis allows this to be rearranged into the form

$$\frac{T_W}{(\rho/2)\omega^2 r^5} = \frac{f(\rho\omega r^2)}{\mu} \tag{2}$$

where

$$\frac{T_W}{(\rho/2)\omega^2 r^5} = C_M \text{ (torque coefficient)}$$

$$\frac{\rho\omega r^2}{\mu} = \text{Re (Reynolds number)}$$

Therefore,

$$T_W = C_M \frac{\rho}{2}\omega^2 r^5 \tag{3}$$

and

$$C_M = f(\text{Re})$$

which is a relationship between two dimensionless quantities that can be plotted on one graph. This is a powerful simplification, and its validity depends on all the situations to which it refers being geometrically similar. Note that for each configuration, the length variable r in effect stands for all the lengths in the system, both rotor and housing.

What has been established here is that if a model of a proposed spinner is made, and tested in such a way that the Reynolds number is the same as that for the "real" rotor, the calculated value of torque coefficient will be the same as that in the "real" case. The assumption of an incompressible fluid is valid for windage, as the gas pressures generated are low and compressibility effects can be neglected.

### 3.2. Model Testing of Spinners

The real beauty of model testing as applied to high-speed spinners is that, for the same Reynolds number, a model can be run in *liquid* at much lower speeds. The following advantages of this method greatly simplify, and therefore speed up, the gathering of this information:

1. The reduced dynamic unbalance forces allow larger tolerances on parts and the use of longer reach spindles to hold the rotor. These simplify the test rig design.
2. The range of speeds is likely to be readily available on an external drive (e.g., a pillar drill).
3. It is difficult to instrument a model running at high speed in gas, especially in a sealed container required for reduced or helium atmosphere, and it is very difficult to incorporate configuration modifications. Testing in liquid eliminates these problems.

A model-testing rig which has been successfully used is shown in Figure 2. This is fairly self-explanatory, but note that the windage reacts back on the housing (labeled "Pot") and thence is applied directly to the torque transducer.

To test the validity of the model testing approach a number of direct measurements were made, at Ferranti, a windage on "real" spinners rotating in gas, and these were compared with predictions from model tests. Agreement was obtained to an accuracy of ±12% in the case of most of the points. Figure 3 shows an example of this, using the rotor shown in Figure 4.

The use of model testing to predict whether the windage of any proposed rotary scanner will fall within acceptable limits can be done in two ways. The first is to model the rotor and housing as exactly as possible, and the second is to do a number of tests modeling different rotor and housing features, to establish a database from which to infer the likely windage of a new spinner and housing design.

The first approach is valid for one particular configuration and will give an accurate prediction of windage. However, a scanner design is rarely fixed at an early stage, and where there is likely to be an optimization process, the second approach is

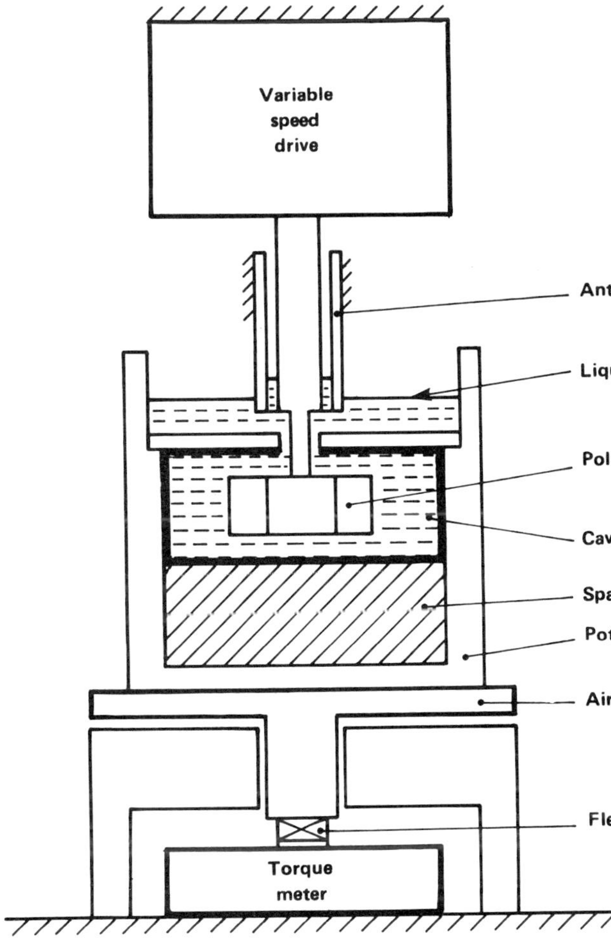


Figure 2 Schematic of windage model test rig.

probably the more useful, in that it predicts trends as well. The latter approach has been the basis of the information presented in this chapter. There is, unfortunately, not enough space to give more than a flavor of the subject, and to point out a few trends that may be of interest to designers wishing to keep rotor windage to a minimum.

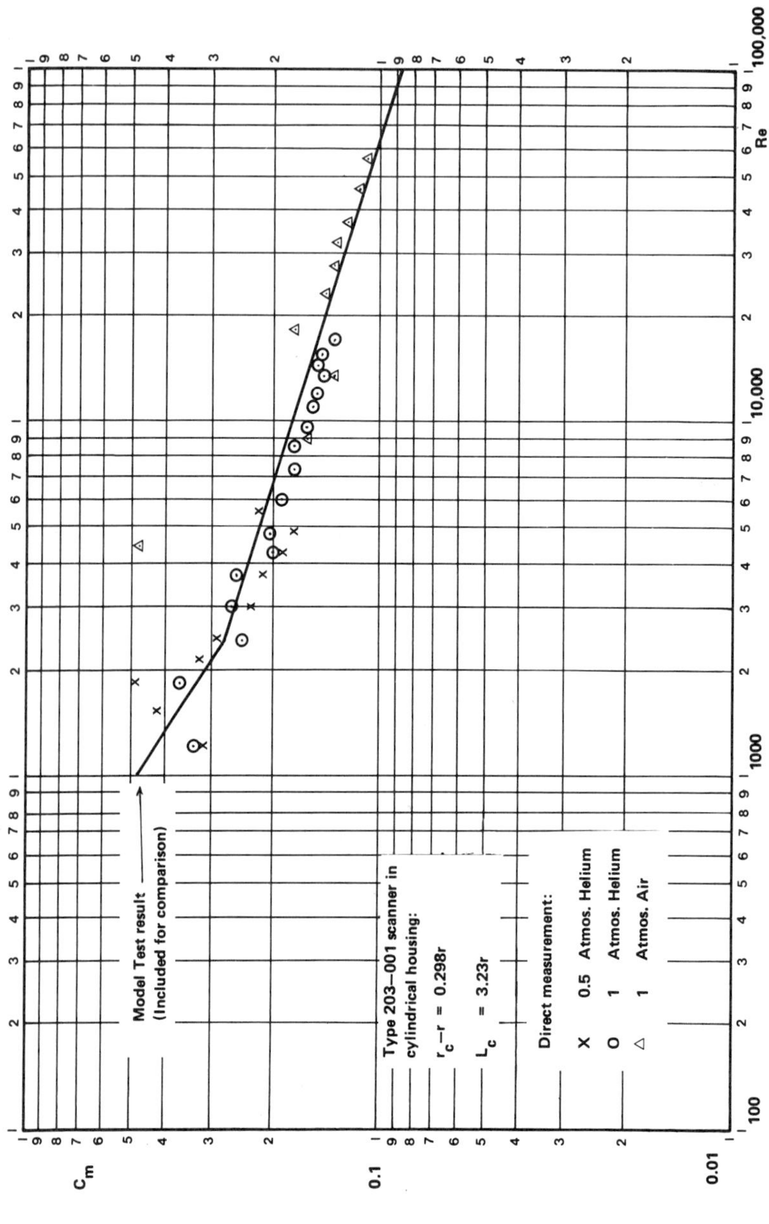


Figure 3 Comparison of model test with direct measurement.

Figure 4 Type 203-001 scanner.

## 4. RESULTS OF MODEL TESTS PERFORMED ON ROTATING POLYGONS

These tests involved only regular polygonal rotors with between 4 and 12 facets, parallel to the spin axis, but since only the broad trends of the results can be given here, the information given may be applicable to other rotor types as well.

### 4.1. Shape of Curves of $C_M$ Against Re

Curves of $C_M$ against Re were typically (although there were some exceptions) found to have the form

$$\log_e C_M = \log_e A - B \log_e \text{Re} \tag{4}$$

The constant B was found to have values between 0 and 0.83, the lower values of B indicating a relatively more disturbed flow (i.e., windage increasing more rapidly with speed). This applies over the range of Reynolds numbers used in the tests (2000-20,000), and in those cases where high values of Re (up to 100,000) were

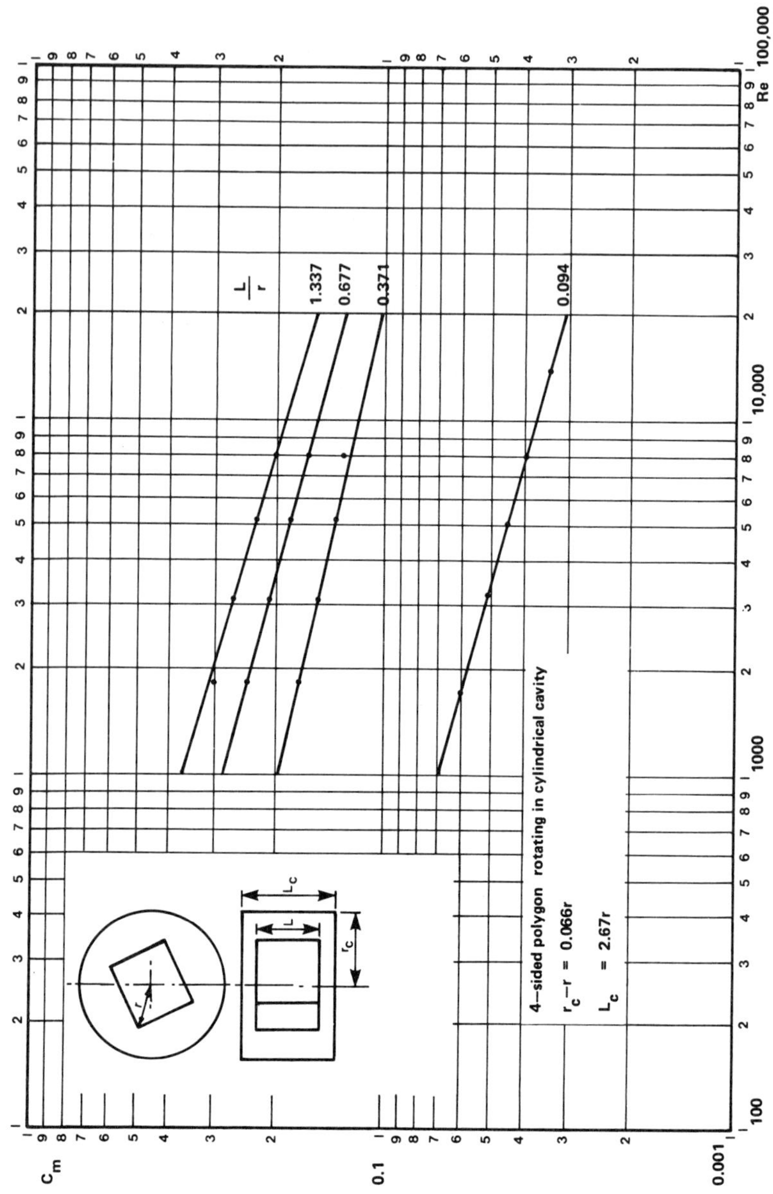


Figure 5 Typical model test results.

tested, the equation remains valid to ±10% approximately. Figure 5 shows an example, a test on a four-sided polygon rotating in a plain cylindrical cavity.

Equation (4) can be rearranged to bring it into a more useful form:

$$C_M = A\,Re^{-B} \tag{5}$$

Substituting

$$\frac{T_W}{(\rho/2)\omega^2 r^5} = A\left(\frac{\rho\omega r^2}{\mu}\right)^{-B}$$

and rearranging, we obtain

$$T_W = \frac{A}{2}\left[\rho^{(1-B)}\right]\left[\omega^{(2-B)}\right]\left[r^{(5-2B)}\right]\mu^B \tag{6}$$

or

$$P_W = \frac{A}{2}\left[\rho^{(1-B)}\right]\left[\omega^{(3-B)}\right]\left[r^{(5-2B)}\right]\mu^B \tag{7}$$

where $P_W$ is the windage power (= $T_W\omega$).

It was found that A and B vary in a complicated manner with the dimensional parameters of the system. Table 1 gives examples of the value of B for various rotors in cylindrical housings, but it should be noted that the values given will be completely upset by features causing gross disturbances to the fluid flow, such as window holes, obstructions, and baffles. Indeed, in one series of tests, on a five-sided polygon at Re = 20,000, a variation in windage of over 500% from best to worst case was obtained, merely by changing the housing shape and by using side cheeks (disks attached to the end faces of the rotor) to smooth the flow. The results of these tests are summarized in Table 2. This shows that not only do the actual values of windage vary very widely for the same rotor at the same speed, but the slope of the graph $C_M$ against Re is very variable. This means that the higher the value of Reynolds number, the more the windage is affected by obstructions to the smooth flow of fluid around the rotor.

Table 1 Values of the Parameters A and B for a Number of Different Polygons Rotating in Plain Cylindrical Housings

| Number of facets | Length-to-radius ratio, L/r | Radial clearance, $r_c - r$ | Curve parameter A | Curve parameter B |
|---|---|---|---|---|
| 4 | 0.094 | 0.066r | 0.46 | 0.272 |
| 4 | 0.371 | 0.066r | 0.90 | 0.220 |
| 4 | 0.677 | 0.066r | 1.62 | 0.255 |
| 4 | 1.337 | 0.066r | 2.68 | 0.287 |
| 5 | 0.367 | 0.062r | 0.83 | 0.228 |
| 6 | 0.374 | 0.066r | 0.74 | 0.226 |
| 7 | 0.373 | 0.065r | 0.75 | 0.251 |
| 8 | 0.425 | 0.109r | 1.03 | 0.288 |
| 12 | 0.366 | 0.066r | 3.3 | 0.454 |
| ∞ (cylinder) | 1.337 | 0.026r | 252 | 0.829 |

## 5. DISCUSSION OF THE EFFECTS OF THE VARIOUS PARAMETERS ON WINDAGE

### 5.1. Facet Size

Given a fixed scanning rate in facets per second, efforts should be made to reduce the facet size as much as possible. This is because, for any one configuration, windage is approximately proportional to size raised to the power 4.5 [see Eq. (7) and Table 1]; that is, if a spinner is scaled down by 10%, windage is reduced by nearly 40%.

### 5.2. Housing Design

Housing design is important from the point of view of windage. The best housing is a plain cylinder with the radial clearance adjusted to give a minimum value of windage. Real housings will depart from this ideal, however, and any additions to the housing in the form of window cutouts, subsidiary mirrors, or other

Table 2 Change in Windage with Changes to Rotor and Housing

| Number of facets | Length-to-radius ratio, L/r | Radial clearance, $r_c - r$ | Comments | Curve parameter A | Curve parameter B | Windage at Re = 20,000 |
|---|---|---|---|---|---|---|
| 5 | 0.68 | 0.023 | Cylindrical housing (see Fig. 6) | 1.90 | 0.298 | 100 (nominal) |
| | | 0.023 | Side cheeks (see Fig. 7) | 8.3 | 0.494 | 59 |
| | | 0.038 | Hole in side of housing (see Fig. 8) | 0.88 | 0.156 | 168 |
| | | 0.369 | Obstruction (see Fig. 9) | 0.66 | 0.110 | 204 |
| | | 2.34 | Radial baffle (see Fig. 10) | 0.70 | 0.070 | 318 |

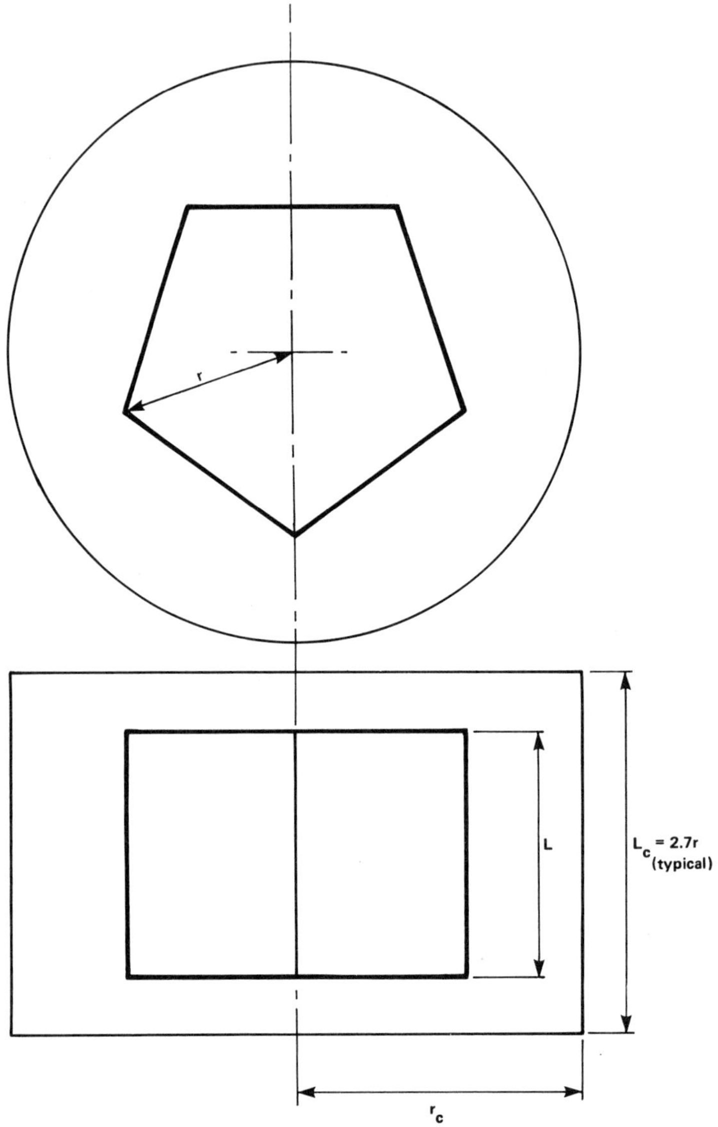


Figure 6 Important dimensions of polygon model and plain cylindrical cavity (for results given in Table 2).

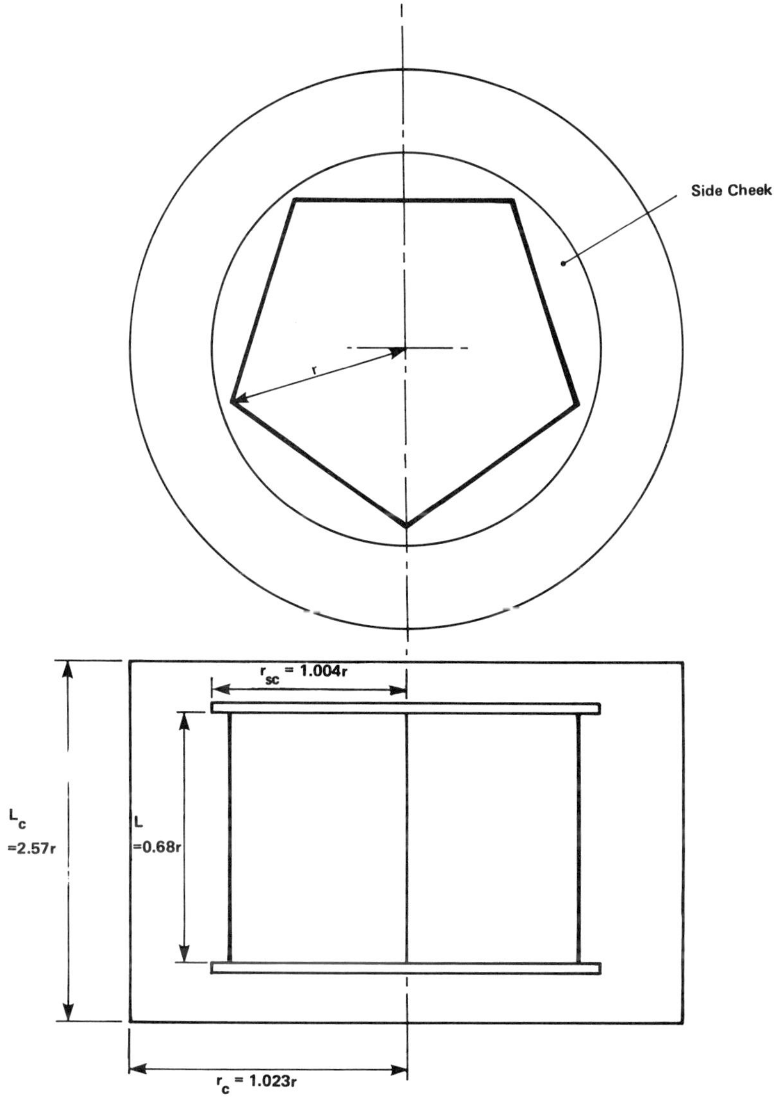


Figure 7 Important dimensions of polygon model and cavity, including side cheeks (for results given in Table 2).

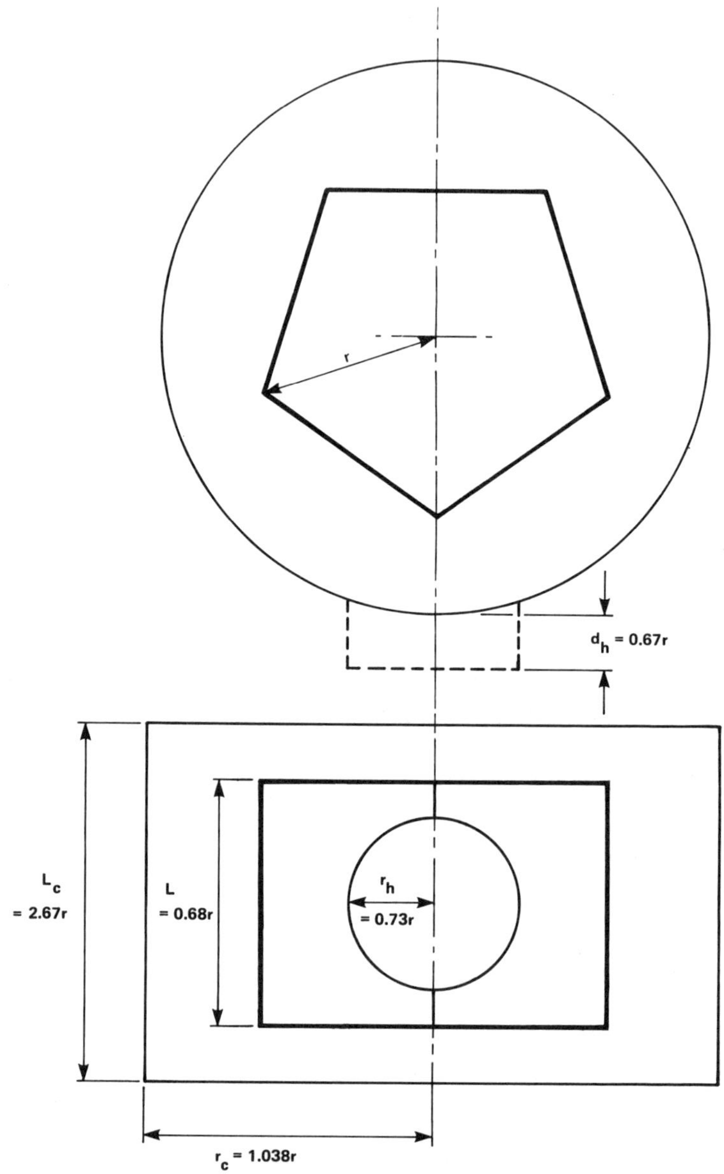


Figure 8 Important dimensions of polygon model and cavity, including window hole (for results given in Table 2).

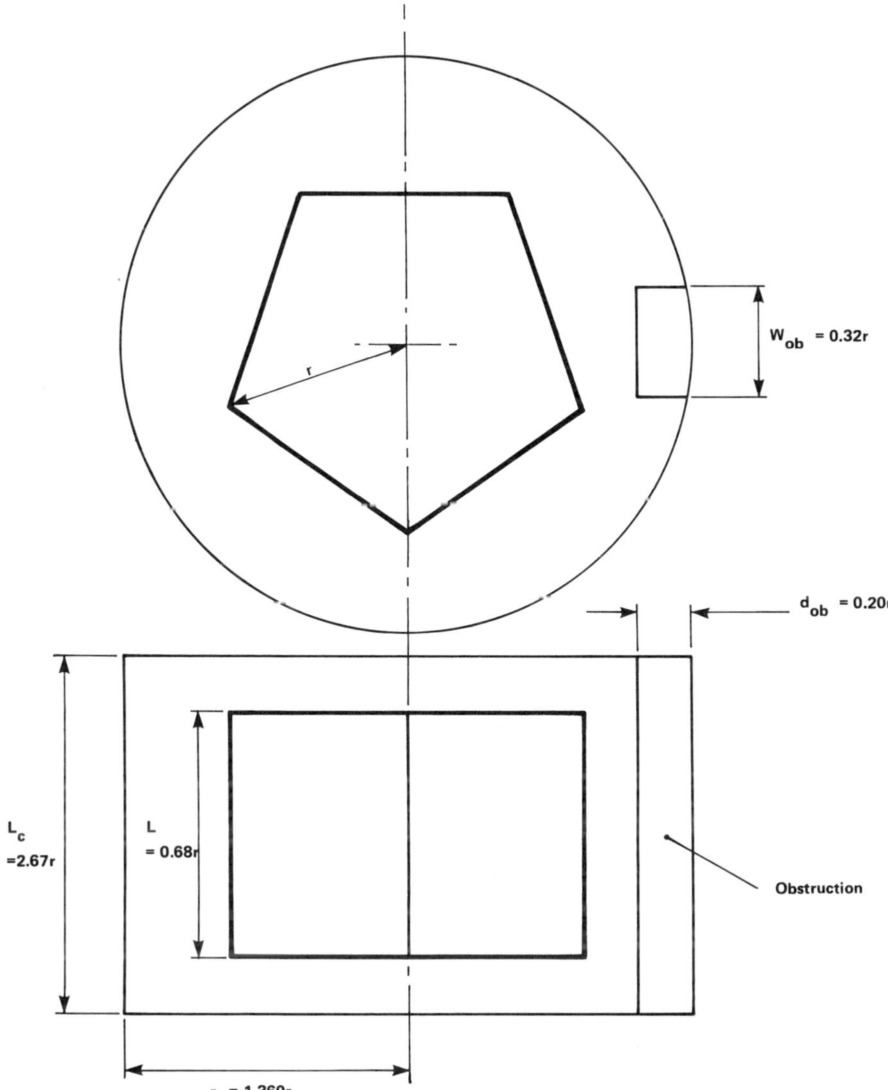


Figure 9 Important dimensions of polygon model and cavity, including obstruction (for results given in Table 2).

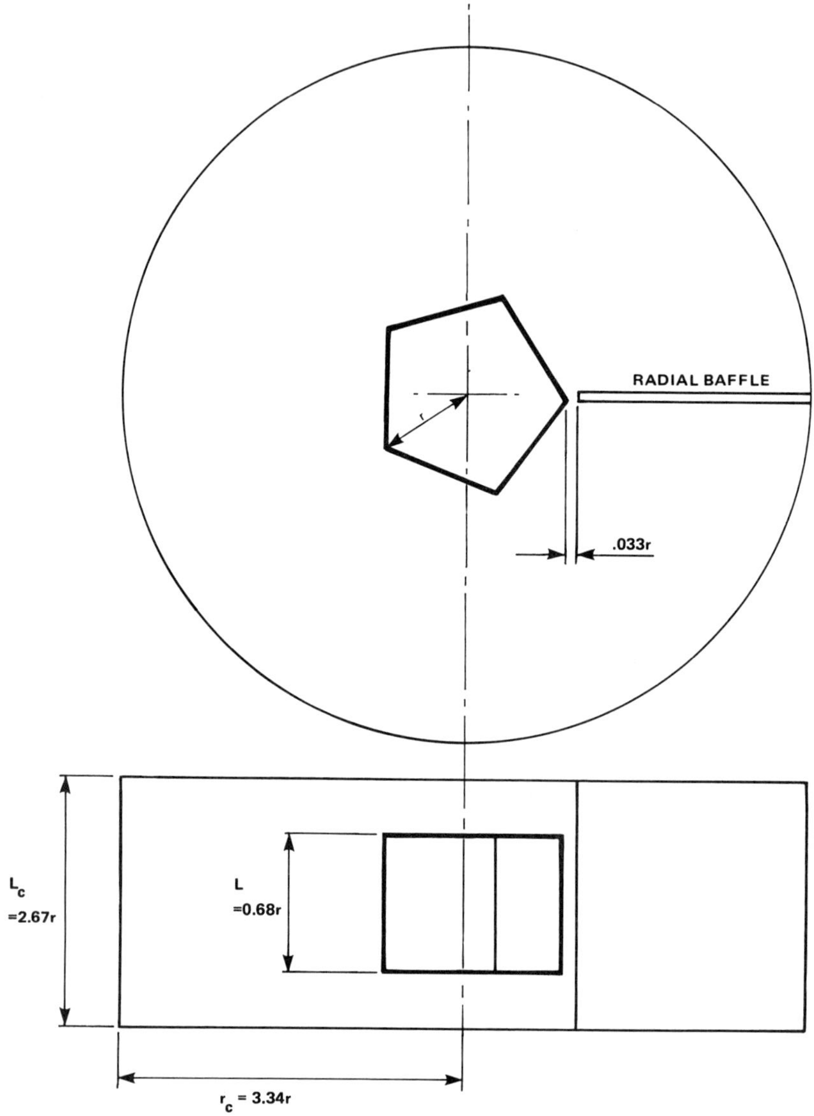


Figure 10 Important dimensions of polygon model and cavity, including baffle (for results given in Table 2).

obstructions may cause large increases in windage. These effects can be mitigated by careful design: keeping obstructions and baffles as far away from the rotor as possible, smoothing window cutouts, and keeping the cutouts as small as possible. In general, disruptions to the flow have relatively less effect on windage power for large, slow rotors. It should be noted that running a polygon rotor in a large cavity can increase windage by up to 80% compared with running the same rotor in a cylinder of the ideal diameter.

## 5.3. Atmosphere

The gas atmosphere in which the rotor turns has an important effect on windage. The choice of atmosphere may be largely a consequence of the structure of the scanner; one might suggest three classes of seal between the polygon rotor and the outside world:

1. Dust seal only; this will require the use of 1-atm air only.
2. Polygon rotor running without a special sealed housing inside the machine, but with good sealing between the scanner and the outside world; this will allow the use of an inert-gas atmosphere, usually helium, at 1 atm pressure.
3. Polygon rotor in separate sealed housing; reduced atmosphere (e.g., $\frac{1}{2}$ atm) helium can be used.

Class 1 is very severe from the windage point of view. In this case the designer has to be careful to take every step to reduce windage by smoothing the airflow and by making the polygon facets as small as possible.

In classes 2 and 3, the use of helium, with its lower density, decreases the Reynolds number by a factor of 7 (1 atm He) or 14 ($\frac{1}{2}$ atm He), which leads to windage power being reduced by a factor of 4.4 or 7.3, respectively, given a typical value of $B = 0.25$.

## 5.4. Polygon Modifications

Reductions in windage can be obtained by the use of modifications to the rotor, typically rounding of the rotor corners or the fitting of side cheeks (see Fig. 7). These devices have more effect with lower numbers of facets and can give useful reductions in windage, especially at higher Reynolds numbers. They can be optically restrictive, however.

## 6. CONCLUSION

The windage of rotary scanners can be affected by many interdependent variables of the mirror polygon and its surroundings. To estimate the effect of these variables using actual scanners would be extremely costly. Model testing enables this work to be carried out much more easily, so that (1) models of complicated scanner arrangements can be checked at the design stage, and (2) a databank of windage for a wide range of configurations can be readily built up.

Although the quantity of information required to cover most likely configurations is too great to be presented in a short chapter, it should be possible for most engineering institutions to generate this information readily using the techniques outlined. A proper analysis of windage can allow substantial savings in scanner power, reduced heat dissipation, more compact designs, higher speeds, lower costs, and decreased risk with prototypes. These benefits should greatly enhance the market potential of scanner systems.

## 7. APPENDIX: WINDAGE EQUATIONS

There are many empirical windage equations in use by rotary scanner designers to predict windage; these are generally derived from dimensional analysis in a similar manner to Eq. (3), namely:

$$T_W = C_M\left(\frac{\rho}{2}\right)\omega^2 r^5$$

In this the torque coefficient $C_M$ is a function of Re and includes the effect on windage of both polygon features and, where applicable, features of the surroundings. This approach was taken in the belief that polygon features and features of the surroundings can often interact in an unpredictable way to affect drag. Plots of $C_M$ against Re must therefore be obtained for a wide range of conditions.

Hayosh has taken a different approach. He assumes that polygon features can generally be separated from features of the surroundings. An equation based on this premise is in a paper given by him in 1976. This has been amplified in recent correspondence with Hayosh and his equation can be summarized as follows:

$$T_W = C_D \left(\frac{\rho}{2}\right) N\omega^2 (r^*)^3 A$$

where

$\rho$ = density of the surrounding medium

$r_{min}$ = polygon-inscribed radius

$r_{max}$ = polygon-circumscribed radius

$r^*$ = effective radius of the paddle due to the facet cross-sectional area; $r^*$ is the distance to the center of percussion of the scanner projected area ($r_{max} > r^* > r_{min}$)

$\omega$ = rotational speed

A = projected cross-sectional area of each facet
$A = (r_{max} - r_{min})L$

L = length of polygon

N = number of facets

$C_D$ = drag coefficient for the scanner configuration

Hayosh suggests that $C_D$ is dependent on Reynolds number and boundary conditions, but is independent of features that affect scanner design (i.e., N, L, r, and $\omega$).

Model test results indicate that $C_D$ and its slope with Re do, in fact, vary with both the number of facets and the clearance, with some interdependence between these variables. It may be possible to produce a best line of $C_D$ against Re, with a fairly wide tolerance, which would cover a limited range of Re for open conditions (large clearance) and another for small clearance. However, when disruptions to the flow are present (e.g., windows, lenses, etc.), these would also interact with clearance and number of facets to affect $C_D$.

An equation such as that suggested by Hayosh, together with simplified best-line charts of $C_D$ against Re, could well be useful to a scanner designer for a quick approximation of windage. To produce these charts and to establish limits to their application, a comprehensive range of windage results for likely polygon clearances, numbers of facets, and obstructions would still be necessary. These results would best be obtained by the methods outlined in this chapter.

## 7.1. Definitions

| | |
|---|---|
| Configuration | A scanner design in which the overall size of the machine may be scaled up or down without the configuration being changed. The restriction is that all the dimensions of the system must remain in a fixed relation to one another. This scaling property is fundamental to the validity of model testing. |
| Rotor | In this chapter, *either* identical to "spinner" *or* a rotating polygon that models a spinner for the purpose of model testing. |
| Scanner | The mechanical part of an optical scanning system, including the spinner and its housing. |
| Spinner | The rotating part of an optical scanner. |

## 7.2. Nomenclature

| Symbol | Meaning | Dimensions | SI units |
|---|---|---|---|
| $C_M$ | Torque coefficient $= \dfrac{T_W}{(\rho/2)\omega^2 r^5}$ | – | – |
| L | Length of polygonal rotor | L | m |
| $L_c$ | Internal length of housing cavity | L | m |
| N | Number of facets | — | — |
| $P_W$ | Windage | $ML^2T^{-3}$ | W |
| r | Tip radius of polygon | L | m |
| $r_c$ | Internal radius of housing cavity | L | m |

| | | | |
|---|---|---|---|
| Re | Reynolds number = $\frac{\rho\omega r^2}{\mu}$ | – | – |
| $T_B$ | Bearing drag (torque) | $ML^2T^{-2}$ | N-m |
| $T_m$ | Motor torque | $ML^2T^{-2}$ | N-m |
| $T_W$ | Windage (torque) | $ML^2T^{-2}$ | N-m |
| $\mu$ | Dynamic viscosity | $ML^{-1}T^{-1}$ | kg/m/s |
| $\omega$ | Rotational speed | $T^{-1}$ | rad/s |
| $\rho$ | Fluid density | $ML^{-3}$ | $kg/m^3$ |

## REFERENCES

Daily, J. W., and Nece, R. C. (1960). Chamber dimension effects on induced flow and frictional resistance of enclosed rotating disks. *J. Basic Eng., Trans. ASME, 82*, 218.

Dorfman, L. A. (1963). *Hydrodynamic Resistance and the Heat Loss of Rotating Solids*. First English ed. Oliver & Boyd, London.

Hayosh, T. D. (1976). Motors and control systems for rotating mirror deflectors. *SPIE, 84: Laser Scanning Components and Techniques.*

# 4

# Bearings for Rotary Scanners

Joseph Shepherd
*Ferranti P.L.C., Edinburgh, Scotland*

## 1. INTRODUCTION

The bearings for rotary mirror scanners must be capable of establishing an extremely accurate rotational axis. They must do this repeatedly within acceptably low limits of deviation, throughout a large number of startups, for a maximum life, and while sustaining specified static or dynamic loads.

A rotary scanner bearing specification will depend on the scanning system requirements. For example, speeds up to 500,000 rpm, rotational center accuracy to millionths of an inch, a wide temperature range, low cost, silence, and cleanliness may be required. Fortunately, the extremes of these are not often all required at once, and different types of bearing give special advantages.

It is a tribute to the work of many engineers and scientists that high-accuracy mechanical scanning is at all possible. Their work over the last two decades has improved the reliability and accuracy of both gas and ball bearings. Many of the advances in bearing technology have been achieved through experience gained in the design of gyroscopes for inertial navigation systems; in such systems the need for the ultimate in spin axis, accuracy and reliability, freedom from contamination, and long life has much in common with that of rotary scanners.

In all types of bearings the principal aim is to produce true fluid film separation of the bearing surfaces when rotating, thereby achieving contactless running which will ensure long life. It is hoped that the designer of systems incorporating rotary scanners, who has little previous knowledge of bearings, will gain from this chapter an appreciation of the advantages and limitations of the various types of spin axis bearings available. While not intended to be a guide to bearing design, a designer should be able to approach the manufacturers of bearings and rotary scanners with a working knowledge of their products.

The bearing types most likely to be considered for rotary scanners are: ball, pressurized gas, and self-acting gas. These have all been covered as fully as possible in this chapter. Further information on ball bearings is well documented and readily available, whereas information on gas bearings, especially self-acting types, is much more restricted. To help to redress this, more illustrations and design references have therefore been included for gas bearings than for ball bearings.

### 1.1. Historical Background

A bearing carries or supports the load of an object while allowing relative movement, preferably with low friction. There are two general classes, linear bearings and rotating bearings.

Many forms of bearings occur in nature, for example, animal joints, which are beautiful examples of sealed low-friction rotating bearings; or snails slithering on their mucus, and glaciers sliding on ice water melted by pressure, both examples of linear bearings. Early man, having slid on scree or slipped on fat or perhaps even a banana skin, soon realized that by placing a "bearing" under a heavy object it became easier to move. Early examples from history are the use of tallow or oil to lubricate the runners of sledges used to convey the massive stones used in the pyramids (fluid film bearings), or a number of round tree trunks placed under heavy weights to transport them (rolling element bearings). The wheel is often regarded as one of the most important advances in history, but a wheel needs a bearing and preferably lubrication to reduce friction.

Plain journal bearings such as those used on early wooden wheels worked much better when lubricated; however, this was usually boundary lubrication without true separation of the bearing surfaces. Such bearings existed, with little change, until advances in manufacturing during the nineteenth century led to more accurate bearing surfaces, which greatly reduced friction and wear of plain cylindrical (journal) bearings. The reduction in friction, with better surface accuracy, was not, however, seen in flat thrust bearings. This was a puzzle, especially to naval designers working on the development of ships' propellers, as this was being held back because of the high losses on the shaft thrust bearings.

The first understanding of a possible reason for this was by Tower, who in 1883 published data showing that the pressure distribution in a plain journal bearing was not uniform around the side of the bearing opposing a load, but increased considerably at points near the convergence in the bearing gap. The importance of this work was recognized by Reynolds, who in 1886 was able to explain Tower's results when he published his now famous theory for hydrodynamic fluid film lubrication.

Air is a fluid, although with some significant differences from liquids as described later, and journal bearings with air as the lubricant were demonstrated within about 10 years of Reynolds' work.

As mentioned previously, rolling element bearings utilizing round logs were used by early man, as were smooth spherical stones, which were the forerunners of ball bearings. There is evidence of a type of ball bearing being used by the Romans, in the form of a thrust bearing incorporating spherical wheels. However, it was many centuries later, during the Industrial Revolution, that ball bearings began to be produced more regularly as machining standards improved. The first true instrument-size ball bearings available commercially were probably produced by the Swiss firm RMB in 1936, as a diversification from precision watchmaking.

## 2. GAS BEARINGS

In many forms of bearing, surface separation results from the pressure generated, by viscous effects, when a fluid is induced to flow in a narrow gap. When the fluid is a gas, the effect is much less because of its relatively low viscosity—which has some disadvantages, but also some unique advantages, as described later. Because of those advantages, gas bearings are now used in a wide range of fields from ultraprecision self-acting bearings for inertial navigation gyroscopes to pressurized bearings used to support elements of machine tools weighing several tons.

Gas bearings are particularly suited to low-noise, high-speed, high-accuracy bearing applications, which makes them attractive for many rotary mirror scanners. In many instances it is the availability of gas bearings, providing a high-accuracy, high-speed spin axis, which has broadened the scope for mirror scanning applications.

They can generally be divided into two groups, *self-acting* bearings, which can generate their own internal pressure by virtue of the relative motion of converging surfaces, and *externally pressurized* bearings, in which the pressure is generated by an external source and the bearing needs no relative motion for lift. There is an intermediate group, *hybrid* bearings, which are a cross between the other two and are basically externally pressurized bearings which have been designed to take account of their self-acting capability whne running at speed, thus producing improved load capacity or stability. All types can be designed to take axial or radial loads, either by separate thrust and journal bearings or by conical or hemispherical bearings which can give both axial and radial load capacity.

### 2.1. Background of Gas Bearings

Gas bearings were a fairly natural derivative of the early work on hydrodynamic fluid film bearings, when it was realized that air could carry loads like any other fluid. The first known experimental self-acting air bearing was produced by Kingsbury as long ago as 1897. Other early workers in this field were de Ferranti, who in 1909 patented an air bearing spindle for textile machinery, and Harrison, who in 1913 put forward an approximate theory for induced fluid film lubrication which took into account the compressibility of gases.

Those early designs were not very successful because of their low load capacity, and it was only in the early 1950s that concentrated effort was applied to improving the design of gas bearings, when it was realized that bearing problems in several advanced fields could best be solved by utilizing the unique advantages of gas bearings. Some of the requirements were for bearings to run at greater extremes of temperature, higher speeds, and better accuracies than was possible with the bearings then available. Other needs were for bearings that did not cause lubricant contamination of the local environment, could withstand nuclear irradiation, and had minimum drag. Much of this work was necessary for the gas-cooled nuclear reactor program in the United Kingdom and for high-speed power plant and inertial gyroscope development in the United States. The importance of these programs stimulated a period of intense theoretical and experimental activity on gas bearings, which has been summarized in several textbooks published in the early 1960s. Those by Gross (1962), and Grassam and Powell (1964) cover the period well and cite comprehensive references. Work on refining the theory and improving the technology of gas bearings has continued, until the present, when they can be designed and used with confidence, useful references being the Mechanical Technology, Inc., manuals, *Design of Gas Bearings*, 1972, Volumes 1 and 2 (Wilcock, 1972).

A range of self-acting gas bearing rotary scanners for thermal imaging and laser scanning is shown in Figure 1. These have been run at speeds up to 90,000 rpm and with facet-to-spin axis accuracy of less than 10 arcsec.

## 2.2. Unique Attributes of Gas as a Lubricant

The drag power of a plain journal bearing is approximately given by

$$P_D = K_1 \frac{\mu\omega^2 R^3 \ell}{c} \tag{1}$$

and the load capacity by

$$L = K_2 \frac{\mu\omega R \ell}{c^2} \tag{2}$$

where

$\mu$ = dynamic viscosity

$\omega$ = speed

$R$ = outer radius

$\ell$ = length

$c$ = bearing clearance (gap)

$K_1$, $K_2$ = constants of proportionality

Gas bearings can be run at much higher speeds than liquid bearings because of the very low viscosity of gas compared to that of liquids. The dynamic viscosity of a typical instrument bearing lubricating oil is several thousand times that of air at 20°C. From Eq. (2), for a similar load capacity, a gas bearing would therefore have to be much larger than a liquid bearing, the gap would have to be smaller, or it would have to be run at higher speed. A larger bearing is not necessarily a disadvantage, however, as it allows larger-diameter stiffer shafts to be used, which can be beneficial in increasing the shaft critical speed, often a limiting factor in high-speed operation.

Dividing Eq. (2) by Eq. (1) shows the load capacity per unit of drag power as some measure of bearing efficiency as

$$\eta \propto \frac{1}{\omega C}$$

which indicates that at first sight, this factor is independent of fluid viscosity, improves as gap decreases, and worsens as speed increases. For a gas bearing, there is a practical lower limit on the gap size; as this in turn is limited by the accuracy attainable on the bearing components, but this is steadily improving.

Another important and exclusive advantage of gases as lubricants is their ability to accommodate wide variations or extremes of temperature. Under normal conditions, gases, unlike liquids, do not change state, or degrade with temperature and their viscosity varies only slightly over the normal range of ambient conditions experienced. This is illustrated in Figure 2.

Finally, and importantly, a gas bearing is surrounded by its lubricant, the atmosphere. The gas is drawn or pumped from this readily available source into the bearing and returned, normally without contamination of the surroundings. If the atmosphere is clean, there will also be no contamination of the bearing. There is therefore no need to seal the bearing as may be necessary in liquid or ball bearings to avoid contamination of adjacent optical surfaces by traces of lubricant.

### 2.3. Self-Acting Bearings

These are also known as self-pressurized, aerodynamic, or hydrodynamic bearings. Their load capacity comes from pressure generated by the "wedge effect" produced when gas is induced to flow through a convergence in the bearing gap, by the motion of the bearing surfaces.

An understanding of this effect can be obtained by considering two parallel plates in a fluid. If one of the plates is moving with velocity U, flow will be induced between the plates by viscous shear, the velocity varying linearly from U to zero (Fig. 3a). When one of the plates is tilted, continuity of mass flow through the bearing dictates that the velocity must increase as the gap narrows, so that the area of the velocity profile is constant

**Figure 1** Gas bearing rotary scanners.

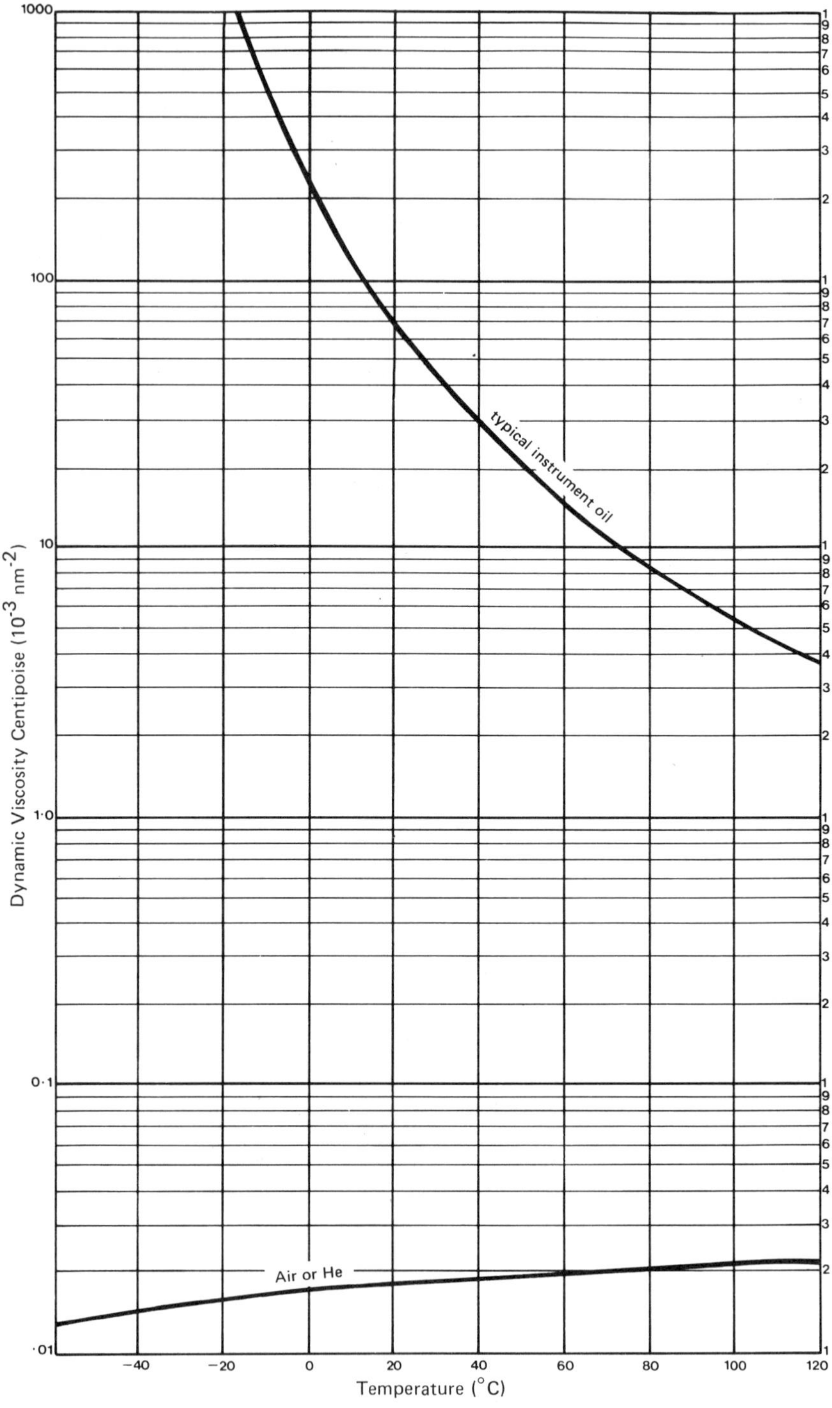


Figure 2 Comparison of gas and oil viscosity change with temperature.

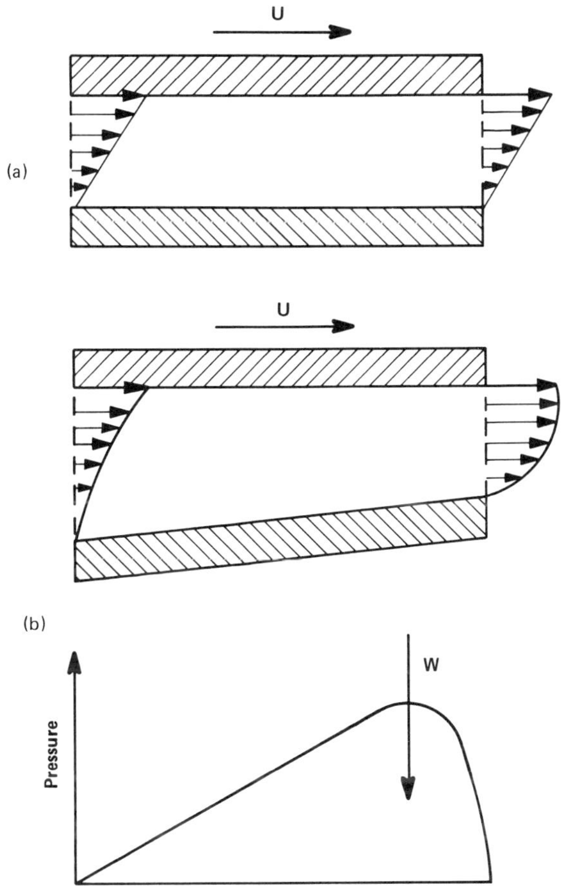


Figure 3 Induced flow between flat surfaces.

(Fig. 3b). This increased velocity must be accompanied by an increase in pressure above ambient, which produces load capacity in a pressure profile as shown.

If the fluid is a gas, the pressure generated will also compress the gas, thereby increasing its density. Consequently, the velocity increase in the constriction will not be as great, resulting in some loss in load capacity. This effect will be slight at low speed but will become more significant as speed is increased.

A plain journal bearing (Fig. 4) develops a wedge effect only when the shaft is deflected from its central position by an applied

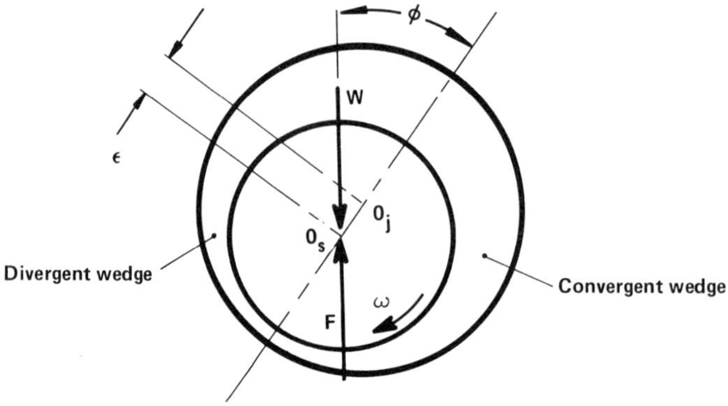


Figure 4 Journal bearing: wedge effect with rotation.

load. It will deflect until the pressure rise balances out the load. This will be accompanied by a sideways movement as a result of pressure rise above ambient in the convergent wedge and a pressure drop below ambient in the divergent wedge. The shaft thus takes up a position at an angle $\phi$ to the load direction, known as the attitude angle, with an eccentricity $\varepsilon$. When running centrally there is no wedge effect in the bearing, and as no restraining forces are generated, it is basically unstable.

A plain thrust bearing with its surfaces normal to the axis of rotation has no wedge effect, even when deflected. This was the problem seen in early ships' thrust bearings described earlier.

To overcome these faults, the surfaces must deviate from true surfaces of revolution, in some way, to achieve a built-in wedge effect. There are two methods of achieving this:

1. By allowing one of the surfaces to move or deflect. (Types include pivoted pad bearings, resilient conforming bearings, and foil bearings.). These bearings have an advantage in their ability to remain stable over a wide speed range because they adjust their shape as speed and load changes. But this adjustment can also be a disadvantage in rotary scanners, where as little movement as possible of the optical axis is required. These types are also generally less efficient as load carriers than fixed geometry bearings.
2. By fixed geometry bearings where one of the mating surfaces is permanently altered from a surface of revolution

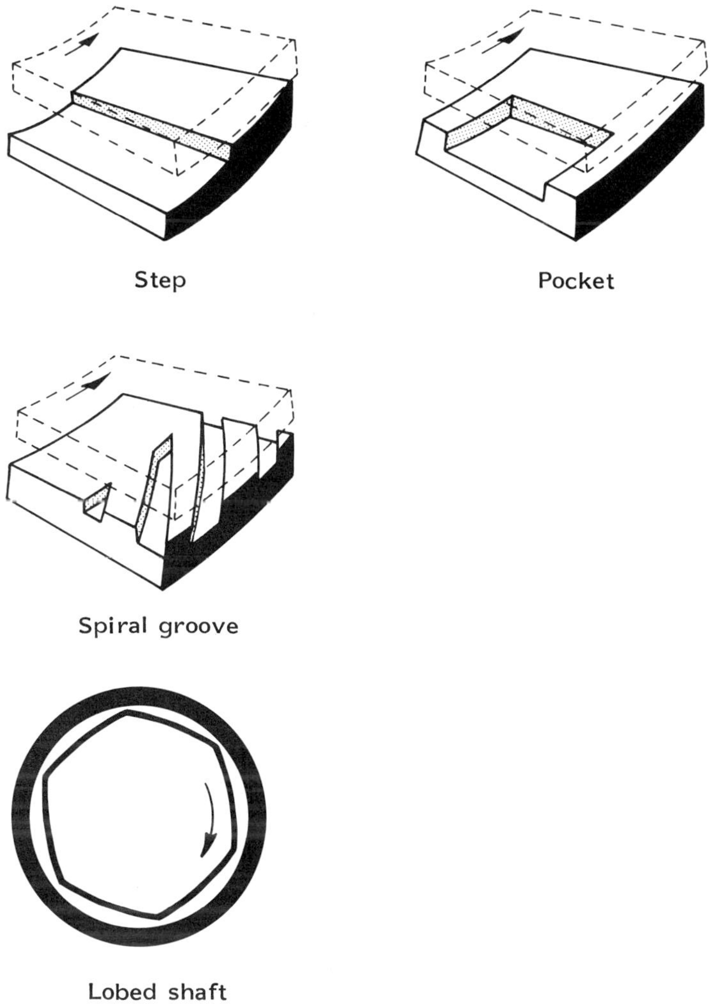


**Figure 5** Built-in wedge effect features.

to produce a built-in wedge compression effect. This effect can be achieved by adding surface steps, pockets, or grooves, which can be applied to both thrust and journal bearings, or by adding lobing to journal bearings (Fig. 5). Of these, spiral groove bearings have several advantages in performance, which make them the most widely used self-acting bearings for precision spin axis applications.

### 2.3.1. Spiral Groove Bearings

In these bearings one of the surfaces is grooved such that when the surfaces move relative to each other, gas is drawn into the grooves by viscous shear. When the flow thus induced is throttled at the end of the grooves, a pressure rise is produced to give load-carrying capacity.

Much of the original work on these bearings was carried out at the U.K. Atomic Energy Authority, and Whipple published the first theoretical analysis of them in 1949; this and other reports from the same source were not published until much later because of government restrictions. They were later adopted as the ideal spin axis bearing for precision gyroscopes, and a general analysis on the application of spiral groove bearings to gyroscopes was carried out by Vohr and Pan (1968), followed by Faddy, who in 1969 produced a design analysis for a conical configuration.

### 2.3.2. Configurations of Spiral Groove Bearings

The spool or H-form bearing (Fig. 6a) was the first type to be produced commercially because the techniques of manufacture (finishing cylinders and flats) were already well established in other fields. The thrust plates have spiral grooves and the journal bearing can be plain, lobed, or grooved.

Two other types available are the opposed conical (Fig. 6b) and the opposed hemispherical (Fig. 6c) bearing. These have more demanding manufacturing problems, but their advantages have encouraged a small number of manufacturers to persevere, and scanners based on these are now also available commercially. Although both types need more specialized techniques to produce accurate geometry, they have the advantage of being able to take radial and thrust loads on one surface. It is thus possible to produce compact standard designs (Fig. 7) which are extremely adaptable for rotary scanner applications. Hemispherical bearings also have the advantage of being inherently self-aligning, which can greatly simplify the required accuracy of assembly.

If the projected apexes of opposed conical bearings coincide (Fig. 8), materials of differing coefficient of expansion can be used for rotor and stator without change in bearing gap with temperature. This can be a major advantage, as lightweight rotors are often used to reduce mass or inertia effects together with heavier or stiffer stators. The gap of conical bearings can also be easily adjusted by moving one of the cones axially; thus the same standard bearings can often be used for a range of loads and speeds. For the same maximum diameter, minimum diameter, and length, the theoretical bearing performance characteristics of both conical and spherical types have been shown to be similar by Vohr and Pan in 1968.

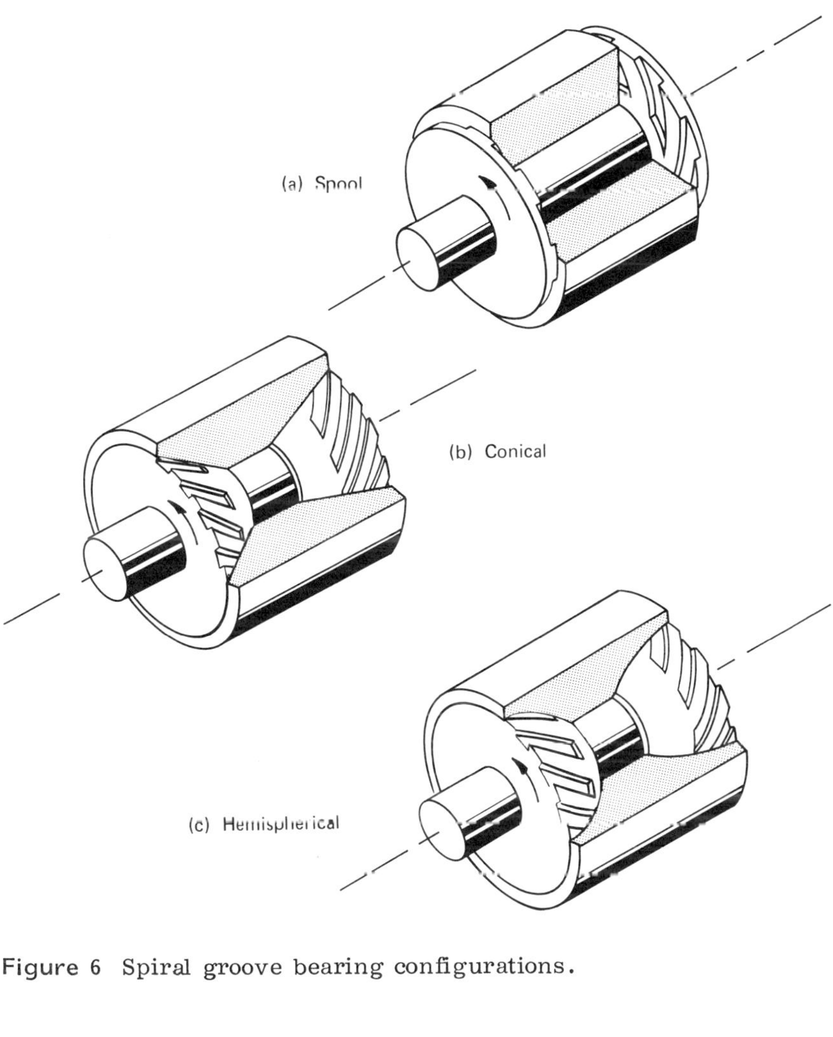


Figure 6 Spiral groove bearing configurations.

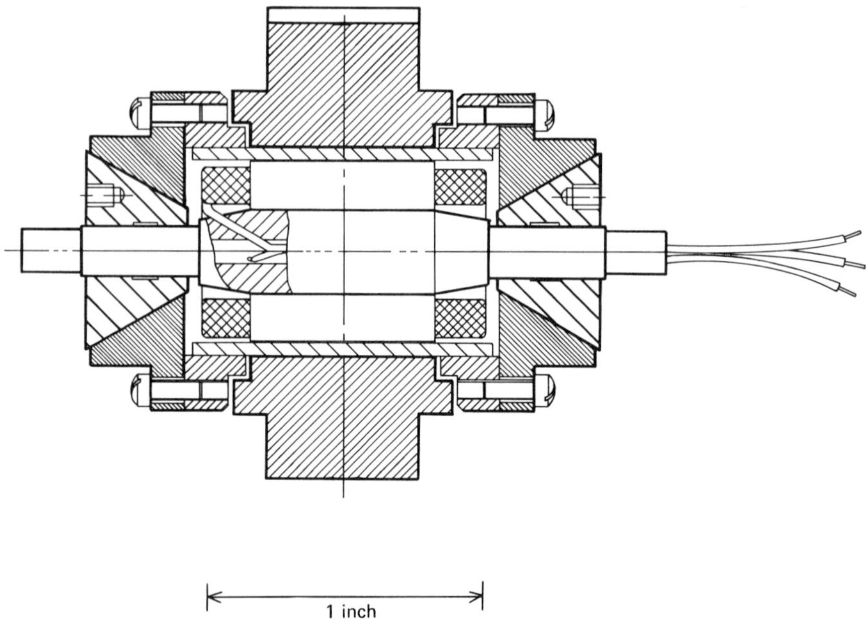


Figure 7 Scanner with conical gas bearings.

2.3.3. Other Characteristics of Self-Acting Gas Bearings

A self-acting gas bearing requires relative movement of the surfaces to generate pressure, and rubbing of the surfaces takes place at startup. To minimize wear, bearing surfaces are normally manufactured from, or coated with, hard materials such as ceramics or cermets; these materials are also more easily finished to the

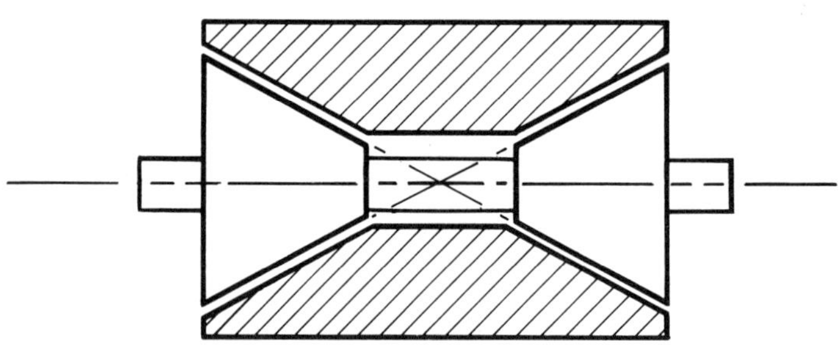

Figure 8 Coapex conical bearing.

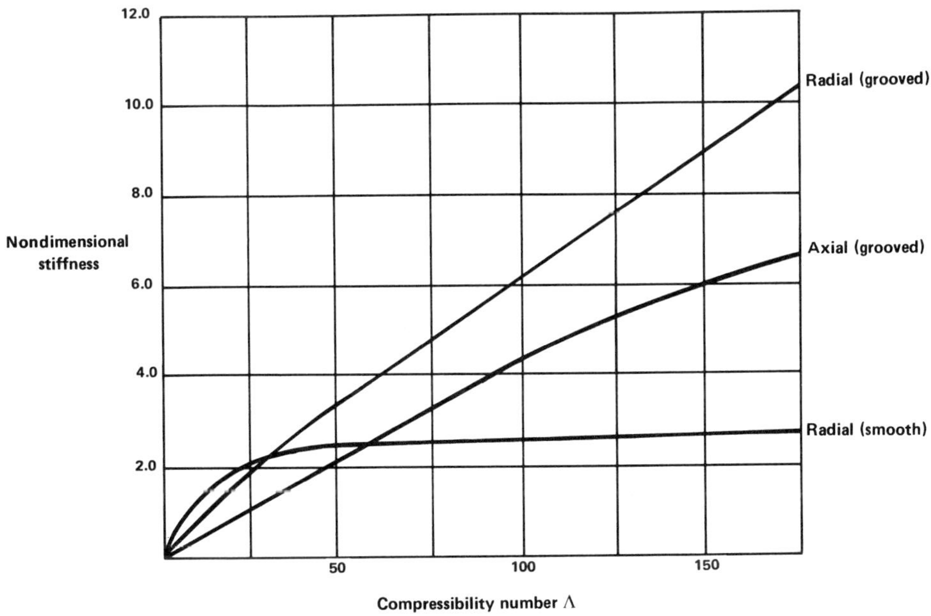


Figure 9 Comparison of grooved and smooth gas bearing stiffness.

extremely high accuracies required for self-acting gas bearings. Boundary lubrication with fatty acids has also been employed to reduce starting friction and wear. The use of these techniques has been highly successful, so that there should be no real problem in achieving many thousands of starts and stops, without damage.

In a plain journal gas bearing, load capacity initially increases linearly with bearing speed [see Eq. (2)]; however, at high speeds the linear increase no longer applies and there is some loss of load capacity, as described previously (see Sec. 2.3). It has been shown by Faddy that this effect is not present in spiral groove bearings (Fig. 9) within the range of compressibility numbers likely for rotary scanners. The compressibility number $\Lambda$ is the dimensionless group:

$$\frac{6\ \mu\omega R^2}{P_a c^2}$$

where

$\mu$ = dynamic viscosity

$\omega$ = speed

R = radius

$P_a$ = ambient pressure

c = bearing clearance

and is a measure of the pressure generated in the bearing relative to ambient. A dimensionless group is a gathering of interdependent variables into groups whose dimensions (length, mass, and time) cancel out. In a complex situation, such groups greatly simplify the interrelationship of the variables.

Two further factors that can influence the performance of self-acting gas bearings, and whose effect should be minimized at the design stage are:

1. Centrifugal effects, which can cause distortion of the bearing surfaces when running at high speed, can become a significant proportion of the small gaps used in self-acting bearings.
2. A loss of load capacity, which was observed by Burgdorfer in 1959, occurs when the film thickness of a self-acting gas bearing becomes less than about 100 times the magnitude of the mean free path of the ambient gas. This is especially relevant when running at reduced pressure or when using low-density gases to reduce windage on the polygon.

### 2.3.4. Instability of Self-Acting Gas Bearings

Instability in self-acting gas bearings can be a complex phenomenon. The following is a simplified description of two of the main factors.

1. Mass/stiffness resonance. This is the well-known type of resonance which occurs in every spring-mass system. However, in self-acting gas bearings, stiffness will change with speed and will not necessarily be linear with deflection. Bearings should therefore be designed such that this resonance lies outside the frequency range of any likley external vibration and the speed range is below this resonance. If this is not possible, sufficient damping must be built in to limit the vibration amplitude.

2. Half-speed whirl. The average induced flow in a plain journal bearing running with no eccentricity is in the same direction as the rotation but at half the speed. If the shaft is deflected from its central position, it will experience a centering force from the wedge effect and a backwards torque from shear forces. These forces will cause the shaft to whirl in a direction opposite to the direction of rotation. When this whirl speed reaches the speed of flow around the bearing (i.e., half speed), there will be no relative flow through the gap and therefore lubrication forces will collapse. Bearings with a built-in wedge effect, such as spiral grooving, are not usually troubled by half-speed whirl, as the effect is much reduced: however, it will still be there and should be considered in any design.

Obviously, if a bearing is run at a speed of two times its resonant frequency, it will tend to whirl at its resonant frequency, which is almost certain to cause bearing failure if it is held at this speed, but if damped, it is often possible to run successfully through this resonance to higher speeds.

For a pair of bearings, instabilities can occur in cylindrical and conical modes. External vibration at specific frequencies, or poor balance, can stimulate these instabilities. The rotor should therefore be balanced as well as possible and external vibrations taken into account.

## 2.4. Externally Pressurized Gas Bearings

Other names for these are hydrostatic or aerostatic bearings. In these, the pressure to produce load support is provided from an external source, avoiding any need for relative motion of the bearing surfaces. They can therefore be used to define an accurate axis from zero up to very high speeds. At zero and low speed they have virtually zero friction torque and many instrument designs make use of this. They have the advantages of the use of gas as a lubricant, described earlier (see Sec. 2.2).

The bearing gap is generally about an order of magnitude higher than that for self-acting bearings, so manufacturing and assembly tolerances can be more relaxed. These tolerances can be met by most precision machine shops, and this has led to fairly wide use and availability of pressurized bearings.

Pressurized bearings should have long life in spin axis applications, as, because the bearings are pressurized to separate the surfaces prior to running, no wear should occur. However, it is still desirable to have wear resistance surfaces, to withstand accidental turning when unpressurized, or overload touchdown.

The need for an external pressure source is often a major disadvantage; if this is not already available, a local compressor will be necessary, which may be bulkier and noisier than permissible. There is also a need to exhaust the gas supplied, which precludes their use in sealed systems. If these conditions are acceptable, the pressurized system has an advantage, since if the supply is filtered and dried, continuous flow through the bearings can keep them cool and clean even when working in relatively dirty atmospheres. It is necessary to use dry air, as the pressure drop across restrictors (see Sec. 2.5.2) can cause ice crystals to form in moist air.

Early pressurized bearings were made with relatively large clearances, which made them easy to manufacture, but also made them comparatively inefficient, with large gas consumption and low stiffness, and also more liable to pneumatic hammer instability (see Sec. 2.5.4). As the demands on these bearings have increased and the design knowledge improved, the clearances have been reduced. This reduced clearance also means that when the bearings are used for high-speed applications, there is an appreciable self-acting effect, which takes them into the hybrid bearing class; both effects must be allowed for in any design.

### 2.4.1. Journal Bearings

The principle of operation of these can be explained with reference to the bearing as shown (Fig. 10). Supply pressure $P_s$ feeds a chamber around the bearing. From this, gas passes onto the shaft through restrictors, which limit the gas flow, causing a pressure drop across the restrictor, to pressure $P_g$ in the gap. There will be a further pressure drop to ambient as gas flows to the ends of the journal. With the shaft central and no load applied, the pressure $P_g$ in the gap will be equal around the shaft and will vary along the shaft as illustrated (Fig. 10a). If a load is now applied to the shaft, it will be displaced in the direction of the load. Because the gap has increased on the load side, flow will increase through the restrictors and the pressure drop will increase, lowering the pressure. Similarly, the decreased gap on the opposite side will produce an increase in pressure, and the resultant pressure difference will balance the load.

The lifting efficiency of a pressurized bearing is normally measured by its

$$\text{Load coefficient } C_L = \frac{W_{max}}{P_s A}$$

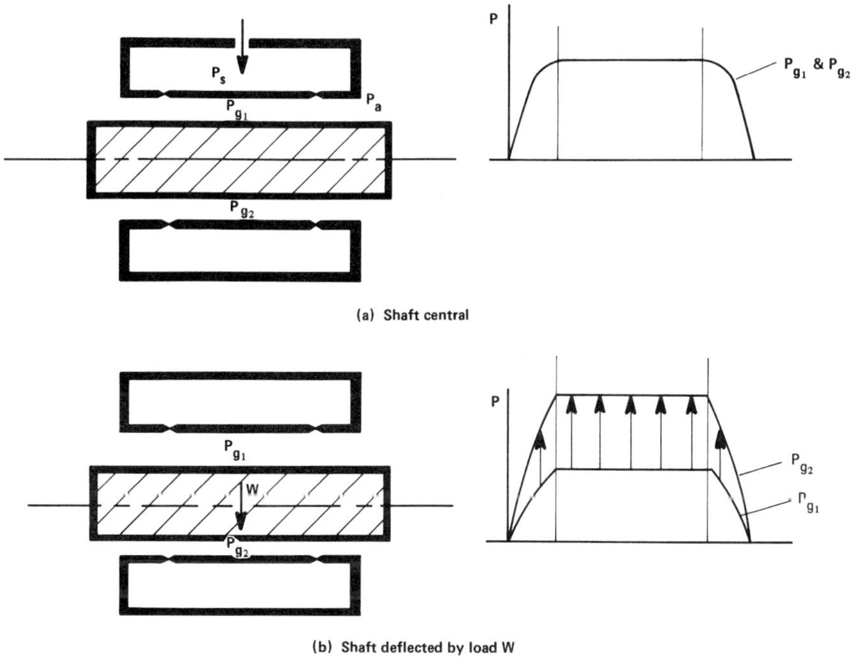


Figure 10 Externally pressurized journal bearing.

where

$P_S$ = supply pressure above ambient

$W_{max}$ = maximum load

A = bearing projected area

The load coefficient varies with bearing deflection. At a deflection of half the bearing clearance the load coefficient is normally about 1/3; however, this can be increased, up to a maximum of about 2/3, for high-efficiency bearings, but usually only with additional manufacturing complication.

As described earlier, the lifting capacity of a journal bearing is determined by the pressure difference created between the two sides of the shaft when it is deflected under load. Any factor

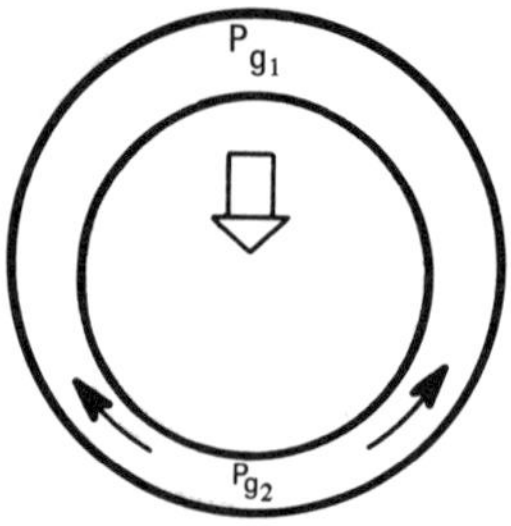


(a) **Circumferential flow**

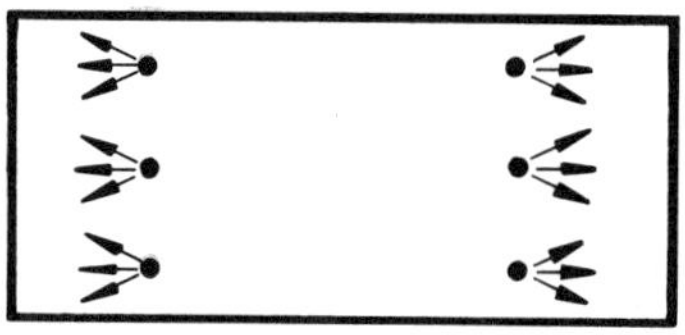

(b) **Dispersion effect at jets**

Figure 11 Factors reducing load capacity in externally pressurized bearings.

that reduces this difference will also reduce the lifting capacity. Typical factors are:

1. The flow of gas circumferentially round the bearing from the high-pressure side to the low-pressure side, limiting the area of the high-pressure zone (Fig. 11a).
2. The effect of dispersion of the gas as it flows from discrete feedpoints (Fig. 11b), resulting in local reductions in pressure; this can be reduced by increasing the number of outlet points.
3. The length of the reduced pressure zone between the feedpoints and the ends of the bearing; this length can

be decreased by moving the outlet points nearer to the ends of the bearing, but only at the expense of increased gas consumption.

4. The restrictor characteristics, which control the maximum pressure differential generated across the shaft when deflected. These characteristics depend on the restrictor type and dimensions.

   There are several basic types of restrictors, with many variations on these. New designs are regularly being introduced with claimed advantages in some aspects of performance. Some of the basic types are outlined below.

### 2.4.2. Types of Restrictor

*2.4.2.1. Jet or Orifice.* There are two main types, the plain jet and the pocketed jet, illustrated in Figure 12. In the plain jet, flow is regulated by the area of the annulus surrounding the jet $\pi dh$ (if $\pi dh > \frac{1}{4}\pi d^2$); this is often described as *annular orifice regulation.* For the pocketed jet, flow is regulated directly by the area of the orifice $\frac{1}{4}\pi d^2$, known as *simple orifice regulation.* The regulation of flow is important in limiting gas consumption and providing control over the bearing response to load fluctuations. Generally, pocketed jets have higher stiffness than plain jets, because of reduced dispersion losses, while plain jets have better stability characteristics.

*2.4.2.2. Porous Pad.* In these, the jet is replaced by a porous pad or, more commonly, a complete porous journal is fitted (Fig. 13). With a complete journal, an almost infinite number of point source outlets will be created, so that dispersion effects will be decreased and load capacity increased. Porous bearings are

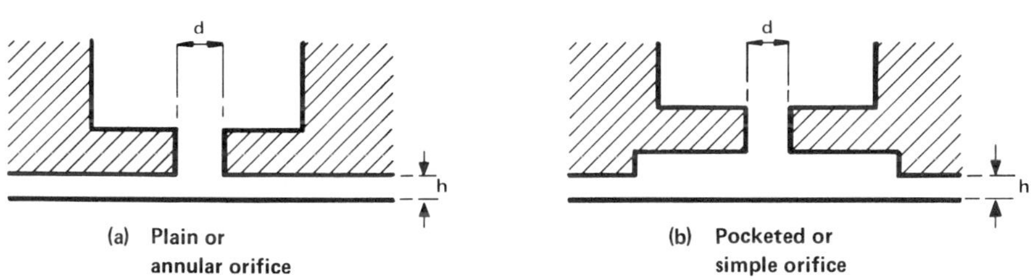


Figure 12 Types of jet restrictor.

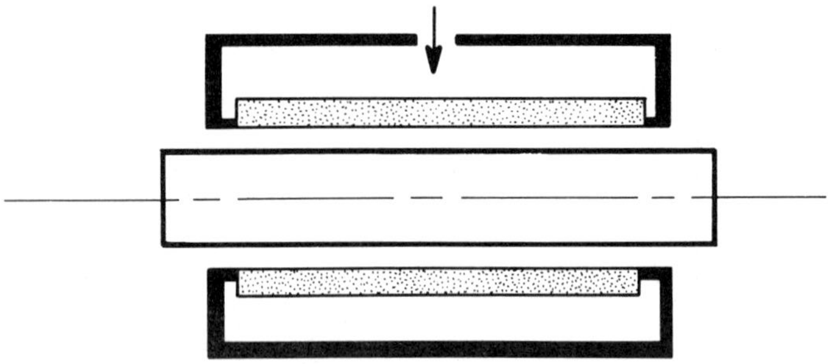

Figure 13 Porous journal bearing.

also more tolerant of contamination in the supply, which can have a disastrous effect on a jet-fed bearing if one or more jets become plugged.

The main disadvantages are that special microporous materials are needed to obtain the correct flow characteristics; these materials must machine cleanly to tight tolerances and should have excellent rubbing qualities, without smearing, to withstand accidental damage before lift-off or during overload touchdown.

*2.4.2.3. Slot Entry.* These normally have a circumferential slot, often around the entire circumference of the bearing (Fig. 14). While these have a lower theoretical stiffness and load

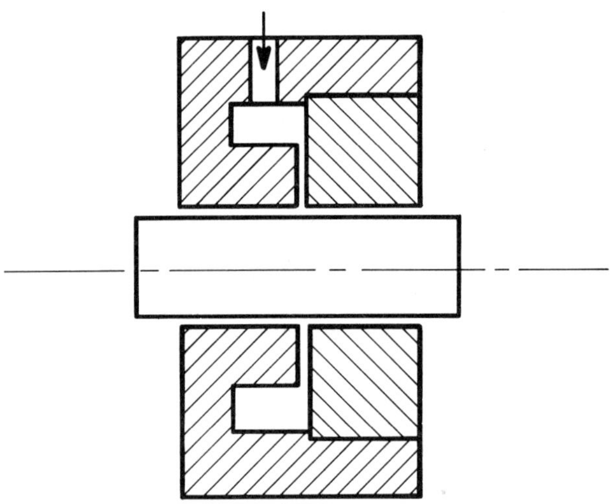

Figure 14 Slot entry bearing.

capacity than similar jet-fed bearings, the line source of supply eliminates dispersion effects which reduce actual performance in jet-fed bearings. Slot entry bearings therefore have an advantage for short, low length-to-diameter bearings, where flow dispersion has the greatest effect on jet-fed types. In addition, the optimum design for slot entry bearings is independent of the fluid characteristics and supply pressure. This feature makes them extremely versatile as "standard" manufactured bearings, which will give good performance at any available supply pressure, and which can even be used with liquids.

### 2.4.3. Other Design Features

For a mirror scanner of given mass and inertia, the bearing load capacity limits the maximum input rates that can be applied before touchdown of the bearing surfaces. To leave a margin of safety it is desirable to design for sufficient load capacity when the bearing gap is closed by only about half. If this is not possible, the surfaces should be hard coated as a precaution against high-speed touchdown.

Spin axis positional displacement under load will be determined by the bearing stiffness, which can be increased by reducing the mean radial clearance; this also greatly reduces gas consumption. Angular displacement will also be controlled by bearing stiffness and the distance between bearing support points.

A journal bearing of cylindrical form will support only radial loads. If axial load capacity is also required, it is usual to provide this on a flat circular or annular thrust bearing surface, normal to the spin axis, which is provided with restrictor outlets. If these axial loads are likely to be small; there are advantages, in simplicity, in using types of combined journal and thrust bearing where no outlets are present on the thrust face. Some axial load capacity is produced by allowing the pressurized exhaust from the journal to pass through the thrust faces to ambient pressure.

Conical or hemispherical pressurized bearings are also sometimes used in special designs, where the advantages of these configurations are considered worth the probable extra cost of manufacture. Their advantages are the ability to take axial and radial loads on one surface and the other attributes described previously for these configurations (see Sec. 2.3.2).

### 2.4.4. Instability of Pressurized Bearings

The instabilities associated with externally pressurized bearings are caused primarily by the resonance and half-speed whirl effects already outlined for self-acting bearings. Where instabilities do occur within the speed range required, a well-established method of overcoming the problem is to mount the stationary member on soft rubber O-rings, which contribute sufficient damping to limit the amplitude. This may be a disadvantage in some accurate spin axis applications, as soft mounting will contribute to the total axis deflection under load. If this is important, other methods of damping out the oscillations will have to be built in, such as the use of stabilizing cavities, as described by Sixsmith in 1959.

The other instability associated with externally pressurized bearings is that known as pneumatic or air hammer, where the rotor vibrates in a self-sustained oscillation similar in principle to that of a pneumatic hammer. This condition occurs independently of any rotation of the bearing, and can often be heard as a low-frequency moan at standstill. It is more likely to occur in thrust bearings than in journal bearings. An analysis of the problem of self-exited vibrations and suggestions for overcoming it were presented by Licht and Elrod in 1960.

## 3. BALL BEARINGS

Small instrument ball bearings were first produced commercially about 50 years ago. They were immediately successful because of their low friction and were widely used on military and commercial equipment. With their increased usage more specialized machinery was introduced to improve output and standards. Even with these improved standards, many bearings suffered from unexplained failures and limited life. The problem continued until the late 1950s when a theoretical understanding of the true nature of the lubrication process was sought and research and development on the influence of ball bearing components was carried out. Much of this work was initiated at the Charles Stark Draper Laboratory (then the Instrument Laboratory of MIT), stimulated by the demand for better performance from spin axis bearings for inertial navigation gyroscopes.

The theoretical work on elastohydrodynamic (EHD) lubrication by Archard and Cowking, published in 1965, showed that the lubricant film thickness in a typical instrument bearing was less than 10 millionths of an inch. This knowledge highlighted the need to control all factors that could affect this extremely thin film; these factors include complete raceway geometry and finish, ball standards, cage material and type, cage geometry, lubricant type and quantity, steel quality, preloading, environment, and freedom from contamination. Control over all these factors is neither easy, or inexpensive, but if attained it will ensure extremely good consistency and accuracy, together with long life.

## 3.1. Description and Terminology

Instrument ball bearings are usually either *radial deep groove* bearings or *angular contact* bearings. For accurate spin axis applications, such as rotary scanners, free movement (play) in the bearings should be eliminated by using them in pairs loaded against each other (preloaded pairs). The load displaces the balls to the side of the race into an angular position (contact angle). Radial bearings can be used in this manner, but there are some advantages in using the more specialized angular contact bearings described later.

A ball bearing consists of an inner race and an outer race, separated by a set of balls held in position by a cage (retainer, separator) (Fig. 15). The outer and inner races have grooves (raceways) machined in them in which the balls run. Generally,

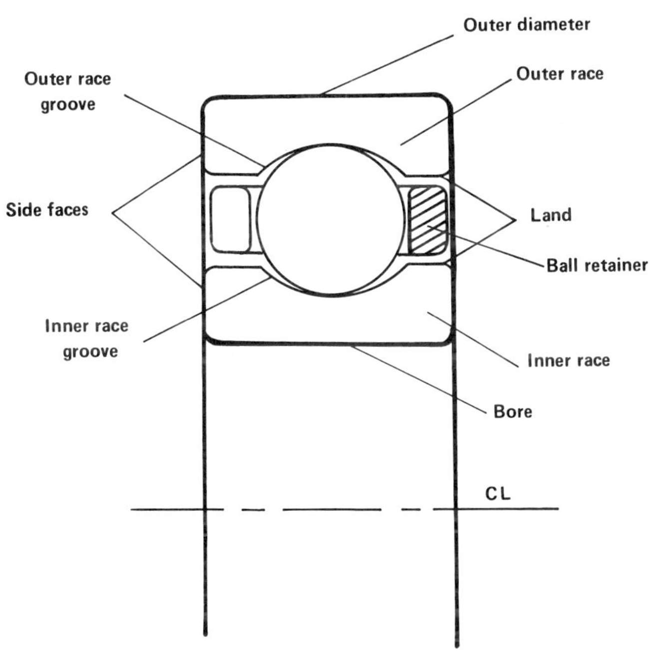


Figure 15 Ball bearing nomenclature.

these raceways are finished to very high geometric accuracies relative to their locating faces by grinding and honing, often finished by lapping and polishing to improve the surface finish further.

The type of bearing shown is known as a *radial deep groove* bearing. The balls have a slight clearance in the raceways, which allows them to deflect to the sides of the groove to sustain axial loads. Radial loads push the balls to the center of the groove, the load being taken by an arc of balls opposite the load. They are assembled by displacing the inner race to one side within the outer and filling this gap with as many balls as can be accommodated. The balls are then spaced evenly round the race and a snap-on or two-piece riveted cage is placed over them. This procedure has certain disadvantages in that the raceways can be only about half filled with balls, limiting the load capacity, and the cage must be one-sided or two-piece, limiting high-speed performance.

These disadvantages can be overcome by using angular contact bearings, which are commonly used for accurate, high-performance spin axis applications. Angular contact bearings (Fig. 16) have one side of either the outer or inner race machined away, allowing more balls and more rigid cages to be fitted, which improves their performance. If only enough of the shoulder is removed to allow the balls to clear when the outer race is heated and the inner race is cooled, a *nonseparable* angular contact bearing is produced, or if one side is removed completely, this gives a *separable* angular contact bearing.

In these bearings, the contact angle is the deviation of the contact line from the radial position. An increase in contact angle gives greater end load or thrust capacity. At large contact angles the life of the bearing can be affected, because of the greater deviation from true rolling contact to that with a component of spinning of the balls. The maximum satisfactory contact angle for high speed will be just above 30° when axial and radial stiffness are about equal.

For elimination of shaft play and to produce an accurate spin axis, bearings must be used as preloaded pairs. The preload (Fig. 16) can be introduced by an axial spring, or by solid preloading using differing length spacers between the two outer and the two inner races, to produce fixed compression. For critical applications the preload should be measured and adjusted to give optimum performance and life.

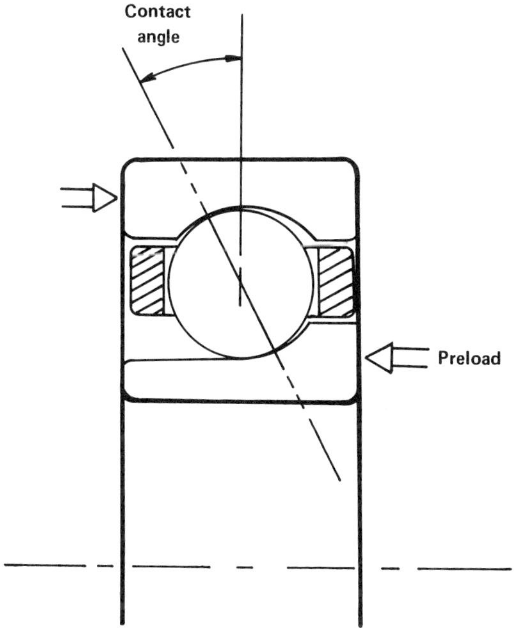


Figure 16 Separable angular contact bearing.

### 3.2. Ball Bearing Standards

Ball bearings bought as separate items from a manufacturer will usually be supplied to ABEC (American Bearing Engineers Committee) or ISO (International Standards Organization) standards. These give limits for the errors measured on the locating faces of either race, when it is rotated, while the other race locating faces are held in a fixed position. Typical figures are shown in Table 1a.

Although the ABEC or ISO standards give some measure of race quality, they omit many important factors. The highest standards of accuracy, life, and performance can only be ensured by (1) buying "specials" (i.e., bearings custom made against a design specification), or (2) quality checking and, if necessary, finishing all bearings immediately prior to instrument assembly by methods similar to those described by Holmes and Shepherd in 1973.

Table 1 Tolerance Errors for Ball Bearings

(a) Precision Ball Bearing Tolerance[a] (in. $\times 10^{-4}$)

| | Bearing class | Radial run-out outer | Bore out of round | Outside diameter out of round | Size bore | Outside diameter |
|---|---|---|---|---|---|---|
| Normal | ABEC 5P | 2.0 | 1.0 | 1.0 | 2.0 | 2.0 |
| Precision bearings | ABEC 7P | 1.5 | 1.0 | 1.0 | 2.0 | 2.0 |
| Special | ABEC 9P | 0.5 | 0.5 | 0.5 | 1.0 | 1.0 |

[a]For shielded bearings, tolerances are worse.

(b) Estimated Bearing Mounted Tolerances (in. $\times 10^{-4}$)

| Optical axis/bearing mounting diameter | Radial run-out | Out of round | Size |
|---|---|---|---|
| Precision | 1.0 | 1.0 | 2.0 |
| Ultraprecision | 0.5 | 0.5 | 1.0 |

The geometrical errors that must be controlled include:

1. Balls: diameter, sphericity, surface finish, and diameter matching of the ball set
2. Outer and inner raceways: raceway pitch circle diameter, roundness, cross track profile (groove radius accuracy), concentricity and squareness of raceway to mounting diameter, squareness of the end face to the mounting diameter, and surface finish.

For the highest accuracy, all these parameters should be controlled on every bearing, and this has become general practice on ultraprecision spin axis instruments such as gyroscopes for inertial navigation. This is a costly procedure and can be carried out only where facilities exist for assembling the bearings after checking.

### 3.3. Spin Axis and Optical Axis Errors

In addition to the bearing run-out errors (Table 1a), the following factors must also be taken into account in assessing spin and optical axis errors:

1. Any clearance fit used in mounting the bearings
2. Distortion of the raceways where an interference fit is used
3. Lack of squareness of parallelism of the races to the mounting surfaces, which will increase their run-out
4. The manufacturing accuracy of the mounting surfaces to the optical axis

Estimates of these figures are shown in Table 1b. When these errors are combined with those from Table 1a, it is obvious that there are built-in limits on the accuracy of optical spin axis that can be achieved with ball bearings. If, for example, the required maximum optical axis error is ±0.1 mrad (±20 arc sec), then for a 1.5-in. length between the bearings, the maximum allowable error at each end would be $0.75 \times 10^{-4}$ in. This could easily be exceeded when the bearing tolerances in Table 1a are added to the mounting tolerances in Table 1b. These errors can be reduced but never entirely eliminated by special manufacturing techniques and by the selection and matching of bearings, which will of course raise costs.

### 3.4. Cages

Cages are also referred to as retainers or separators. The purpose of the cage is to keep the balls in their correct angular position and to provide lubricant as required. They may be located in three ways: (1) on the balls (ball riding), (2) their outer diameter may ride on the outer land (outer riding), or (3) the inner diameter may locate on the inner land (inner riding).

In general, for higher speeds and longer life, one-piece cages of phenolic-impregnated cotton, phenolic paper, porous polyimide, or porous nylon are used. Porous cages can be partly filled with lubricant to provide a reservoir of fresh oil and act as a filter to trap degradation products; they increase bearing life but are more expensive.

While raceways and balls are finished to microinch accuracy, the machining accuracy of plastic cages is necessarily several orders lower. Geometric or surface finish variations inherent in plastic cages are often the source of inconsistent performance in bearings.

To avoid migration of the lubricant to adjacent optical surfaces, the cages should be vacuum impregnated and then centrifuged to an acceleration level higher than they are likely to sustain while running. This step is also employed to reduce the risk of uneven distribution of the lubricant, which could cause running problems.

### 3.5. Lubricants

The choice of oil or grease will affect bearing life and temperature range and the possibility of oil contamination of the optics. Some properties likely to be important are the following.

1. *Volatility* (or vapor pressure at a specific temperature) will affect the condensation of vapor on adjacent optical surfaces. Low-volatility (or high-vapor-pressure) lubricants are available, but even with these, vapor can still be released if lubricant degradation occurs.

2. *Viscosity*, and its associated viscosity indexes (temperature and pressure), determine the film thickness and therefore influence the running life. The change in viscosity with temperature is important if wide variations in temperature are expected; the pressure viscosity index also affects the film thickness because of the extremely high pressures at the ball contact zone.

3. *Oxidation resistance* is directly related to the effectiveness of lubricants in that failures of lubricants in bearings are commonly caused by oxidation of the lubricant. Oxidation produces resins, sludges, and acids, which cause an increase in temperature when running and therefore more oxidation. Partial control is possible by using oxidation inhibitors or by using an inert atmosphere (N, He), or by changing to an oil with greater oxidation resistance (see also Sec. 3.6).

4. *Lubricity* is important but, like a good whiskey, it is difficult to analyze what makes a good lubricant. While running with a true fluid film separation of the surfaces, only the viscosity affects performance, but under adverse conditions, such as touchdown or shock loads when boundary lubrication takes place, good lubricants protect bearing surfaces and themselves usually by beneficial chemical or physical reaction with the surfaces, forming

protective layers. This action is usually too complex for full analysis, and the best way to prove a lubricant is to test run it in instrument bearings. Additives to improve the performance are also often tried by empirical testing.

5. *Hydrolysis* is the reaction of a lubricant with water. Moisture is present in air and can react with lubricant molecules, causing decomposition products (sludges, acids). Hydrolysis increases with temperature, as can the possible moisture content of air.

6. *Surface tension* controls the spread of a lubricant over surfaces. The optimum surface tension is a compromise between the need to "wet" the bearing surface, to ensure a continuous oil film, and the need to keep the lubricant in the bearing. Surface tension depends on both the lubricant and the surface condition (cleanliness). Barrier films have been used to control the spreading of lubricants to nonbearing surfaces; these should be used with great care to avoid contamination of the bearing surfaces, making them nonwettable.

### 3.5.1. Types of Lubricant

The choice of lubricant for precision instrument bearings is probably limited to mineral (petroleum), synthetic diester, or synthetic hydrocarbon (SHC). The poor lubricating properties of silicone oils and greases probably rules them out except where their ability to withstand extremes of temperature or their extremely low volatility are necessary, and limited life is acceptable.

Mineral oils have been used with great success in instrument bearings and have excellent lubricating properties. Synthetic oils have lower volatility, less change in viscosity with temperature, inherent oxidation resistance, and can be used over a wider temperature range. These improved properties could make them attractive for rotary scanners.

Each class of synthetic comprises a wide range of lubricants with differing properties. In general, SHC's have better lubricating properties than diesters and are much more compatible with rubbers and elastomers.

Grease consists of an oil with a thickener; thus it can be packed into a bearing to make more oil available. Common lubricating oils are also often available in grease form. They have some advantages when running in dirtier atmospheres, as the grease also acts as a seal, but they are less suitable at higher speeds, because of higher friction losses.

### 3.6. Life

The life of a ball bearing depends on the establishment and continued existence of the thin EHD fluid film to separate and protect the bearing surfaces. A great many factors can affect this, such as preload, cage type, temperature, atmosphere, lubricant type and quantity, contamination, loads applied (including vibration and shock), raceway geometrical accuracy, and raceway surface finish and speed. Bearing manufacturers seldom have control of these factors and cannot therefore guarantee life; however, a close liaison between manufacturer and user is always beneficial.

Fatigue life is usually quoted in manufacturer's literature, but this is seldom a factor in precision instrument bearings. In the author's experience of many thousands of precision gyroscope bearings, fatigue has never been the cause of failure, even after life tests exceeding 50,000 hr running.

Bearing failure is often seen as lubricant degradation (or polymerization); this is an increasing thickening and darkening of the lubricant. The breakdown is often progressive, as the thickening causes increased drag, higher temperatures, and so further degradation, often ending in complete failure of the bearing to run. Failure can usually be traced to:

1. Inadequate specification or control of the bearings or lubricant, resulting in poor geometry or surface finish, wrong bearing type, incorrect contact angle, unsuitable cage, unsuitable lubricant, and so on.
2. Mishandling or poor assembly methods, resulting in contamination of lubricant or bearing surfaces, excessive preloading, brinelling, surface damage, and so on.
3. External influences, such as extremes of temperature, shock, slew, vibration, and so on.
4. The cage, which needs special mention, as it is the most difficult component to control adequately (see Sec. 3.4): lubricant failure is often initiated at the cage rubbing surfaces.

Despite this catalog of potential troubles, where ball bearings are properly selected, handled, and installed, there is no reason why true fluid film separation of surfaces should not be achieved, ensuring long life and consistent performance. There will be a few where statistical chance works against this, and it is advisable to test run all bearing assemblies for a time to weed these out.

Bearing life decreases as speed increases above a certain level, where dynamic forces on the cage, ball set, and lubricant cause such increased frictional drag that overheating and lubricant breakdown occur.

## 4. SPIRAL GROOVE LIQUID BEARINGS

Plain oil bearings were available long before either ball or gas bearings, but it is only since the emergence of the spiral groove bearing in the early 1960s that such bearings using oil have become contenders for precision spin axis applications. Spiral groove oil bearings are similar to those using gas described in Section 2.3, but because of the much higher viscosity of oil, they will be much smaller for the same load capacity. This makes them especially suitable for small-diameter, high-load-capacity instruments.

Unlike gas bearings, spiral groove oil bearings do not normally have the advantage of being surrounded by their lubricant supply, and the oil must therefore be retained within the bearing. When rotating, the pumping action of the spiral grooves does this, but when stationary the oil would normally leak out; this has been a major problem in the development of these bearings. Another problem is that voids can occur in the fluid because of cavitation or air entrapment at the inlet.

Methods of preventing leakage when stationary include (1) the use of barrier film coatings; (2) using ferrofluids, which are held in the bearing gap by magnetic force; and (3) using special greases. Good results on the use of special greases to overcome both leakage and voids have been reported by Muijderman et al. in 1980, on the application of spiral groove journal bearings to video equipment spindles. Another method to overcome the problems of these bearings described by Molyneaux in 1982 is to completely surround the bearing with a chamber, attached to the rotor, filled with a grease developed to stay in position, but to have low shear viscosity.

All these methods, while successful, have some limitations (e.g., in size, speed, or design adaptability). Further development is probably required before spiral groove oil (or grease) bearings become generally available for use in rotary scanners.

The simpler construction of these bearings makes them potentially more accurate than ball bearings, but they will probably have similar limitations, in speed, temperature range, and because of oil migration to ball bearings. They should have greater shock capacity than ball bearings, since the load will be distributed over a larger area.

Within their range of application, they should cost less to manufacture than spiral groove gas bearings; the larger bearing gap will ease tolerances. They are also likely to be inherently stable, because of the grooves and the damping of the oil (see Sec. 2.3.4).

## 5. PERFORMANCE COMPARISON

Any comparison of the types of bearing covered (gas, ball, liquid) must necessarily be generalized because of the large number of variables outlined in previous sections. There are general characteristics, however, which make one type more suitable for certain applications. Table 2 summarizes some of the most likely requirements, together with the suitability of the types of bearing considered to meet these requirements.

It is only where more than one type would be suitable for a particular set of requirements that a choice is necessary. To aid this choice, the advantages and disadvantages of all four types are listed in more detail in Section 5.1.

Spiral groove liquid bearings are not generally available and only limited information has been published on their performance. They are most likely to compete with ball or gas bearings in the lower-speed, higher-accuracy range.

Externally pressurized bearings are most likely to be used in less compact applications, where their excellent performance over a wide speed range, or their ability to work in dirtier atmospheres, is needed and where an external pressure supply can be provided.

The most likely area of contention is the choice between angular contact ball bearings and spiral groove gas bearings, for compact, medium-speed (up to 25,000 rpm) applications. A more detailed discussion on the merits of these two types is therefore given in Section 5.2.

This comparison cannot be comprehensive over such a wide possible field; magnetic, squeeze film, and ferrofluid bearings could also have been considered, as could hybrids between these and the other types. Magnetic and squeeze film bearings are more specialized and complicated than any of the others, and would probably be considered only if none of these were suitable. Ferrofluid bearings, as such, have low load capacities and are most likely to be used, as seals, to extend the performance of the other types, for example, to prevent oil migration from ball bearings, or to allow self-acting gas bearings to be used in dirty atmospheres, or to hold lubricant in place in spiral groove oil bearings. Finally, it should be emphasized that this comparison is for guidance only, as all the bearing types considered are capable of

Table 2 Summary of the Performance of Likely Bearing Types[a,b]

| Feature | Bearing type | | | |
|---|---|---|---|---|
| | Ball | Pressure gas | Self-acting gas | Self-acting liquid |
| Speed (rpm) | | | | |
| 0-1000 | L | G | P | L |
| 1000-20,000 | G | G | G | G |
| 20,000-30,000 | L | G | G | L |
| Above 30,000 | P | G | G | P |
| Spin axis error (arcsec) | | | | |
| < ±10 | L | G | G | G |
| > ±10 | G | G | G | G |
| Life (hr) | | | | |
| 0-1000 | G | G | G | G |
| 1000-20,000 | L | G | G | G |
| Above 20,000 | P | G | G | L |
| Low noise | L | L | G | G |
| Noncontaminating | L | L | G | L |
| Use in dirty atmosphere | L | G | P | L |

[a]This is a general comparative guide only; special designs could improve on these ratings.
[b]G, Good performance; L, limited performance; P, poor or no performance.

improvement, by special design or manufacturing, to take them outside the limits of performance suggested.

## 5.1. General Comparison

### 5.1.1. Ball Bearings

*Advantages*

Can be run in either direction

Will usually continue to run after shock loads (but performance and life will be affected)

Can be run in relatively dirty atmosphere

Inexpensive for low-performance scanners

Readily available

More compact than gas bearings

Lower starting torque than self-acting gas bearings

*Disadvantages*

Limit on spin axis accuracy

Possibility of oil migration onto optics

Limit on speed

Limit on life

Susceptible to damage by shock loads especially when stationary

### 5.1.2. Self-Acting Liquid Bearings (Spiral Groove)

*Advantages*

High-accuracy spin axis possible

Very compact

Should have long life (limited information available)

Low noise

Fairly easy manufacturing tolerances

*Disadvantages*

Possibility of oil migration onto optics

Very limited availability

Limit on speed

Small size may limit shaft stiffness

### 5.1.3. Self-Acting Gas Bearings (Spiral Groove)

*Advantages*

Extremely high spin axis accuracy possible

High speed

Almost unlimited life in clean atmospheres

No contamination of optics

Compact

Wide temperature range

Very low noise

*Disadvantages*

Run in only one direction

Need reasonably clean atmosphere

Less widely available than ball bearings

Higher starting torque than ball bearings

Tight machining tolerances

### 5.1.4. Externally Pressurized Gas Bearings

*Advantages*

Extremely high spin axis accuracy possible

Good performance from zero to high speed

Very high stiffness possible

Long life with clean air supply

Wide temperature range

Fairly easy manufacturing tolerances

*Disadvantages*

Need separate air supply

Fairly bulky

Can be liable to instabilities

Sealed system impossible because of exhaust gas

## 5.2. Choosing Between Ball Bearings and Self-Acting Gas Bearings

Generally, ball bearings have been used in lower-speed, lower-accuracy systems, while for higher-speed, higher-accuracy systems it is necessary to use gas bearings. It is in the medium-accuracy, medium-speed (up to 25,000 rpm) range that the choice is not so clear cut and would depend on the particular specification. Some of the factors likely to be considered are reviewed here.

### 5.2.1. Bearing Power

Figure 17 shows a comparison between measured running power for spiral groove conical bearings and bearing manufacturers' data for ball bearing dynamic lubricant drag. The ball bearings shown are R2 (3/8 in. nominal OD), R3 (1/2 in. nominal OD) and R4 (5/8 in. nominal OD), which cover the likely range of sizes for rotary scanners. The gas bearings considered are of roughly similar load capacity.

The effect of lubricant viscosity dominates ball bearing drag torque, as illustrated in Table 3, where a 40°C fall in lubricant temperature produces a fourfold increase in drag. In practice, a viscosity should be chosen to have sufficient elastohydrodynamic film thickness (say, 6-12 μin.) to ensure a safety margin over raceway surface roughness at maximum operating temperature. Gas bearing running power is proportional to gas viscosity, which changes little with temperature.

### 5.2.2. Spin Axis Errors

Gas bearings are inherently more accurate because they have only two surfaces separated by a gas film, whose compressibility averages out roundness errors of these surfaces. Errors should therefore be an order of magnitude lower than for ball bearings.

As described in Section 3.3, a spin axis accuracy of better than ±0.1 mrad (±20 arc seconds) is much more difficult to achieve with ball bearings. However, a few manufacturers are now quoting ball bearing axis errors of less than half this figure.

### 5.2.3. Life

The life of an instrument ball bearing is usually terminated by lubricant degradation (Sec. 3.6), which depends on factors such as speed, load, atmosphere, bearing quality, and lubricant. The life of a gas bearing, if it is run in a clean atmosphere, is almost certainly superior to that of a ball bearing.

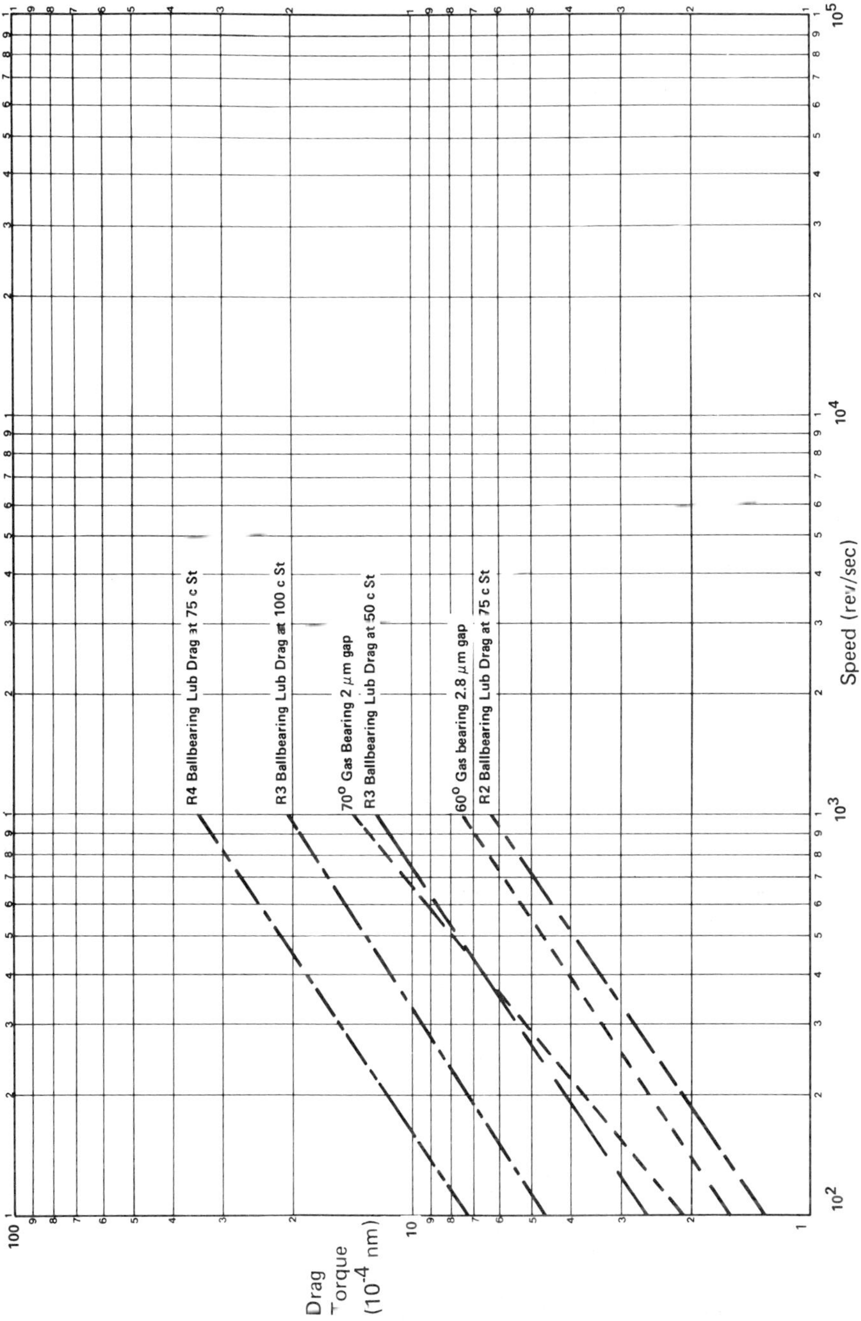


Figure 17 Comparison of ball bearings and self-acting gas bearing: drag torque versus speed.

Table 3 Effect of Lubricant Viscosity on Ball Bearing Drag and EHL Film Thickness[a]

| Lubricant | | Minimum EHL film[b] thickness (m × $10^{-6}$) | | Drag torque (N-m × $10^{-4}$) | |
|---|---|---|---|---|---|
| Temp. (°C) | Viscosity ($10^{-3}$/N-m/$sec^2$) | 100 rev/sec | 250 rev/sec | 100 rev/sec | 250 rev/sec |
| 63 | 50 | 0.1 | 0.2 | 3 | 5 |
| 47 | 100 | 0.18 | 0.33 | 4 | 8 |
| 21 | 430 | 0.48 | 1.0 | 11 | 20 |

[a]Typical results for an R3 bearing.
[b]Minimum elastohydrodynamic film thickness between balls and inner race.

### 5.2.4. Shelf Life

Ball and gas bearings should both have adequate shelf life under suitable storage conditions.

### 5.2.5. Vibration and Shock

1. Bearings stationary. This is the worst condition for a preloaded ball bearing because of the high local stresses (hertzian stresses) already present in the ball contact zone and the absence of an oil film between the surfaces. On the other hand, a gas bearing will take higher loads because of the larger areas and close conformity of bearing surfaces.

2. Bearings rotating. In this case the elastohydrodynamic film and the distribution of the load stresses lead to a large increase in the ability of ball bearings to withstand vibration. When running, gas bearings have been shown to withstand similar levels of vibration to ball bearings.

### 5.2.6. Contamination from Bearings

With ball bearings, the lubricant may be gradually centrifuged from the bearing, especially at higher speeds, or may migrate as a vapor as temperature increases. Suitable shields or seals may be required to protect adjacent optical surfaces. Gas bearings produce no significant contamination.

### 5.2.7. Number of Starts

There is no normal limit to the number of startups with a ball bearing. For a gas bearing, where surface rubbing occurs during startup, some wear occurs, and special, hard (antiwear) surfaces are used. With suitable surfaces, gas bearings have been shown to be capable of tens of thousands of starts/stops without degradation in bearing performance.

### 5.2.8. Speed

It is probable that the top speed of ball bearing scanners should be between 25,000 and 30,000 rpm. Higher speeds can be achieved but only at the expense of life, even with the highest-quality bearings. In this case the cost would be similar to gas bearings, but their performance would be worse.

Self-acting gas bearing designs have been run successfully at speeds from about 1000 rpm to 500,000 rpm. In practice, the upper speed limit will normally depend on overall design considerations, such as the mass and inertia of the rotor, dynamic forces generated, and external forces applied.

### 5.2.9. Cost

Experience has shown that for high-accuracy systems the cost of ownership of gas bearing rotary scanners can be similar to or less than that for equivalent ball bearing versions. For low-accuracy, low-speed systems ball bearing scanners should cost less.

### 5.2.10. Atmosphere

Ball bearings can be run in very low atmosphere with special lubricants. They will have longer life in a clean inert-gas atmosphere. Gas bearings require a clean dry gas atmosphere for maximum life. Some designs have been run in an office and laboratory atmosphere with good results.

## REFERENCES

Archard, J. F., and Cowking, E. W. (1965). Elastohydrodynamic lubrication at point contacts. *Proc. Inst. Mech. Eng.*, *180*, Pt. 3B, 47-56.

Burghdorfer, A. (1959). The influence of the molecular mean free path on the performance of hydrodynamic gas lubricated bearings. *J. Basic Eng., Trans. ASME, 81D*, 94-99.

de Ferranti, S. Z. (1909). *Air Bearing for High Speeds.* U.S. Patent 930,851.

Faddy, D. (1969). The theoretical performance of statically loaded pairs of conical gas bearings. *U.K. Royal Aircraft Establishment Report 69174.*

Grassam, N. S., and Powell, J. W. (1964). *Gas Lubricated Bearings*, Butterworth, London.

Gross, W. A. (1962). *Gas Film Lubrication*, Wiley, New York.

Harrison, W. J. (1913). The hydrodynamical theory of lubrication with special reference to air as a lubricant. *Trans. Camb. Philos. Soc., 22*, 39-59.

Holmes, J., and Shepherd, J. (1973). *International Ball Bearing Symp.*, Charles Stark Draper Lab., Inc., Cambridge, Mass.

Kingsbury, A. (1897). Experiments with an air lubricated journal. *J. Am. Soc. Nav. Eng., 9*, 267-292.

Licht, L., and Elrod, H. (1960). A study of the stability of externally pressurized gas bearings. *J. Appl. Mech., Trans. ASME, 82*, 250-258.

Molyneaux, A. K. (1982). Theoretical and empirical evaluation of spiral grooved grease bearings. *Proc. U.S./U.K. Fluid-Film Bearing Meet.* Southampton University, U.K., pp. 62-76.

Muijderman, E. A., Remmers, G., and Tielmans, L. P. M. (1980). Grease lubricated spiral groove bearings. *Philips Tech. Rev., 39*(6-7), 184-198.

Reynolds, O. (1886). On the theory of lubrication and its application to Mr. Beauchamp Tower's experiments including an experimental determination of the viscosity of olive oil. *Philos. Trans. R. Soc. London, 177*, 157-234.

Sixsmith, H. (1959). The theory and design of a gas lubricated bearing of high stability. *First Int. Symp. Gas-Lubricated Bearings*, Washington, D. C., pp. 418-434.

Tower, B. (1883). First report on friction experiments. *Proc. Inst. Mech. Eng.*, (a) 1883, pp. 632-659; (b) 1884, pp. 29-35. Also, Second report on friction experiments. *Proc. Inst. Mech. Eng.*, 1885, pp. 58-70.

Vohr, J. H., and Pan, C. T. H. (1968). Design data of gas-lubricated spin axis bearings for gyroscopes. *Mechanical Technology, Inc., (New York) Technical Report M.T.I.-68 TR 29.*

Whipple, R. T. P. (1949). Herringbone pattern thrust bearing. *U.K. Atomic Energy Authority Report AERE-T/M-29.*
Wilcock, D. F., ed. (1972). *Design of Gas Bearings*, Vols. 1 and 2, Mechanical Technology, Inc., New York.

# 5

# Galvanometric and Resonant Low Inertia Scanners

Jean Montagu
*General Scanning, Inc., Watertown, Massachusetts*

## 1. INTRODUCTION

Low inertia scanners have long been used by science and industry in optical scanning. These scanners were used in the early electrocardiograph recorder and in the first oscilloscope, as well as in the making of film sound track. Advances in computer and laser technology have stimulated their evolution, and they are now found in complex precision instruments.

The term "low inertia scanner" (LIS) designates any scanner that deflects light with the aid of a moving mirror and which can achieve full scanning speed in less than about 1 sec. There are two fundamentally different types of low inertia scanners: galvanometric and resonant.

Galvanometric scanners can produce a steady-state deflection and follow with considerable fidelity sinusoidal as well as nonsinusoidal wave forms such as sawtooth, triangular, and random access. Resonant scanners can operate at only one frequency or its immediate vicinity, and they can oscillate only in harmonic motion; they are of interest because much higher scanning angles can be achieved at high frequencies by taking advantage of resonant amplitude magnification.

This chapter will begin with a description of typical applications of LIS, summarize the performance provided by galvanometric and resonant scanners, and finally show how such performance is achieved by design.

## 2. APPLICATIONS OF LOW INERTIA SCANNERS

Like most industrial products, the low inertia scanner (LIS) has evolved to meet the needs of its users. Literally hundreds of different applications have appeared in industry, medicine, military hardware, and communications. Some of these are artificial horizon displays for aircraft cockpits, scanning laser acoustic microscopes, silicon wafer annealers, bar code readers, birthmark erasers, lead bonders, microfilm equipment for computer output, color separators for printing, component insertion designators, laser communication systems, image digitizers, dye laser tuners, facsimile transmitters and receivers, film thickness monitors, surface flaw detectors, wire-harness wire-path designators, interferometers, infrared scanners, laser trimmers, laser cutting machines, laser markers, laser printers, liquid crystal large-scale displays, light and sound shows, 35-mm movie converters for television, night vision systems, laser ophthalmoscopes, printed circuit board exposure and inspection equipment, phototypesetters, prepress proofers, Q-switches, ring lasers, range finders, scanning infrared microscopes, optical disk readers, robotic vision systems, and war game simulators.

Although the market for each application is different, applications tend to include common technological disciplines. Therefore they can be grouped in specific categories.

### 2.1. Visual Communication

Visual communication is one of the most demanding applications of the LIS. The human eye has not only an astounding resolution, but also very fast and accurate autocorrelation capabilities which permit the detection of minute image imperfections. Physiologists have shown that the human visual system readily detects pixel position errors as small as 1/1000 in. and that it is even more sensitive to the "periodic scanning artifacts" which consist of periodic shading variations, generally called a venetian blind defect. The numerous applications of visual communication can be classified under the following categories:

1. Print plate exposure
2. Laser printers
3. Phototransmission
4. Photo enhancement and manipulation

5. Phototypesetting
6. Computer output on microfilm
7. Large-scale image projection

### 2.1.1. Print Plate Exposure

The use of the hot lead linotype process began to diminish shortly after World War II when the first phototypesetters were introduced. At present, newspaper composition is largely automated. Text, headlines, line art, photographs, and advertisements are paginated and edited into final form via computerized systems.

As computer input devices for image data, or digitizers, scanners convert text and graphics into picture data elements which can be manipulated digitally. Capabilities range from processing letter-sized documents to full newspaper pages with monochromatic or full-color content. As output devices, or plotters, scanners are capable of scanning a full printing plate (typically 20 in. × 24 in.) as well as a single photograph or poster for enlargement or reduction. The scanners must be fast, reliable, versatile, and economical, with resolution compatible with the industry. Originally drum scanners were used. Flatbed scanners are now accepted because they are simpler, faster to operate and can handle nonflexible paste-ups.

Flatbed scanners share a common layout. The work to be digitized or the photographic film to be sensitized is translated at constant speed under a beam scanned along the other axis. Optical elements are usually located near the work, assuring field flatness and normal incidence at all points. Pixel position is usually determined with the help of a grating. In order to achieve small spot size, the scanning mirror is greater than 1 in. in diameter.

As a digitizer, the focused laser beam is absorbed or reflected by the work, and the reflected light is collected by a wedge or bundle of fiber optics located in close proximity to the trace and then guided to a photomultiplier for further processing. As a plotter, a modulator interrupts the beam to create the desired image. In order to correct for line straightness, a small high-speed linear scanner corrects the laser beam prior to its entering the beam expander. Figure 1 is a schematic presentation of such a unit built around the SX resonant scanner of General Scanning Inc. (GSI), and Figure 2 is a photograph of the scanner itself.

Table 1 shows the range of the performance needed for system continuity in the printing industry.

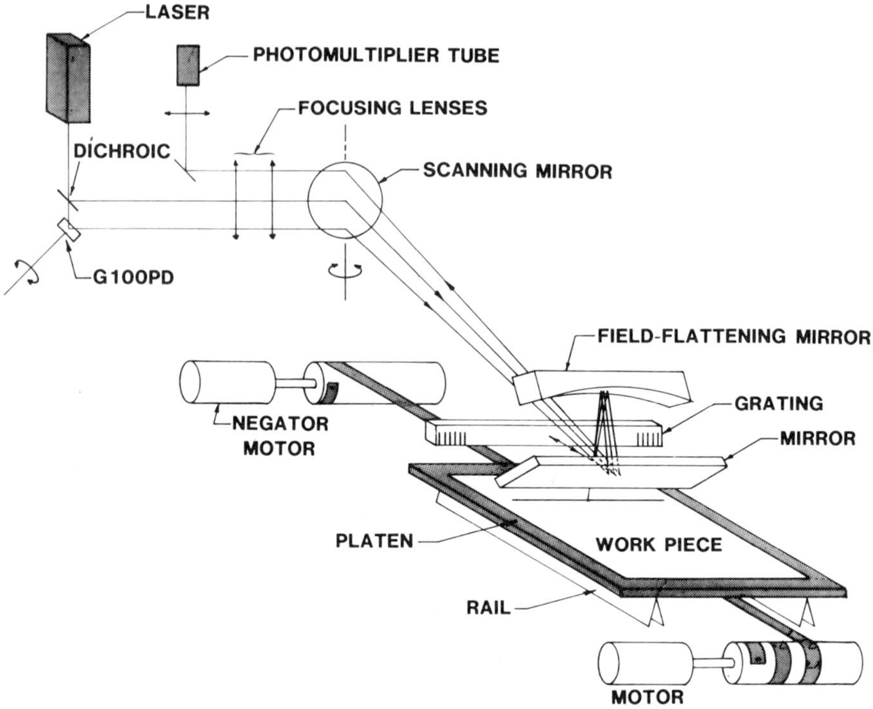


Figure 1 Flatbed scanner system. (Courtesy of Optronic International, Chelmsford, Massachusetts.)

Printing quality varies as widely as applications do. The requirements shown in Table 1 are typical of those for scanners and plotters used in the printing industry. The reader unfamiliar with graphic art reproduction may find these unnecessarily demanding; however, the production of sharp, clean pictures can only be obtained with resolution accuracies of the magnitude cited in Table 1. These dictate in turn the use of large mirrors and high-speed scanners with extremely low wobble as indicated in Table 2.

### 2.1.2. Laser Printers

A new generation of computer printers combines electrostatic and laser printing technologies. These are the low-cost versions of the first-generation laser printers such as the IBM 3800 or Siemens

3352. Their output format, letter size $8\frac{1}{2}$ in. × 11 in., replaces the perforated computer printout. Their print quality is comparable to that of a conventional typewriter; in addition, they make it possible to produce graphics and half-tone pictures with eight possible recognizable shades of gray. Though these printers are similar in concept to the plate-making printer of Figure 1, they use a much larger pixel, smaller format, and lower line density. This leads to simpler and more economical constructions, as field-flattening optics and a grating reference are not required.

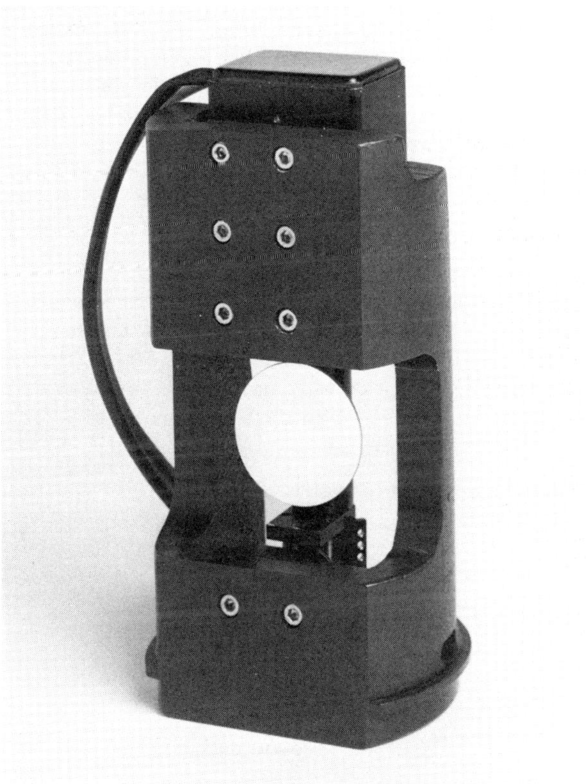

Figure 2 SX scanner. (Courtesy of General Scanning, Inc., Watertown, Massachusetts.)

**Table 1** Flatbed Scanner System Performances

| | |
|---|---|
| Half-tone dot density | 80–200 per inch |
| Scan line per inch | 800–2000 per inch |
| Line spacing position tolerance | 1–2 μm |
| Resolution | 12,000–25,000 lines per inch |
| Maximum plate size | 20 in. × 24 in. |
| System speed | 1–2 min per plate |

The resolution limitations of the xerographic technology and the small format opens the way for scanners which can be produced in large volume at low cost. However, the office environment in which these lightweight machines must operate demands that the scanners be capable of operation under moderate vibration. Table 3 shows expected system performance, and Table 4 shows typical scanner performance requirements for such an application.

### 2.1.3. Phototransmission

In electronic printing or facsimile transmission, images are digitized at one location and transmitted via telephone lines or satellite to another location, where they are reconstructed. Phototransmission handles continuous tone information as well, and unattended operation at the receiving end is possible. The data per image are

**Table 2** Typical Resonant Scanner Specifications for Flatbed Scanners

| | |
|---|---|
| Beam rotation | 32° |
| Resonant frequency | 150 Hz |
| Mirror dimension | 35 mm diameter |
| Mirror inertia | 39 g-cm$^2$ |
| Repeatability, scan/scan | 0.1 arcsec maximum |
| Wobble, scan left to right/ scan right to left | 1 arcsec maximum |
| Life | $10^{11}$ cycles minimum |

Table 3 Laser Printer Performances

| | |
|---|---|
| Line density | 200–300 lines per inch |
| Pixel density | 250–300 pixels per linear inch |
| Pixel overlap | 50% |
| Spot size | $4–6 \times 10^{-3}$ in. |
| Speed | 3–5 sec per page |

considerable because of the gray tone information attached to each pixel.

Laser scanning systems printing on dry silver, heat-developed photographic paper are the only systems now used for newspaper phototransmission. A standard 8 in. × 10 in. picture with a resolution of 5 lines per millimeter requires 4 or 5 min for image transmission. Reproduction speed is low because it is not commercially justifiable to use transmission lines with a greater capacity than the 4800-Hz bandwidth of the telephone lines. Police or military systems which operate without communication bandwidth limitation transmit at a rate of 60–200 scan lines per second for both transmission and reception.

Since such pictures are intended for viewing rather than scaling, avoiding image discontinuities is important, because the human eye readily detects a pixel 0.1 mm out of position. Critical scanner features are repeatability, low wobble, and low jitter.

Table 4 Scanner Performances

| | |
|---|---|
| Mechanical | |
| Oscillating frequency | 600–1200 Hz |
| Mechanical deflection | 20–30° |
| Mirror size | 10 mm |
| Wobble | 7 arcsec maximum |
| Environmental | |
| Vibration | 1 G DC, 100 Hz |
| Temperature | 20–100°C |

Linearity and temperature stability are sometimes required by special applications and environmental conditions.

### 2.1.4. Photo Enhancement and Manipulation

By using lasers of appropriate wavelength, pictures can be digitized and separated according to their basic colors. Typically HeNe and ArKr lasers are used. Image data are either manipulated and printed on the run on color film or paper or stored on magnetic medium. Image sections can be changed or added; color replaced, enhanced, or dissolved; shadows created; and faults removed. Algorithms can retrieve underexposed negatives, compensate for a limited range of photographic printing paper, or even regroup the picture element to create special visual effects. This technology has been developed and refined for the printing and advertising industries as well as for military uses.

The LIS is commonly used for photo manipulation because it requires very simple optics. Rotating polygons are considerably more expensive, and they require color- and pincushion-corrected, field-flattening telecentric diffraction-limited lenses, which are very expensive.

### 2.1.5. Phototypesetting

In phototypesetting the image of a font character generated by a cathode ray tube or a rotating set of photographic negatives is positioned in line on photographic film. Printing plates in relief are created by reimaging and etching on special materials. The LIS is especially well suited for this application. Figure 3 is a schematic presentation of a modern Mergenthaler Company laser typesetter using an LIS scanner.

### 2.1.6. Computer Output on Microfilm

The output of computer data directly on microfilm makes it possible to produce microfiches with user-readable information. Two machines now commercially available are capable of printing over 200 pages a minute, with 64 lines of 132 characters, as well as producing graphics.

The "Komstar," shown schematically in Figure 4, uses a line scanner to position characters on a line, and a page scanner to position lines on a page. Each line scanner moves an array of nine beams. The characters are composed by turning each beam on or

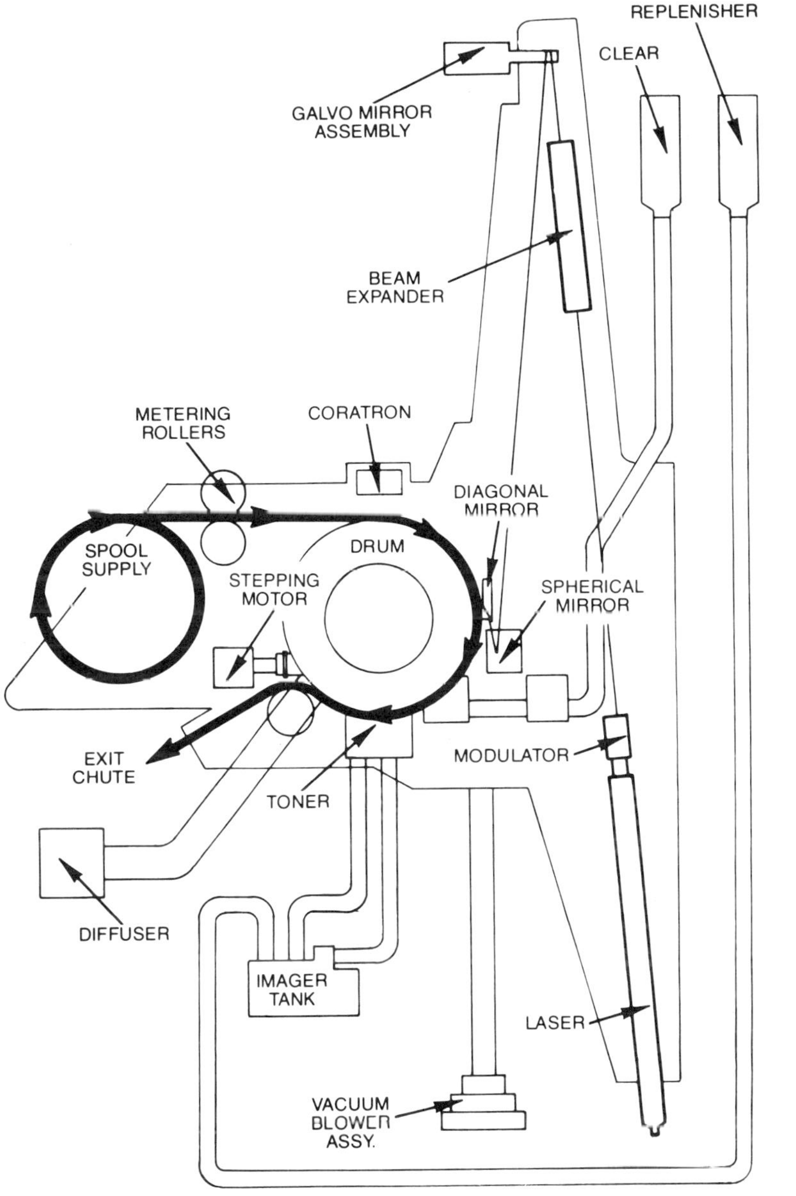


Figure 3 Laser typesetter. (Reprinted from *Photonics Spectra*, May 1982.)

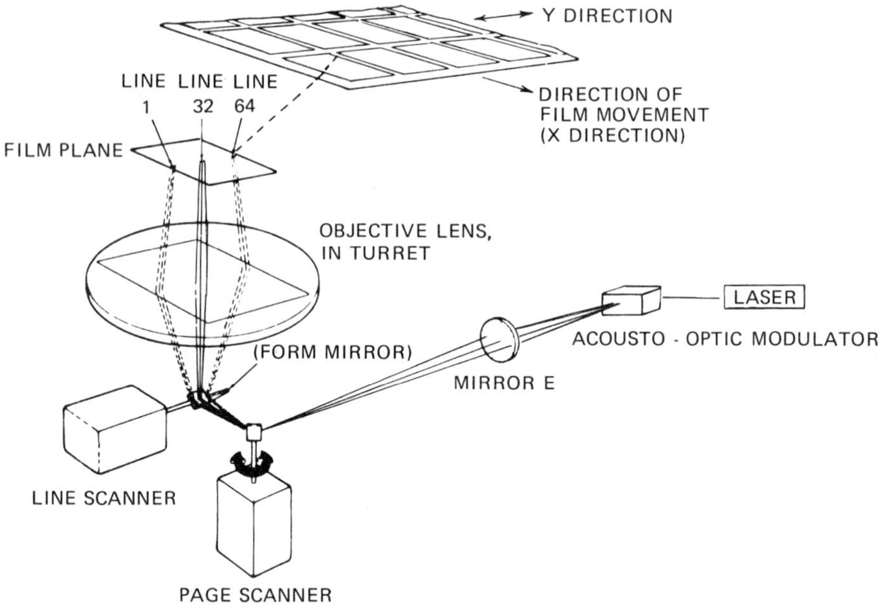


Figure 4 "Komstar." (Courtesy of Eastman Kodak Co., Rochester, New York.)

off at any of the 1500 possible positions on a line. The nine-beam array is created from a single laser passing through an acousto-optic modulator operating at nine different frequencies between 40 and 50 MHz. Each frequency is controlled in amplitude to ensure individual beam output independent of the number of beams sharing the same source at any instant. Of the 4.8-msec cycle of the line scanner, 1 msec is reserved for retrace and settling. The scanner operates with optical position feedback and locates characters with an accuracy of 1 part in 10,000 of line length. The cross-axis wobble of the two scanners and the interaction between them must be below 5 arcsec to produce consistent accuracy.

The ERIS II, manufactured by Datagraphix, Inc., has similar performance characteristics but uses a holographic rotating scanner to position a four-beam cluster to generate symbols. A number of line scans is required to complete a line of characters. Page scan is achieved with an LIS. The scan duration is 150 msec, with an accuracy of 30 arcsec and less than a 5-arcsec wobble.

### 2.1.7. Large-Scale Image Projection

Images with large and changing information content are often displayed by large-scale projection. These types of displays are used in applications such as public address panels; traffic control displays for air, city, and harbor activity; interactive computer-aided design/computer-aided manufacturing (CAD/CAM) output for the design of complex products such as very large-scale integrated (VLSI) chips or hydraulic systems for aircrafts or power plants; and CAD/CAM microfiche or microfilm generation.

A high contrast image is created on a 2-cm to 10-cm square intermediate substrate with a modulated laser scanned in vector, raster, or spiral mode. It is then projected enlarged 10–50 times onto a viewing screen by conventional methods. The intermediate substrate, called a light valve, is either a photochromic film or a smectic liquid crystal cell.

The resolution of these displays exceeds the best cathode ray tube screens, with densities of 2000–64,000 points per axis. Refresh rates often exceed 30 sec; although slow, the refresh speed is consistent with other limitations such as data handling and audience information assimilation. Figures 5 and 6 are schematic presentations of two such displays.

Scanner resolution, wobble, and jitter are critical parameters for these applications. Most systems use LIS for both axes; in those where vector generation is not necessary, rotating polygons may be used for one axis.

## 2.2. Artificial Extension of Vision

Artificial extension of vision systems includes both high-precision machines, such as scanning grating spectrometers, graphic digitizers, and laser gages, and low-resolution systems, designed to extend vision senses into the infrared, ultrasound, or microscopic fields and to display the information on a conventional TV monitor for visual interpretation.

### 2.2.1. Scanning Grating Spectrometer

A conventional spectrometer design can be dramatically improved with a high-precision LIS to support and scan the grating. As each scan is calibrated with a known mercury or other source, long-term drifts can be compensated for. Commonly achieved LIS angular

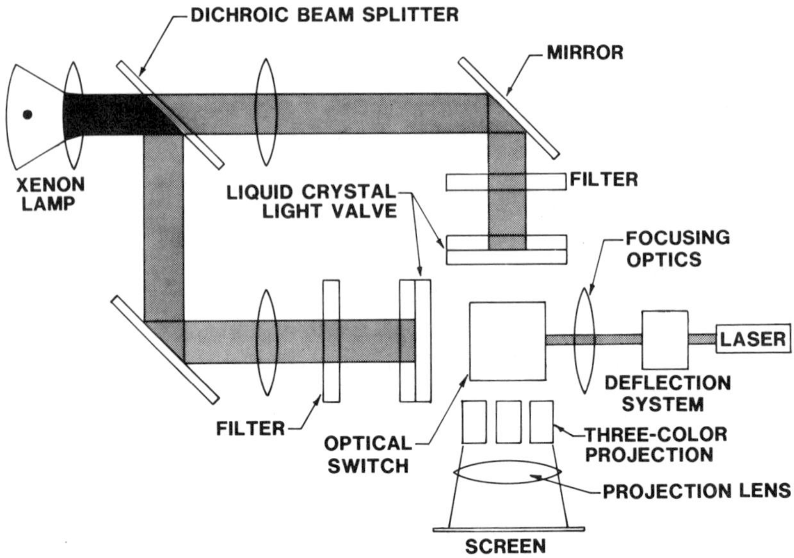


Figure 5 Liquid crystal display. (Reprinted from *Electro-Optics System Design*, May 1982.)

resolutions of 1 part in 262,000 correspond to a spectral resolution smaller than 1/40 Å in the visible range.

### 2.2.2. Forward-Looking Infrared Sensors

Forward-looking infrared sensors (FLIRS) and night vision systems are made for both civilian and military applications. The civilian sector includes applications such as heat loss surveys for homes and pole-mounted transformers, forest fire control, and security. In the military FLIRS are used to protect foot soldiers, as well as tanks, aircraft, and ships.

All objects emit electromagnetic radiation (light). By measuring the intensity and color of this radiation, a thermal image of a field of view can be created. A fairly good representation of the scene is possible even through smoke and vapor, however, the quality of the image is strongly dependent upon the thermal resolution of the system.

A modern FLIRS uses a single or double HgCdTe detector, cooled at liquid nitrogen temperatures to receive a narrow light

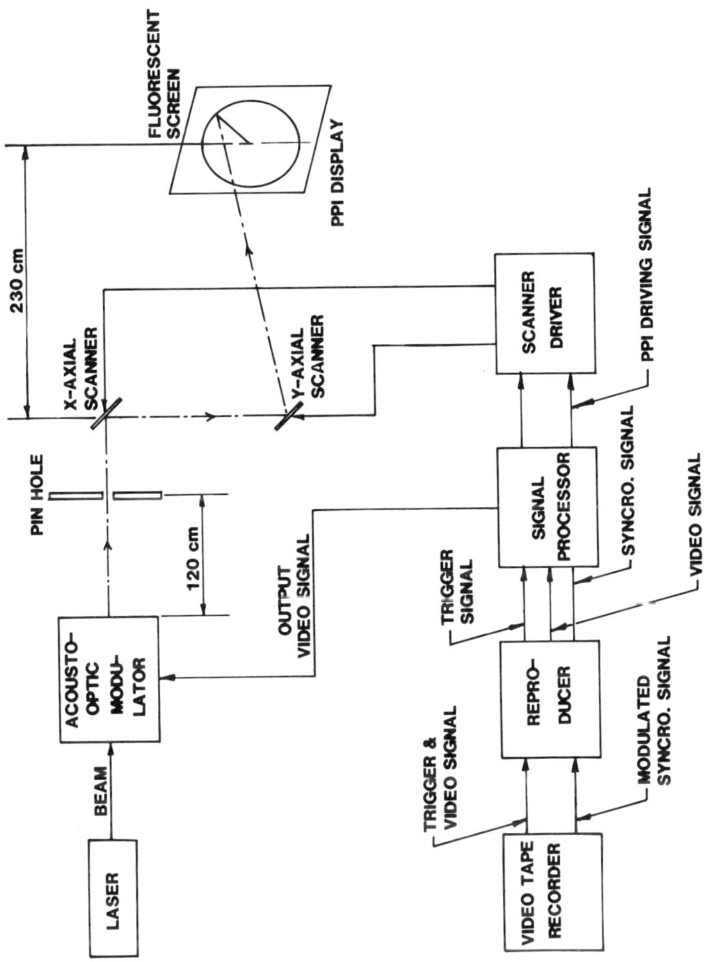


Figure 6 Radar imaging. (Courtesy of Society of Information Display, Los Angeles, California.)

bundle of limited spectral range. A scanner, acting as a television camera in a raster mode, aims the detector to a sequence of points in the object field. Since commercial TV raster scan rates are used, this equipment interfaces with conventional TV equipment.

Figure 7 is a schematic view of a FLIRS using two LIS. The horizontal trace is obtained by a resonant scanner generating 7866 scan lines per second with a viewing angle of 28°. A dual-element detector obtains conventional TV raster density; a position sensor, closed-loop galvanometric LIS generates the frame scan. The thermal resolution of the system is 0.1°C, and image resolution is comparable to that of commercial TV. Temperature is color coded for ease of interpretation.

Because of their long procurement cycles FLIRS presently used by the military incorporate traditional rotating polygon scanners. The advantages of resonant scanners, that is, greater reliability, compactness, less power consumption, and freedom of gyroscopic effect, are now being recognized, particularly by non-U.S. military procurers.

### 2.2.3. Laser Scanning Microscope

Conventional optical microscopes are limited in resolution by the wavelength of light, and their depth of focus at high magnification is inconveniently small. Laser scanning microscopes have similar limitations, but, because images are created by accumulating discrete picture elements, they can be assembled from different focus planes by a tracking dynamic focusing scheme. Even in applications where an image is not required, significant advantage is obtained by employing laser scanning techniques. For example, it is possible to detect light diffusely scattered from surface features much smaller than the focused beam spot. Dust or defect detection scanners for silicon wafers or masks use this property.

An important application in the semiconductor industry is the infrared microimaging capability of this technology. Figure 8 is a schematic description of such a unit with a conventional microscope used for alignment.

### 2.2.4. Scanning Laser Acoustic Microscope

A scanning laser acoustic microscope is a specialized application of the scanning microscope. The work is subjected to standing acoustic waves which create surface deformation conditions representative of subsurface density variations. The scanning microscope digitizes this surface information so that it can be interpreted and

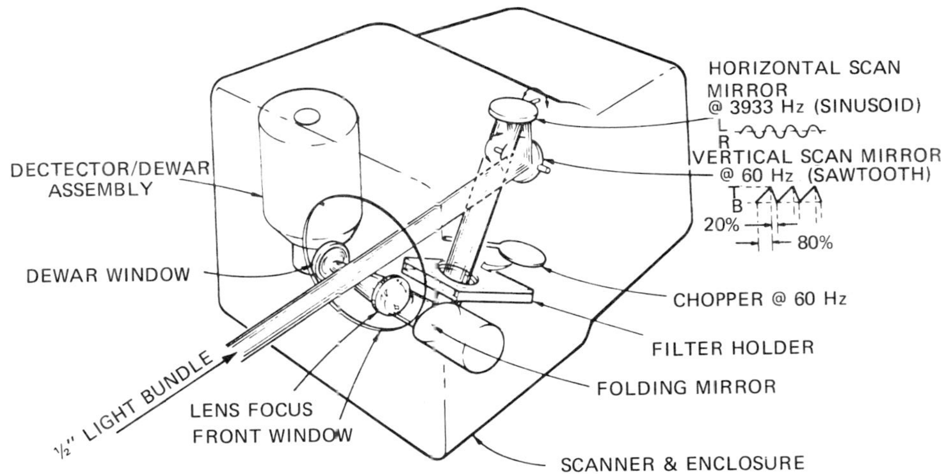


Figure 7 Forward-looking infrared sensors. (Courtesy of Inframetrics, Inc., Bedford, Massachusetts.)

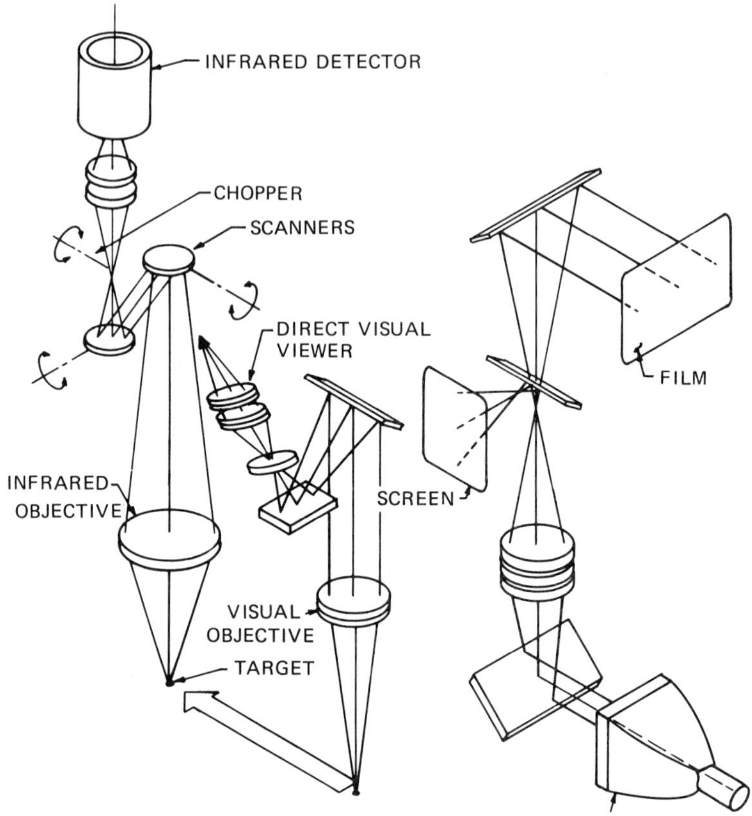


Figure 8 Infrared microscope. (Courtesy of Barnes Engineering, Inc., Stamford, Connecticut.)

displayed on a monitor. Figure 9 shows a system where the line scan is obtained from an acousto-optic scanner and the frame rate with a LIS. Acousto-optic scanners will not be discussed here.

## 2.3. Medical Applications

The LIS has found its way into a wide number of medical applications, notably ophthalmology.

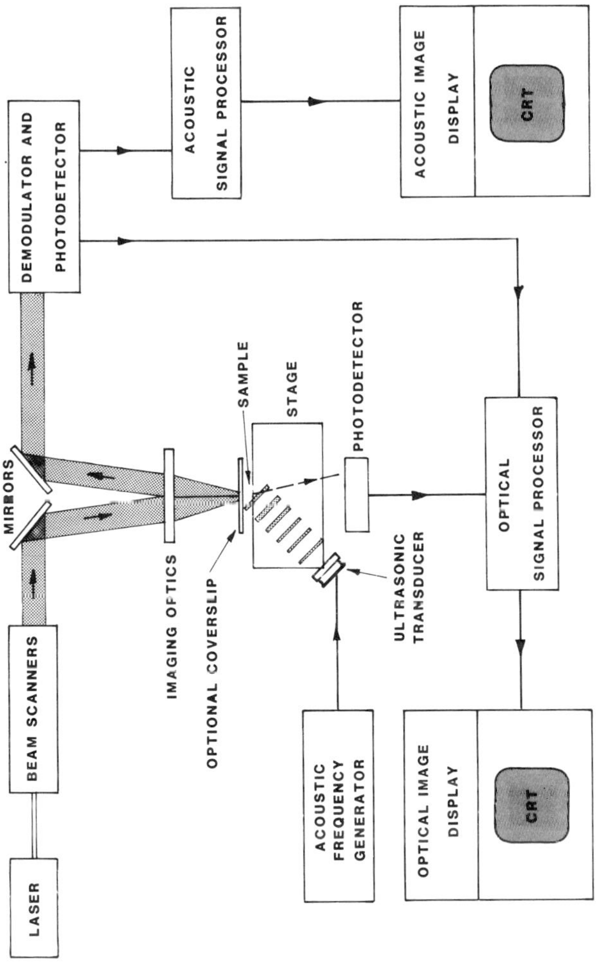


Figure 9 Scanning laser acoustic microscope. (Courtesy of Sonoscan, Inc., Bensenville, Illinois.)

### 2.3.1. Computerized Electro-Oculographic System

Computerized electro-oculographic system techniques are used to gain information about patients with disturbed sensations of balance. Such information is acquired by observing the patient's eye movement during certain kinds of visual and vestibular stimuli. A two-dimensional LIS scanning system is used to present such stimuli in the patient's field of vision.

### 2.3.2. Three-Dimensional Visual Stimulus Deflector

This deflector was made for general vision experiments such as the study of saccadic movements. A typical instrument is shown

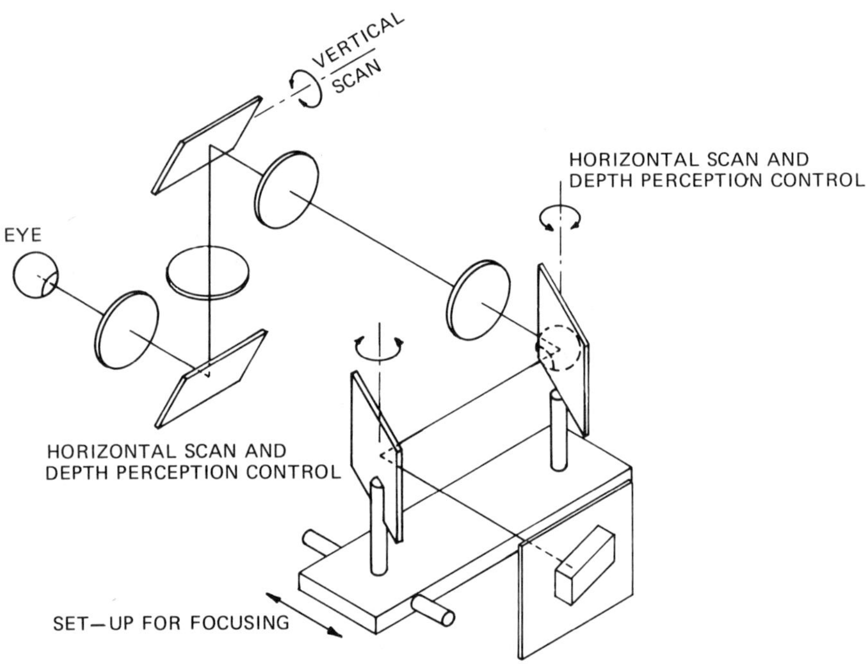


Figure 10 Three-dimensional visual stimulus deflector. (Courtesy of H. C. Crane and Michael R. Clark, S.R.I. International, Menlo Park, California.)

in Figure 10. The speed of eye response (0.005 sec) and visual acuity (0.3 mrad) determine the required performance of the scanners.

### 2.3.3. Scanning Laser Ophthalmoscope

The scanning laser ophthalmoscope is used to view the fundus oculi, the retina, and its supporting structures, including blood vessels, nerve bundles, and underlying layers. The instrument, built by R. H. Webb and G. W. Hughes at the Retina Foundation in Boston, is a TV ophthalmoscope. It uses a flying spot scanner to achieve high light collection efficiency and reduce the light needed for viewing and recording. A resonant scanner provides the horizontal scan at the commercial TV rate and a galvanometric scanner provides the 60-Hz frame rate. The viewing field is 55° peak to peak.

Figure 11 is a schematic description of the scanning laser ophthalmoscope's optical path. The instrument uses 2% of the pupil opening to create the raster scan on the retina, and 98% for collection of reflected light. It resolves 10 μm on the retina with 5 mW of laser power. A HeNe and an ArKr laser are used to cover the visual spectrum. The power density on the retina does not exceed 50 $\mu W/cm^2$. Continuing developments will lead to greater resolution, and eventually performing laser surgery while viewing the retina may be possible.

### 2.3.4. Diagnostic Eye-Tracking Instruments

Diagnostic eye-tracking instruments to screen reading disabilities have been built, and miniaturized versions are used in pilot training.

## 2.4. Material Processing

Industry has accepted high-power lasers as working tools. Where it is necessary to displace the beam, the LIS offers excellent features of programmability, random addressing, speed, precision, and power handling. Many industrial applications exist for welding, cutting, stitching, annealing, drilling, and inducing local chemical reactions.

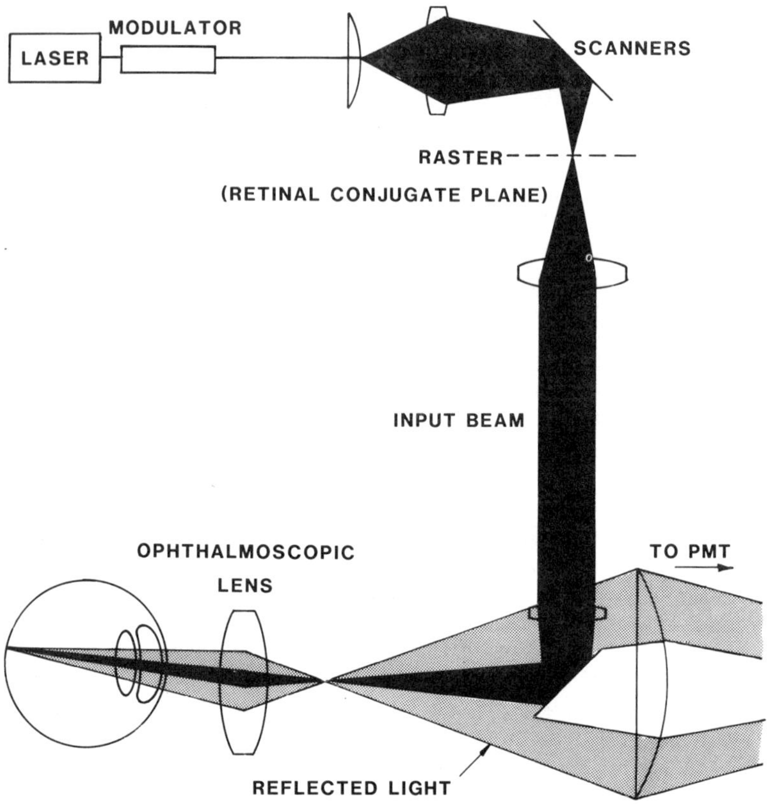


Figure 11 Scanning laser opthalmoscope. (Courtesy of Dr. R. H. Webb and G. W. Hughes, Eye Research Institute of the Retina Foundation, Boston, Massachusetts.)

### 2.4.1. Laser Trimming

Laser trimmers are used to adjust by ablation the value of resistors formed by thin-film or thick-film technology. With the growth of the semiconductor industry, system speed, accuracy, and reliability have grown more demanding and the LIS has become the standard solution to beam positioning. The main optical element of these systems is a very precise laser flying spot scanner as shown in Figure 12. A YAG laser is used for cutting, and a colinear

HeNe Laser is used for programming by observation through a TV camera. A high-quality, multielement, flat-field focusing lens ensures uniform spot size over the entire work area. Performance of such a beam positioning head is shown in Table 5.

The success of these trimmers has lead to the development of new machines capable of operating directly on integrated circuit chips in order to correct faulty elements or to customize devices. In these machines the beam is focused to a 6-μm spot and positioned to an accuracy of 1.5 μm with a system resolution of 0.16 μm. Position accuracy is 1 part in 64,000 in each axis. Such new trimmers have contributed to solving problems of low yield in 64K and 256K memory chips.

Figure 12 Laser Trim system. (Courtesy of Teradyne, Inc., Boston, Massachusetts.)

Table 5 Beam Positioner Head Performance

| | |
|---|---|
| Work area | 2 in. × 2 in. |
| Repeatability | $1 \times 10^{-3}$ in. |
| Resolution | $0.5 \times 10^{-3}$ in. |
| Trim resolution | $0.025 \times 10^{-3}$ in. |
| Peak acceleration | 800 *G*'s |
| Settling time, maximum | 12.5 msec |
| Spot size | $1 \times 10^{-3}$ in. |
| Depth of focus | $22 \times 10^{-3}$ in. |

### 2.4.2. Laser Marking

Marking products with proper identifying codes or serial numbers can be a difficult manufacturing task, and laser scanning systems have proven to have important advantages over mechanical and inking processes, especially when the marking must be indelible. The first laser markers were introduced more than a decade ago as a natural extension of laser resistor trimming. Now hundreds of systems have been installed. Scanning laser markers are best suited for engraving metal parts, for situations requiring rapid change of marking such as in serialization, and for the marking of large symbols on nonflat surfaces such as on the ends of compressed-air cylinders or pipes.

A typical scanning laser marker is shown schematically in Figure 13. A Q-switched or continuous-wave YAG laser usually supplies the power. More recently shorter-wavelength lasers such as nitrogen or eximer lasers have been used successfully to mark silicon wafers.

Marking applications cover a wide variety of parts, from automobile engine blocks to hypodermic needles. One of the larger-volume applications is wafer, chip, and package identification or serialization in the semiconductor industry. Figure 14 shows typical YAG laser marking patterns.

As marking is expected to be machine readable, location and type must be quite accurate. The optical system has a low F number to reduce the required power and to minimize ablation debris; therefore, field-flattening lenses or focusing systems are desirable to provide a constant spot size. The performance parameters of a typical laser marker are shown in Table 6.

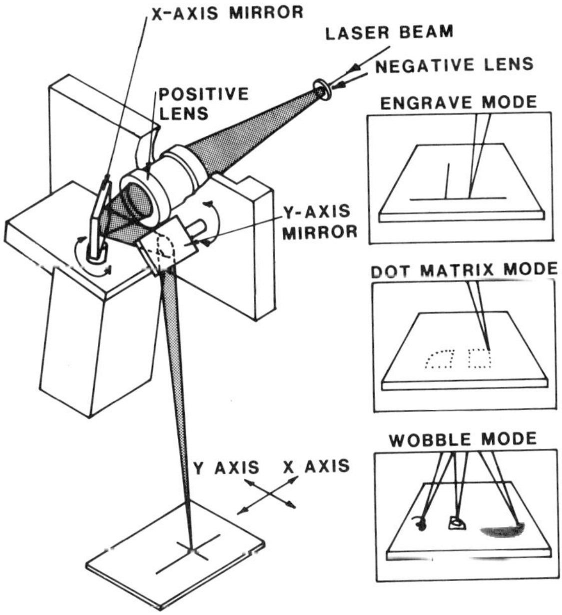


Figure 13 Scanning laser marker. (Courtesy of Quantrad Corporation, Torrance, California.)

Figure 14 Laser marking patterns. (Courtesy of ADE Corporation, Newton, Massachusetts.)

Table 6 Laser Marker Performance Parameters

| | |
|---|---|
| Mirror throughput | 20 mm diameter |
| Scanner response time | 5 msec |
| Beam excursion | 32° |
| Linearity error | 0.3% of peak to peak |
| Null drift | 12 arcsec/°C |
| Marking field | 12 mm × 45 mm |
| Marking rate | 12 characters per second |
| Marking accuracy | |
| Intracharacter | ±.005 mm |
| Intercharacter | ±.012 mm |
| Character font | |
| Modified OCR-A | 5 × 9 dot matrix, continuous |

### 2.5. Entertainment

The use of lasers for entertainment began in the late 1960s, and scanners were incorporated soon after. The first scanners were loudspeakers with mirrors glued to the cone. Later on, diffraction devices, mirrored spheres, and moving coil galvanometers were used. In the early 1970s laser displays, in which modulated Lissajous figures and mandalas typified the esthetics, used all the capabilities of the open-loop scanners then available. In the fall of 1975 lasers with scanners made graphic images. The November cover of *Industrial Research and Development* was a photograph of the magazine's initials laser scanned on the clouds over Boston, Massachusetts. Computer-generated animation, which used the advances in hardware, software, and laser technologies, soon followed.

Scanned laser graphic projections are best known for the planetarium shows and animated "son et lumiére." They also are effective in enhancing meetings and promotional presentations, as well as advertisements and art.

The image is produced by aiming a laser beam at a screen and repetitively deflecting the beam to retrace a controlled path. When the time period is shorter than that of vision decay, the viewer perceives the path as a persistent line. A flicker-free image requires a minimum retrace rate of 15–25 Hz, depending on the color, light intensity, and ambient illumination.

### 2.5.1. Studio Projector

The first entertainment shows were designed as visual representations of music. Suitable signals were filtered out of the sound track and used as source data to drive scanners directly or to stimulate auxiliary drive patterns; they were analog systems. Modern laser entertainment studios create specialized graphic generation systems adapted to the amplitude–bandwidth constraints of low inertia scanners.

Scanner and amplifier dynamic responses determine the complexity and often the message content of the imagery. All other parameters have wide tolerances. The small-angle, 2°, response of 100 μsec is commonly demonstrated with the G120D scanner of Table 9. This is possible because the components, the system, or the dynamic distortions are compensated by the graphic artist as he or she views the imagery being generated. Images with the equivalent complexity of 20 detached cursive letters can be presented in vector mode and flicker free with only one pair of scanners.

The studio has a computer-based system with many editing features which are used to develop animation through scan-to-scan dynamic image deformation. Figure 15 presents the elements of a studio with full graphic capability. A typical program is composed of elements held on magnetic tape, disks, or programmable read-only memories (PROMs) and then stored on sets of PROMs for later use. The playback unit or projector also has manually controlled functions such as interpolation, pan, zoom, rotation, as well as color hue and intensity. The operator can, if needed, follow actors or speakers on stage.

### 2.5.2. Programming

The process of programming is best described by one of its practitioners, Jennifer Morris of the Image Engineering Corporation:

> The most familiar laser graphics effect, widely used in light shows and discotheques, is the lissajous figure. More complex abstract designs, with circular motif, can be achieved by an analogue rotator.

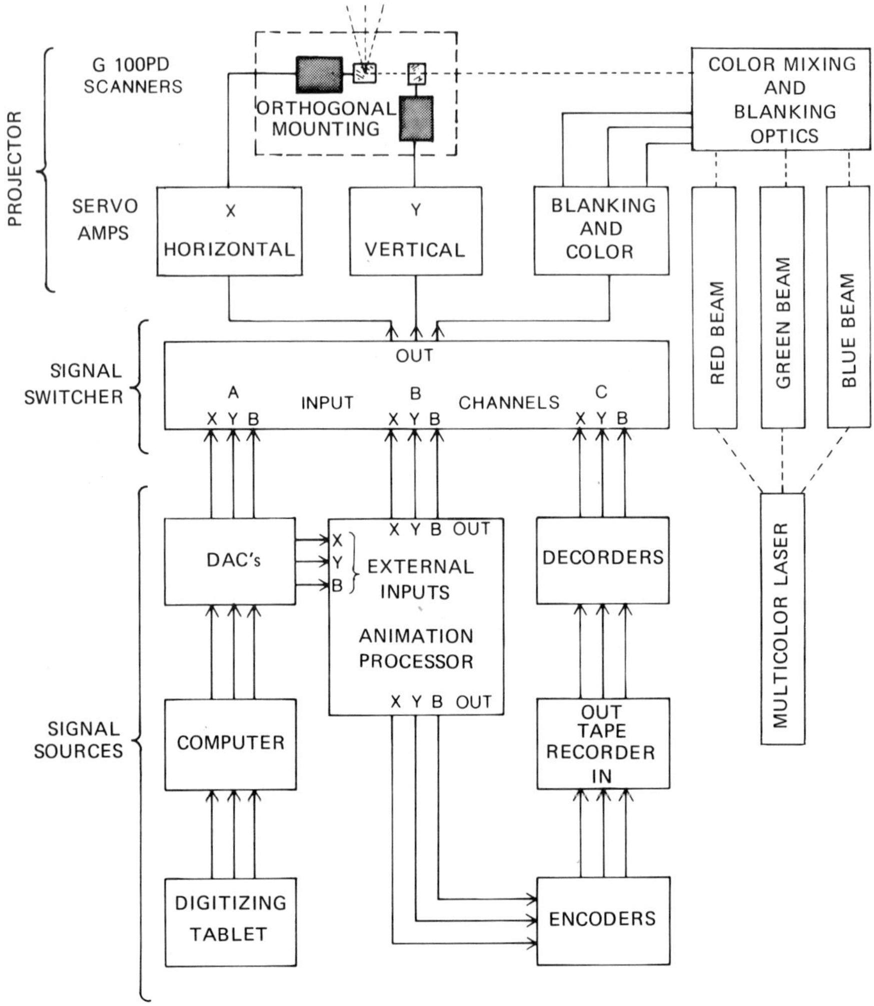


**Figure 15** Studio/projector functions. (Courtesy of Image Engineering, Inc.)

Representational imagery, such as cartoons, messages, logos and other illustrations, are generated by means of an electronic data tablet interfaced to a microcomputer. An artist may then trace a line drawing and have it translated into the X–Y coordinate points stored in the computer. As the images are played back, they may be animated by using special circuits. Animated sequences can also be stored on magnetic tape with FM encoding.

A third signal is required to generate blanking of the laser beam. This makes it possible to display images with discontinuous lines. Blanking may also be used with Lissajous pattern generators to create more intricate dynamic designs.

In a full-spectrum color projector, three blanking signals are required, one each for red, blue, and green. These signals also contain intensity level information. With this system, it is possible to create images with precise, discrete colors, from soft pastels to vibrant primaries, at specific points within the image.

## 3. PERFORMANCE OF LOW INERTIA SCANNERS

The above review of typical applications of LIS defines the range within which LIS devices must often operate. Table 7 recapitulates the typical performance requirements for the galvanometric LIS, and Table 8 does the same for the resonant scanner.

### 3.1. Performance of Galvanometric Low Inertia Scanners

The galvanometric LIS is essentially an actuator of high torque to inertia, capable of precise movement and stable characteristics. Characteristics relating to the accuracy of servo-controlled moving iron scanners have been studied by Brosens (1976) for speed, drift, wobble, and linearity. The data shown in Tables 9 and 10 were obtained with the latest commercial general-purpose scanner and commercial error-integrating drive electronics optimized for both speed and accuracy.

### 3.2. Performance of Resonant Low Inertia Scanners

Performances of resonant LIS vary dramatically with mounting techniques and must be verified on the final instrument base. Improper

Table 7 Parametric Requirements of Typical Galvanometric LIS Applications

| | Retrace or step time (msec) | Frequency (Hz) | Mirror size (mm) | Type of scan |
|---|---|---|---|---|
| Phototransmission | 1–6 | 2–133 | 10 | Linear |
| Computer output microfilm | 1 | 200 | 7 | Linear |
| Phototypesetting | 8 | Asynchronous | 30 | Step |
| Trimming of thick-film resistors | 20 | Asynchronous | 30 | Step |

Table 8 Parametric Requirements of Typical Resonant LIS Applications

| | Retrace or step time | Frequency (Hz) | Mirror size (mm) |
|---|---|---|---|
| TV Display of infrared scans | NA | 3937 or 7875 | 15 or 8 |
| Newspaper plate image transmission | NA | 175 | 35 |
| Intelligent copiers | NA | 1000 | 10 |

| Absolute mechanical position accuracy (μrad) | Mechanical wobble (μrad) | Mechanical scan angle (rad) | Mechanical repeatability (μrad) | Time scan efficiency (%) |
|---|---|---|---|---|
| 2000 | 50 | 0.20 | 40 | 80 |
| 35–100 | 35 | 0.15 | 35–100 | 80 |
| 1000 | 50 | 0.15 | 75 | NA |
| 50 | 50 | 0.10 | 10 | NA |

| Absolute mechanical position accuracy (μrad) | Mechanical wobble (μrad) | Mechanical scan angle (rad) | Mechanical repeatability (μrad) | Angular scan efficiency (%) |
|---|---|---|---|---|
| 2000 | – | 0.25 or 0.20 | 20 | 85 |
| 1000 | 5 | 0.30 | 2 | 75 |
| 2000 | 35 | 0.30 | 5 | 80 |

Table 9 Step Response of Moving Iron Galvanometric Scanners with Closed-Loop Integrating Amplifier

| Scanner type | GSI G120D | | | GSI G325D | | |
|---|---|---|---|---|---|---|
| Mirror | | | | | | |
| Face (mm) | 7 × 7 | | | 26 × 26 | | |
| Thickness (mm) | 1 | | | 3.2 | | |
| Inertia, with mount ($g\text{-}cm^2$) | 0.016 | | | 3.3 | | |
| Rotor inertia ($g\text{-}cm^2$) | 0.023 | | | 3.7 | | |
| Stator | | | | | | |
| Height (mm) | 33 | | | 79 | | |
| Cross section (mm) | 23 × 33 | | | 32 × 45 | | |
| Weight (g) | 107 | | | 435 | | |
| Life (cycles) | $10^{10}$ | | | $10^{10}$ | | |
| Bearing noise (arcsec) | 10 | | | 5 | | |
| Step response (msec) for angular step | a | b | c | a | b | c |
| 1° | 0.50 | 0.5 | – | 3.4 | 2.8 | 4.2 |
| 5° | 0.55 | 0.57 | 1.0 | 3.5 | 3.4 | 4.4 |
| 10° | 0.63 | 0.75 | 1.1 | 3.6 | 3.7 | 4.6 |
| 15° | 0.70 | 0.95 | 1.1 | 4.1 | 4.4 | 4.8 |
| 20° | 0.80 | 1.1 | 1.1 | 4.6 | 4.8 | 5.1 |
| 25° | | | | 4.9 | 5.4 | 5.4 |

Conditions: Closed-loop integrating amplifer
a. Settled accuracy or error: 1% of step angle
b. Settled accuracy or error: 75 arcsec
c. Settled accuracy or error: 0.1% of step angle

Table 10 Drift of Scanner Transducers

| Scanner type | GSI G120D | GSI G325D | State of the art |
|---|---|---|---|
| Test range (°C) | 26–46 | 28–46 | 10–50 |
| Gain drift (%/°C) | 0.03 | -0.16 | 0.007 |
| Zero drift (μrad/°C) | 100 | 100 | 2 |

Table 11 Typical Performance Characteristics of Resonant Scanners

| Scanner construction | Taut band | X-Flexure | Torsion bar | Isogonic | IDS | Double node |
|---|---|---|---|---|---|---|
| Performance range: resonant frequency, mirror diameter/angle (Hz/mm/deg) | 10/26/30 to 800/5/2 | 5/40/30 to 200/40/15 | 750/10/7 to 16,000/2/1.5 | 800/10/30 to 3500/7/10 | 600/10/30 to 4000/7/10 | 4000/15/15 to 18,000/3/4 |
| Typical Performance | | | | | | |
| Resonant frequency (Hz) | 100 | 150 | 1000 | 1000 | 1000 | 4000 |
| Temperature stability (ppm/°C) | 1000 | NS[a] | 500 | 200 | 200 | 200 |
| 2/1 amplitude change (ppm) | NS | NS | NS | 200 | 400 | NS |
| Mechanical scan angle (deg) | 15 | 15 | 6 | 30 | 30 | 15 |
| Temperature stability (ppm/°C) | NS | NS | NS | NS | 100 | NS |
| Mirror dimension (mm) | 15 × 16 | 40, diameter | 10 × 10 | 10, diameter | 10, diameter | 15, diameter |
| Mirror flatness (λ/cm) | ½ | ½ | NS | 1 | ½ | ½ |
| Wobble (μrad) | NS | 5 | NS | NS | 35 | NS |
| Repeatability (μrad) | NS | 0.5 | NS | NS | 5 | 5 |

Table 11 (Continued)

| Scanner construction | Taut band | X-Flexure | Torsion bar | Isogonic | IDS | Double node |
|---|---|---|---|---|---|---|
| Power requirement (mW) | 40 | 1000 | 300 | 500 | 100 | 100 |
| Operating temperature (°C) | -20–40 | NS | -40–40 | -55–77 | 0–77 | NS |
| Weight (g) | 21 | 800 | 56 | 70 | 100 | 50 |
| Vibration sensitivity | Poor | Excellent | Good | Very good | Excellent | Very good |

[a]Not specified.

*Source*: Taut-band and torsion bar data from American Time Product data sheets; X-flexure, Isogonic, IDS, and double-node data from General Scanning, Inc., data sheets.

mounting permits energy transfer with the instrument structure and affects the following features, depending on the type of scanner:

1. Damping and current consumption
2. Scan amplitude
3. Cross-axis motions

Table 11 shows the performances of a number of commercially available resonant LIS. Normally these performances can only be obtained when the LIS are installed on structures with recommended inertia and/or isolation.

## 4. DESIGN OF SCANNING SYSTEMS WITH LOW INERTIA SCANNERS

The selection of a scanning system configuration is usually the result of an evaluation process in which the system designer weighs the relative merits and drawbacks of specific approaches. This evaluation addresses aspects such as the following:

Spot size, quality of spot, and accuracy

Spot travel speed and repetition rate

Spot size variation over the scanned field due to the shape of the work surface, lens field variation, and the chromatic response of the lens

Angle of attack of beam and specular reflection

Motion of the work surface

Technical literature on these subjects is generally available to the designer. Urbach et al. (1982) have reviewed their application to optical scanning with high inertia scanners, while Montagu (1976b) has analyzed the needs of low inertia scanners. Some of the theoretical constraints involved are described below.

### 4.1. Dynamic Considerations

The dynamic behavior of resonant LIS is predictable (Zook, 1974), and manufacturer's data should be used for design purposes.

Special consideration needs to be given to scanners' ability to exchange energy with the frame.

The dynamic performances of nonresonant LIS can be described by the bandwidth and step response of a second-order system. If it is assumed that drive power and control are of secondary importance, Newton's law applies and the limiting condition is as follows:

$$T = I\frac{d^2\theta}{dt^2}$$

where T is the torque required to impart the angular acceleration $d^2\theta/dt^2$ and I is the moment of inertia. To optimize a reciprocating motion system, equal energy and time must be allowed for both acceleration and deceleration of the mirror and armature. The maximum potential mechanical energy W which can be given to a rotating system is expressed as

$$W = \tfrac{1}{2}T\beta$$

where $\beta$ is the total angular displacement (peak to peak). This energy can be expressed as equivalent dynamic energy by the kinetic-energy equation as follows:

$$W = \tfrac{1}{2}I\left(\frac{d\theta}{dt}\right)^2$$

Solving the above three equations for time yields the expression for the minimum stepping time or step response:

$$t = 2\sqrt{\frac{I\beta}{T}}$$

### 4.2. Geometric Considerations

Other aspects of the scanning system design which are less well known relate to the problems encountered specifically in the scanning process by mirror rotation. These have been treated by Pelsue (1983) and others. They are the following:

Focus error in postobjective scanning of flat fields

Pincushion error in two-dimensional postobjective scanning of flat fields

Electronic correction of error.

### 4.2.1. Focus Errors in Postobjective Scanning

Most schemes for focus and pincushion correction are preobjective or postobjective. These terms describe the position of the objective (or imaging lens) in relation to the scanning elements. Therefore a preobjective system is one in which the scanning mirrors act upon the laser beam before the beam is imaged, and a postobjective system is one in which the scanning elements act upon the laser beam after the imaging optics. Diagrams of both types of systems are presented in Figures 16 and 17. Figure 18 shows how the focus error for two-mirror, two-axis flat-field scanning is derived. In this configuration, a is the center of the X mirror, b is the center of the Y mirror, and c is the point at coordinates $(0, Y_i)$; d is the length from b to (0,0) and e is the length from a to b. The optical scanner angles are $\theta_x$ and $\theta_y$, and the coordinate $(X_i, Y_i)$ is any point on the target fields. It can be seen that when $X_i = Y_i = 0$, then $\theta_x = \theta_y = 0°$. The equation that relates $Y_i$ to $\theta_y$ is derived from the triangle of points (0,0), $(0, Y_i)$, and d. Solving for the length (0,0) to $(0, Y_i)$, which equals $Y_i$, we get

$$Y_i = d \tan \theta_y$$

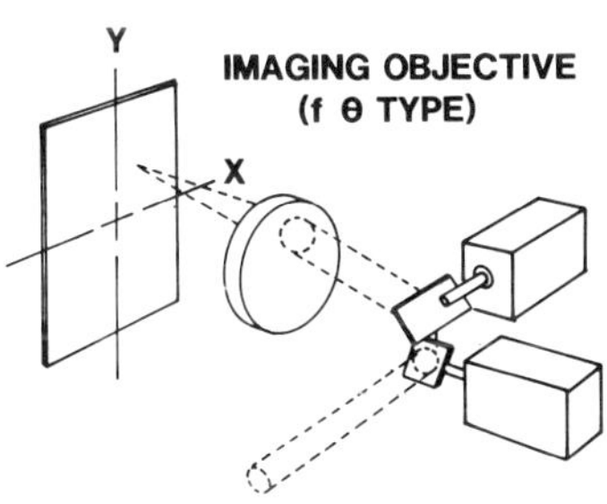


Figure 16 Preobjective scanning.

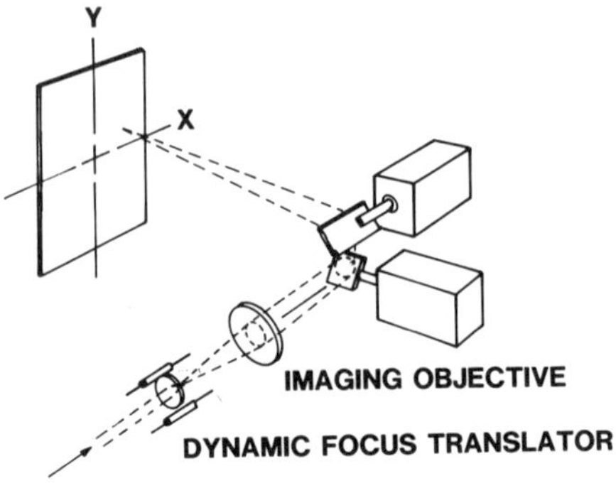


Figure 17 Postobjective scanning.

The determination of the X equation is somewhat more complex and is best illustrated by projecting the target image onto the virtual image position of the Y mirror, as shown by the phantom lines and phantom coordinates $(0,Y_i)$, $(X_i,Y_i)$, and a in Figure 18. By solving the triangle of points a, $(0,Y_i)$, $(X_i,Y_i)$ for the length $(0,Y_i)$ to $(X_iY_i)$, which equals $X_i$, we have

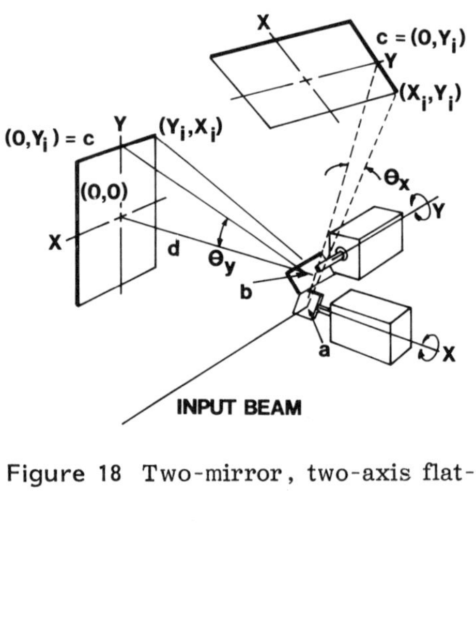


Figure 18 Two-mirror, two-axis flat-field scanning.

$$X_i = ac \tan \theta_x$$

Since $ac = \sqrt{d^2 + Y_i^2} + e$, where $e = ab$, the solution is

$$X_i = \left(\sqrt{d^2 + Y_i^2} + e\right) \tan \theta_x$$

If we solve for the length from a to $(X_i, Y_i)$ we will get the equation for the focus length:

$$f_i = \sqrt{(\sqrt{d^2 + Y_i^2} + e)^2 + X_i^2}$$

The resulting change in focus length for $(X_i, Y_i)$ is

$$\Delta f = \sqrt{(\sqrt{d^2 + Y_i^2} + e)^2 + X_i^2} - (d + e)$$

### 4.2.2. Pincushion Errors in Two-Dimensional Scanning

Two-dimensional scanning can be accomplished with simple two-mirror systems as illustrated in Figure 18, with the use of relay lenses as illustrated in Figure 19, or with a paddle two-mirror system as illustrated in Figure 22.

*4.2.2.1. Pincushion errors in relay lens scanning.* Relay lens scanning makes it possible for the intersection of the scanning beam with the optical axis, or scan pivot point, to be a fixed point in space for all scan angles. This configuration is shown in Figure 19. The scan axis of the first scanner intersects the optical axis at point A, the focal point of the first relay lens with focal length $f_1$. A second relay lens, of focal length $f_2$, is located at a distance of $f_1 + f_2$ from the first. The scan axis of the second scanner is oriented orthogonally to the axis of the first scanner, and it intersects the scan axis at point B. For the case where a fixed beam enters at A and exits at B, the input beam and the scanner are oriented so that the first scanner reflects the input beam into the

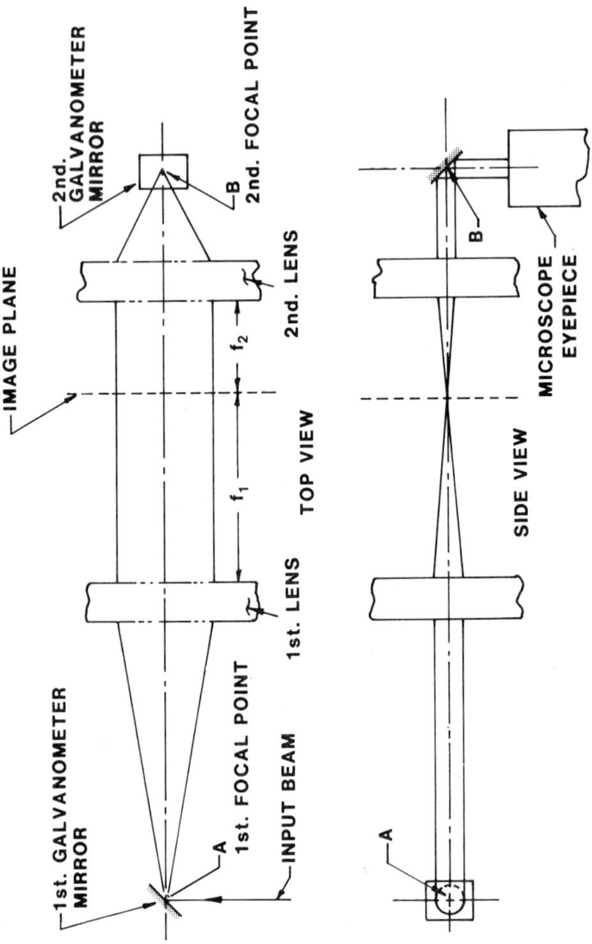


Figure 19 Relay lens configuration.

first lens at all scan angles. The second scanner is oriented so that its mirror scans the desired image area. If a collimated input beam is used and the distance between the relay lenses is maintained at f1 + f2, the exit beam is also collimated.

Figure 20 illustrates the scanning errors introduced by this type of system. Here $\theta_1$ is the angle of incidence at mirror B and is produced by the deflection of mirror A. As mirror B rotates, the reflected beam describes a conical surface in space, with a half angle at the apex of $90° - \theta_1$. However, for any fixed position of mirror B, the rotation of mirror A causes the incident and reflected beams at B to sweep within a plane. As a result, if the beam reflected from B is projected onto a flat plane perpendicular to the beam position when both mirrors are in their neutral position, the scan traces produced by rotation of B while holding A fixed are hyperbolic lines. On the other hand, the scan traces produced by A while holding B fixed are straight, parallel lines. These errors are referred to as one-dimensional pincushion errors.

The magnitude of the scan pincushion error is calculated using the geometric views presented in Figure 20. The error is defined as the ratio of the maximum deviation from straightness e to the peak-to-peak amplitude 2p. From Figure 21 it is evident that

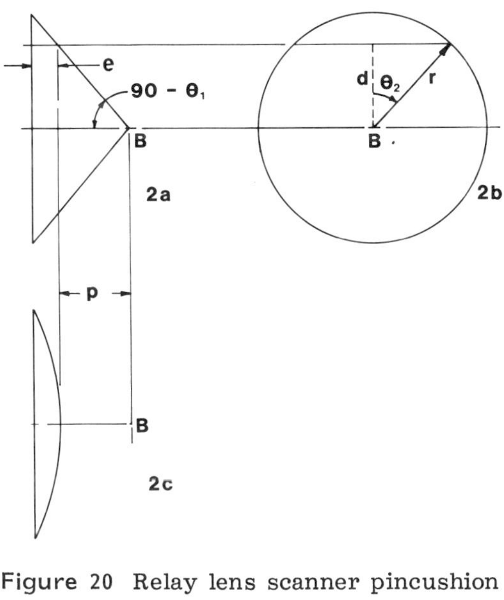


Figure 20 Relay lens scanner pincushion error.

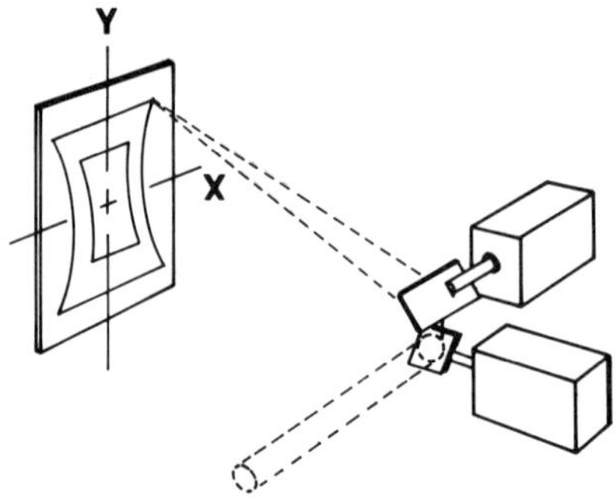


**Figure 21** Simple two-mirror system pincushion error.

$$(e + p)/p = r/d$$

If $\theta_2$ designates the beam deflection produced by mirror B rotation, then

$$r/d = 1/\cos \theta_2$$

from which it follows that the pincushion distortion is given by

$$e/2p = (1 - \cos \theta_2)/2 \cos \theta_2$$

*4.2.2.2. Pincushion errors in simple two-mirror systems.* For the simpler system illustrated in Figure 21, the derivation of pincushion errors is more complicated. Using the terminology introduced with Figure 18, the equation derived earlier,

$$X_i = d^2 + Y_i^2 + e \tan \theta_x$$

can be combined with $Y_i = d \tan \theta_y$ to yield

$$X_i = d(1 + e \cos \theta_y)/\cos \theta_y \tan \theta_x$$

The pincushion error $\varepsilon$ is the ratio of the change in the value of $X_i$ as $\theta_y$ changes from zero to a specified value, to the peak-to-peak amplitude $2X_i$ at $\theta_y = 0$:

$$\varepsilon = \frac{X_{i\theta_y} - X_{i0}}{2X_{i0}} = \frac{1 - \cos \theta_y}{2(1 + e/d) \cos \theta_y}$$

*4.2.2.3. Pincushion error in paddle two-mirror system.* The paddle two-mirror scanning configuration is an alternative to the relay lens method when the attenuation of the lens system is objectionable. This configuration is shown in Figure 22. The two scan axes are orthogonal and coplanar, forming plane x–y. The first

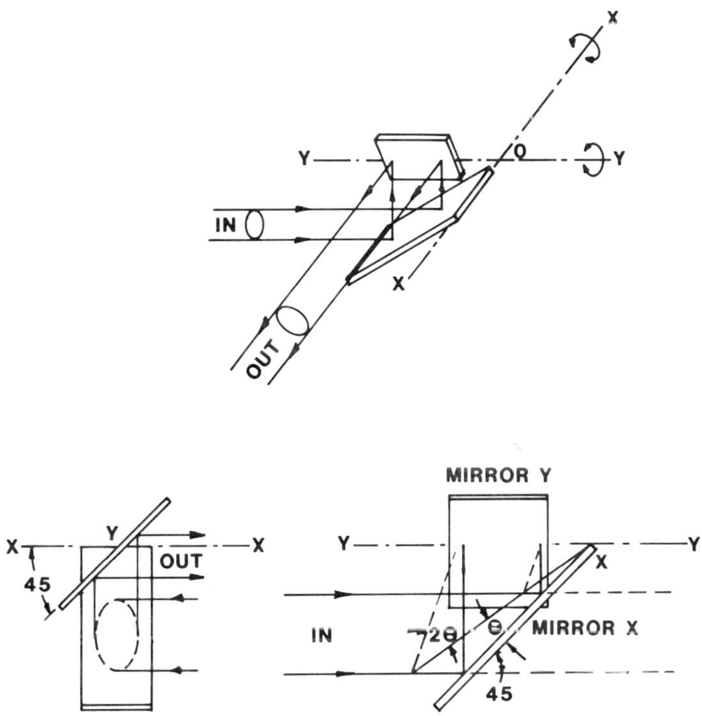


Figure 22 Paddle two-mirror scanning system.

mirror, x, is rotated from its extreme edge and, when centered, the plane of the mirror intersects the x−y plane at a 45° angle. When the incoming beam is parallel to the y axis as a first-order approximation for small scan angles, upon reflection of the x mirror the beam intercepts the y mirror at the same location of mirror x. The plane of the y mirror intercepts the x−y plane at a 45° angle, and the y mirror rotates on its bisector. The output beam is orthogonal to the input beam and, when scanned, appears to pivot on the y mirror in two dimensions like a relay lens scanner. To a first approximation the same errors are found to be present.

### 4.2.3. Cartesian Spherical Mapping with Dedicated Computer

Pincushion and other errors of two-dimensional scanning systems can be electronically corrected within the limits of the repeatability accuracy of the system. Figure 23 is the diagram of a complete electronic mapping system designed to correct pincushion errors of a simple two-mirror scanning system. The pattern to be scanned originates from a digital coordinate generator such as a host computer, a fast vector generator, or a raster generator. The output from the host coordinate generator inputs a stream of X and Y

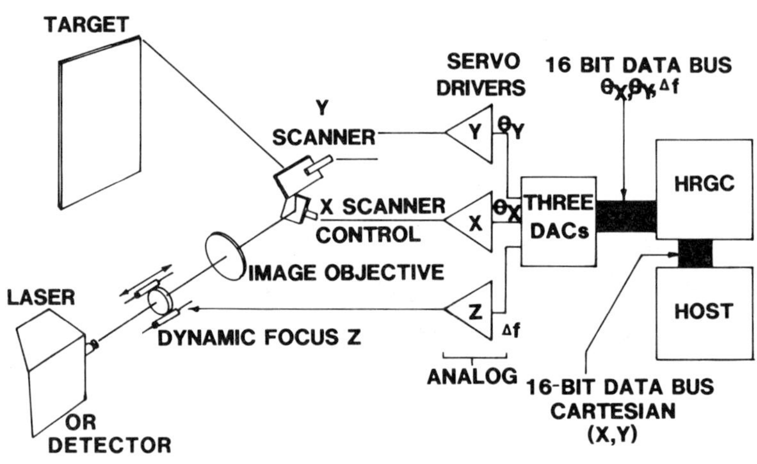


Figure 23 Electronic error correction system.

cartesian coordinates to a high-resolution geometric correction computer for correction. The high-resolution geometric correction computer has the geometry of the scanning system in memory. It uses this information together with the incoming coordinates to compute the exact scanner rotation angles and focus length corrections required to direct the laser beam without error.

The high-resolution geometric correction computer outputs the digital $\theta_x$, $\theta_y$, and $\Delta f$ commands to three digital-to-analog converters, which feed the respective servo amplifiers. The entire process of coordinate and focus error correction computation and digital-to-analog conversion occurs in a few microseconds.

Data processing is to 16-bit accuracy and the corrected analog drive signals are accurate to 15 parts per $10^6$. Since the information is in digital form until converted to drive the scanners, the drift, noise, and adjustment problems inherent in analog circuits are avoided.

In a typical system, dynamic focus correction is accomplished by a device such as illustrated in Figure 24, which is able to

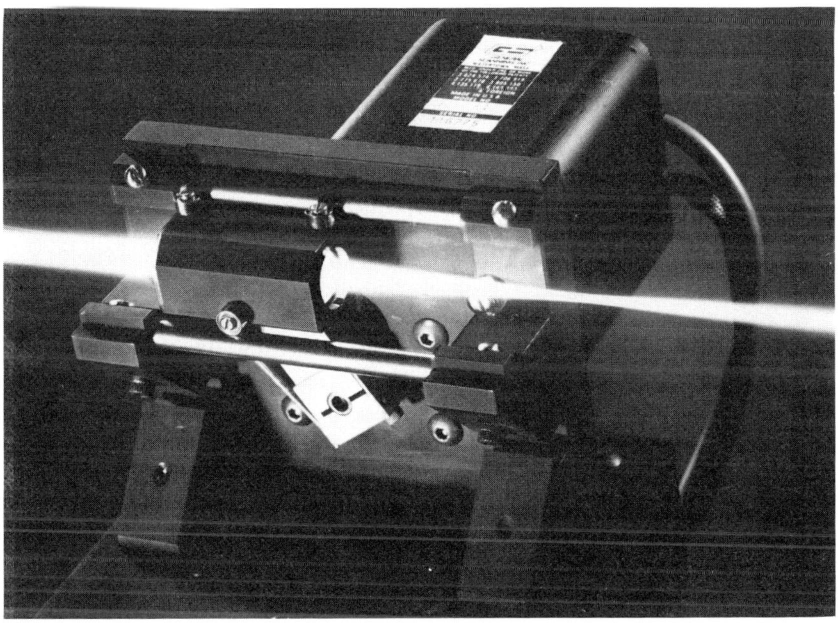

Figure 24 Dynamic focus correction.

translate a 10-mm aperture lens 10 mm in 9 msec. If a larger translation is needed, optical leveraging can be used.

In spite of the accuracy of corrected signals, typical systems exhibit up to a 0.2% linearity error. This error is of mechanical alignment origin and is very repeatable. In a digital correction system it is a simple matter to access an error map to make the necessary corrections. A 64 × 64 coordinate error map is sufficient to provide linearity to 0.0125%.

### 4.3. Mirror-Induced Errors

A general description of mirror definitions is available in the technical literature. Aberrations due to mirror deformation, dimensions, and shape as applied to scanning have been analyzed by Bieser (1974). Dynamic deformations have been studied by Brosens (1972). The following additional mirror-related problems are incurred in the use of scanners:

Mirror mounting

Dynamic deformations

Erosion of mirror surface

Thermal deformations

#### 4.3.1. Mirror Mounting Techniques

To provide precise positioning of a scanning mirror, the mirror must be attached as rigidly as possible to the shaft of the driving scanner. This is usually accomplished by attaching the mirror to a mount which, in turn, can be attached without deformation to the shaft of the scanner. The attachment of the mirror to the mount is normally made in one of the following two ways:

1. The mount is provided with a surface to which a substantial portion of the back surface of the mirror is cemented.
2. An edge of the mirror is cemented into a groove in the mount.

The first method simplifies aligning the mirror surface in parallel with the optical axis, but cement shrinkage tends to introduce mirror distortions. It is not uncommon for a mirror 1 mm thick to pick up a distortion of one-quarter wavelength over the area of the bond. The problem can be minimized by using thin fillets of cement in the area of the bond and by using flexible and high peel strength adhesives instead of hard-curing cements. In some resonant scanners it is also necessary to verify that the bond between mirror and mount will withstand the dynamic loads without breakage.

### 4.3.2. Dynamic Deformations

The accelerating torques imparted to a scanning mirror can produce significant mirror surface distortions. This is particularly true in scanners driven with a sawtooth wave form and in high-frequency resonant scanners. Brosens' (1972) analysis of the deformations induced by accelerating torques yields the approximate formula

$$f = 0.065(s^2T/Eh^3l)$$

where f is the maximum deflection from the original mirror shape, s is the width measured across the axis of rotation, E is the Young's modulus of the mirror material, h is its thickness, l is its length in the axial direction, and T is the applied torque, which is related to the angular acceleration a by

$$T = hs^3lda/12$$

with d the density of the mirror material. Combining the two equations, we get

$$f = 0.0055a(ds^5/Eh^2)$$

This expression points to the desirability of avoiding mirrors of appreciable width. For a glass mirror 1 cm in diameter and 1 mm thick, the resulting deflection at an acceleration of $10^6$ rad/sec$^2$ is about 1/25 of the sodium D-line wavelength. Since the inertia of such a mirror is 0.011 g-cm$^2$, the above acceleration corresponds to a torque of only 11,000 dyne-cm. The mirror deformation actually expected may be smaller when a substantial portion of the width of the mirror is cemented to a mount.

### 4.3.3. Thermal Deformations

When a scanning mirror is exposed to radiation, a substantial part of the radiation that the reflecting surface absorbs is transferred in the form of heat to the rear surface. This heat is discharged by conduction to the mount and by radiation and convection to the surrounding atmosphere. The conduction of heat to the rear surface causes differential expansion and, if the incident radiation is particularly intense, significant distortions can occur. Such distortions can be estimated by assuming one-dimensional heat transfer to the rear surface.

The radius of curvature caused by a uniform temperature gradient is

$$R = \frac{1}{a} \cdot \frac{dt}{dx}$$

where a is the coefficient of linear thermal expansion, and dt/dx is the temperature gradient in the material. From Fourier's law of conduction,

$$\frac{dt}{dx} = \frac{q}{kA}$$

where q is the heat transfer rate, k is the thermal conductivity, and A is the cross-sectional area. The camber assumed by a plate of width s, when it curves with a radius of curvature R, is given to a first-order approximation by

$$e = s^2/2R$$

Combining this equation with the previous expressions, we obtain

$$e = aqs^2/2kA$$

For a glass mirror of width 1 cm conducting 0.1 $W/cm^2$ to the back surface, the resulting camber is 0.5 μm, or about 1 wavelength.

### 4.3.4. Erosion

When a scanner mirror is moved through air at high speed, the collision of dust particles with its surface can cause gradual erosion of its reflective coating. Experience shows that for any coating the process of erosion does not occur below a critical impact velocity. It is believed that surface erosion occurs when the stress developed at the impact interface between the coating and the dust particle exceeds a value that is characteristic of the coating.

The stress developed by the impact of a rigid body against an elastic mass was analyzed by Timoshenko and Goodier (1951). Assuming that echoes from the stress wave generated by impact are of insignificant magnitude, the maximum stress S caused by impact is given by the formula

$$S = E(V/c)$$

where E is the Young's modulus of the substrate, V is the relative speed at impact, and c is the velocity of wave propagation (sound velocity) in the substrate. Experimental evidence shows that AlSiO coatings on fused silica degrade through erosion at all points where the speed of motion exceeds 3 m/sec at any instant during the scan cycle.

Users of high-speed scanners should take precautions to minimize the presence of suspended dust particles near scanning mirrors. Where such protection cannot be provided, hard coatings should be used in combination with substrates of low Young's modulus.

Table 12 Properties of Selected Mirror Substrate Materials[a]

| Fused silica | Beryllium |
|---|---|
| $E = 7.29 \times 10^{10}$ N/m$^2$ | $E = 30.8 \times$ N/m$^2$ |
| $d = 2.2$ g/cm$^3$ | $d = 1.8$ g/cm$^3$ |
| $k = 0.012$ W/cm-°C | $k = 1.57$ W/cm-°C |
| $a = 0.5 \times 10^{-6}$/°C | $a = 12 \times 10^{-6}$/°C |
| $c = 5760$ m/sec | $c = 12{,}870$ m/sec |

[a]Values are from the *American Institute of Physics Handbook*, McGraw-Hill, New York, 1957.

### 4.4. Metallic Mirrors

The problems of dynamic and thermal deformations which are associated with mounting conventional mirrors have stimulated the development of metal unitary mirror mounts. Besides getting rid of bonding difficulties, the unitary construction eliminates the troublesome distortions caused by the differential thermal expansion coefficients of mirrors and mounts. The simpler construction also allows system dynamics to be extended. Metal unitary constructions are commonly of lower inertias, especially when beryllium is used. The bonding interfaces are eliminated, and the metal mirrors can be shaped more easily with ribbed or honeycombed backings. Finally, unitary metal mirrors can be manufactured to exacting alignments and dimensional tolerances. The shaft anchoring design and the mirror surface finish are the salient features of these mirrors.

#### 4.4.1. Shaft Anchor Design

The mirror mount serves three major functions.

1. The mount must anchor the mirror securely to the scanner shaft so that dynamic forces cannot cause slippage. Detailed stress analysis is required when large mirrors are used or when scanning at high-frequency rates (greater than 1 kHz) with resonant scanners. The shaft must be fastened by a strap soldered or epoxied to a neck, mount, or hole. Figure 25 shows two typical examples.
2. The mounting must index the mirror in proper alignment with tight tolerance. The most noticeable consequences of mirror misalignment are the following:

   The beam misses the target.

   The scan pattern is deformed.

   Undamped periodic cross-axis motion appears after a step.

   There is random cross-axis noise.

   There is a different sagittal position during forward and reverse scan.

A 1-mrad mirror alignment tolerance necessitates a 4-mm-wide target band at 1 m. Because other opportunities for misalignment are present in an optical system, mirror construction tolerances of 1 or 2 mrad are common.

(A)

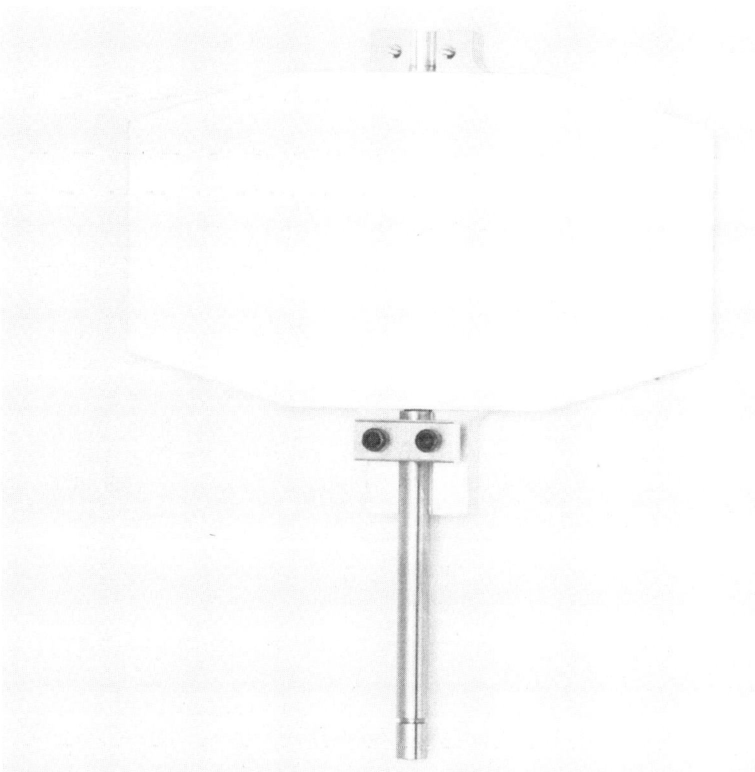

(B)

Figure 25 Mirror shaft mounts.

Curvilinear scan patterns are due to geometric misalignment and can be studied as the intersection of a conic and a plane.

Cross-axis perturbations are due to excitation of shaft bending modes or bearing play when the dynamic forces exceed the radial magnetic or mechanical preload. At worst, the entire rotor can bend and can excite a cross-axis mode through the transducer and amplifier. This renders the system unstable.

3. A unitary mirror mount must satisfy the above conditions and also be free of fastening distortion. Normally the mirror mount is considered an independent entity and is designed to be economically manufactured, handled, aligned, and fastened.

The tolerances of the shaft and mounting bore are large when compared to the wavelength of light; consequently the mount must be designed to prevent any deformation of the mirror surface. Three mounting processes known to prevent mirror distortions are:

Bonding in alignment fixture

Clamping collet buffered from the reflective surface

Clamping with "hydrostatic" compressive stresses

Figure 26 exemplifies these methods. The last two designs are proprietary.

### 4.4.2. Mirror Surface Finish

In earlier millennia only metallic mirrors were available. Glass mirror manufacture was perfected in the past century, but the technology cannot be applied directly to mirror finishing with all metals. To optimize dynamic performances, the light metals, aluminum or beryllium, are normally used for metal unitary mirror mounts; however, the polishing abrasives used on glass become imbedded in them. Three commonly accepted solutions are:

1. The finished mirror, blank after diamond lapping, is plated with 0.002–0.005 in. of hard nickel by electrolytic or electroless deposition. Nickel can be processed to a high-quality surface and then coated by vapor deposition for protection and to get the desired reflectivity.

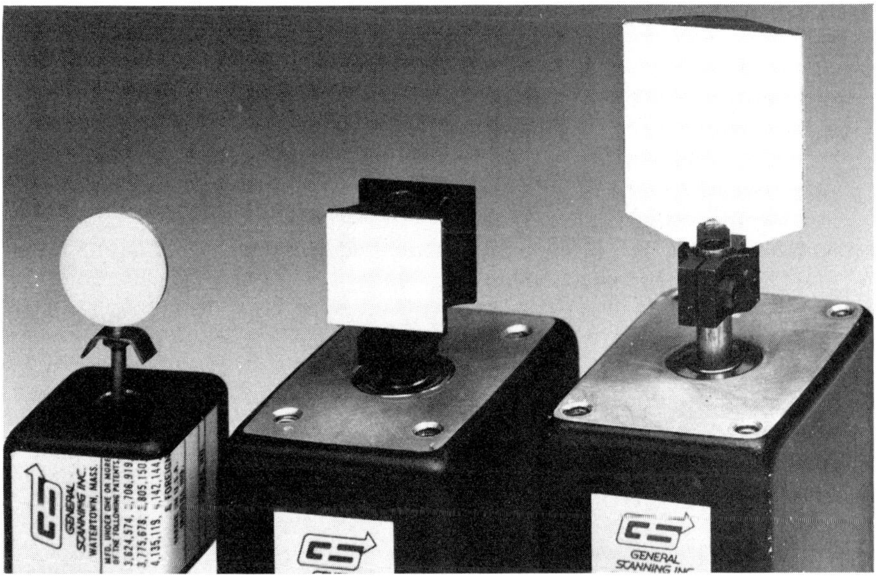

Figure 26 Metallic mirror mounts.

2. Replication is widely used, especially for short runs. The process calls for a tolerance of ½ mrad.

3. The development of diamond machining in the last decade has made possible high-quality mirrors for far- and near-infrared applications. The optical surface is finished directly on the aluminum, copper, or brass. Other substrate materials such as beryllium or steel must be coated by electrolysis with nickel or gold. Machining marks can cause diffraction of visible wavelengths and necessitate additional finishing.

## 5. DESIGN OF LOW INERTIA SCANNERS

Laws (1952), Blitzer (1948), and others have described the basic design of the moving iron galvanometers used in optical applications. These devices were conceived as relays or hydraulic valve actuators. However, an imaginative design evolution took place to meet the demands for speed of response, accuracy, random access, and stability needed for optical systems.

## 5.1. Galvanometric Low Inertia Scanner Design

Prior to 1969 galvanometric LIS were open-loop taut-band suspended moving coil constructions with oil damping. Cross-axis motion was uncontrollable, and damping was a function of repetition rate. The introduction of moving iron LIS took optical scanning out of the laboratory and into the world of industry and office automation.

Modern galvanometric LIS have three critical elements: an electromagnetic drive, a position transducer, and a rotor suspension. Rotor suspension, including ball bearings and flexure bearings, are covered elsewhere and will not be reviewed here.

### 5.1.1. Electromagnetic Drivers

For limited rotation scanning with galvanometric LIS, both positive and negative torques must be applied to the scanning mirror. Similar dynamic performances must be achieved at all angular positions.

Figure 27a shows a typical linear scan cycle with fast retrace, and Figure 27b shows fast random addressing. During the active scan the load (mirror) rotates at a constant velocity under servo control. Retrace and addressing usually must be performed as fast as possible, so the scanner torque is saturated for maximum acceleration and then reversed for maximum deceleration. Servo control for retrace is reestablished at the final resting point.

The dynamics of a transition through an angle $\theta$ are shown in Figure 28. The retrace time is inversely related to the square root of the galvanometer's torque capability, as shown in Section 4.1. Torque is generated on a moving iron vane structure which permits rotation over a limited angle. This structure provides exceptional ruggedness, as only the solid iron rotor moves. The coils are enclosed in the stator, which permits good heat dissipation and eliminates the need for electrical connections to the rotor. Relative freedom from undesirable rotor vibrations make possible a wide servo loop bandwidth.

The magnetic circuit of Figure 29 shows how torque is produced in a moving iron galvanometric scanner. Polarizing flux is created at all four air gaps by the two permanent magnets. The remaining parts of the magnetic circuit are made from soft magnetic iron. The control flux is created in the same air gaps by the two control coils.

The magnetic circuit is analogous to an electric bridge circuit. The coils generate flux in the magnetic circuit such that a polarizing flux at diametrically opposed corners of the rotor section is

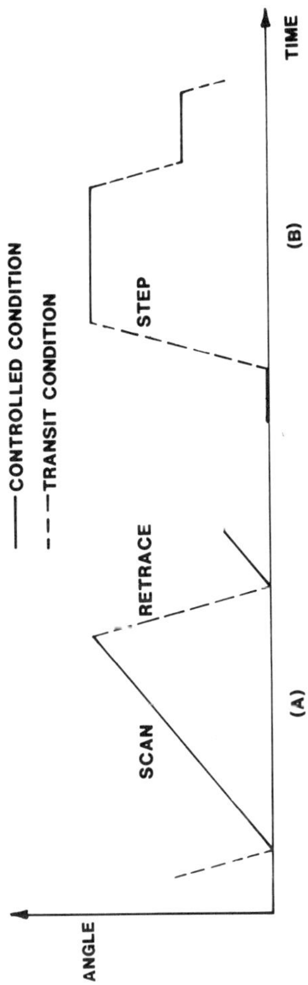


Figure 27 (A) Linear scan with fast retrace and (B) scan with fast random addressing.

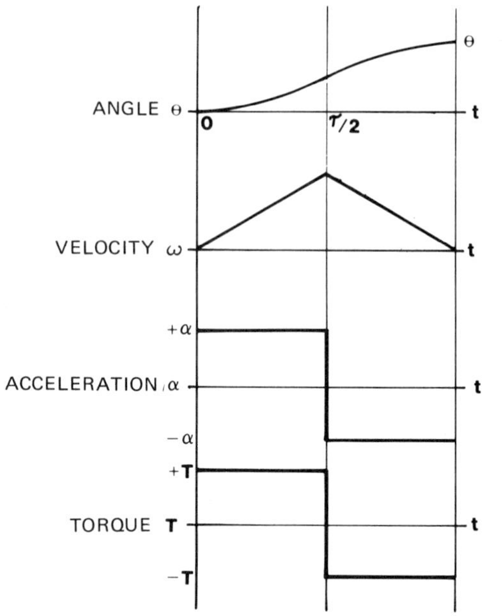


θ MECHANICAL ANGLE ( rad )

t TIME (sec)

α ANGULAR ACCELERATION (rad/sec$^2$)

T TORQUE (dyne - cm)

I INERTIA (g - cm$^2$)

**Figure 28** Dynamic response at constant acceleration and deceleration.

either augmented or diminished. The resulting torque is derived in the Appendix (Sec. 9.1) and expressed by the relation

$$T = BLNID - (2\pi\ 10^{-7} CN^2 I^2 D/g)$$

where T is the torque (N-m), B is the air gap flux density (T), L is the rotor length (m), D is the rotor diameter (m), g is the air gap length (m), N is the number of coil turns, I is the coil

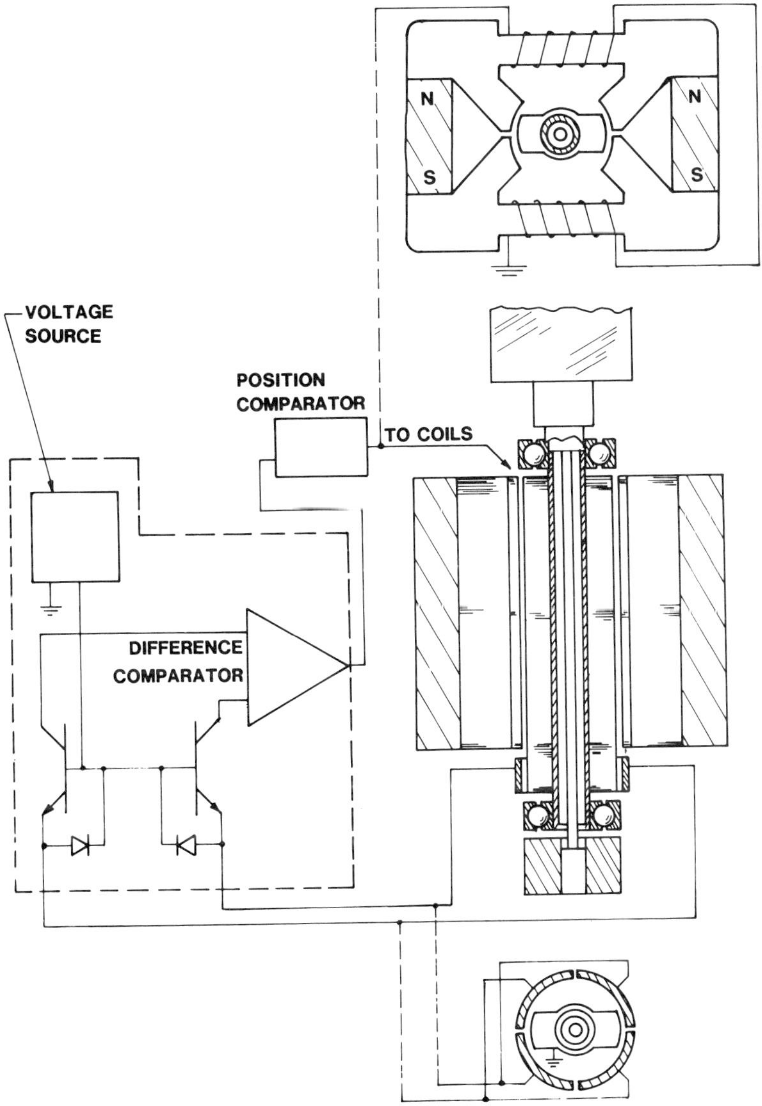


Figure 29 Conventional moving iron galvanometric magnetic circuit and capacitive transducer.

current (A), and C is the ratio of angular to maximum rotation from the center. If optimum performances are to be obtained, the nonlinear term cannot be disregarded, as it affects the stability and speed of response of the system. The consequences of the nonlinear term are that at any one angular position the torque is nonlinear with current value, but also with current polarity. The torque is position dependent, and, consequently, in a position servo the damping and settling time become complex functions of both current value and current polarity.

Scanner designs have been guided by the desirability to have a linear torque—current relation. Montagu (1965, 1971) has described combined geometric and magnetic compensations which minimize the apparent nonlinearity of the torque. Figure 30 shows the angular torque dependence for three galvanometers with a ½-in. bore built with linearity correction schemes. The theoretical torque-angle relation is also given. These correlations are analyzed in the Appendix (Sec. 9.2).

### 5.1.2. Transducers

Applications of galvanometric LIS which necessitate precise angular aim or position sensing are best fulfilled with a device where the sensing is achieved via an angular transducer capable of independent control. Numerous optical, magnetic, and capacitive noncontact techniques are available. The quality of the sensor defines the accuracy of the scanner and the system, and therefore deserves considerable scrutiny. The important parameters are the following:

Accuracy

Temperature stability

Null or reference stability

Gain stability

Signal-to-noise ratio

Resolution

The simplest position sensor is a torsion bar. The torque applied to the rotor determines the deflection, and the necessary current is measured to give the desired position information. This design has good temperature stability but poor resolution, accuracy, and dynamic behavior. It is rarely used without an additional position-sensing loop.

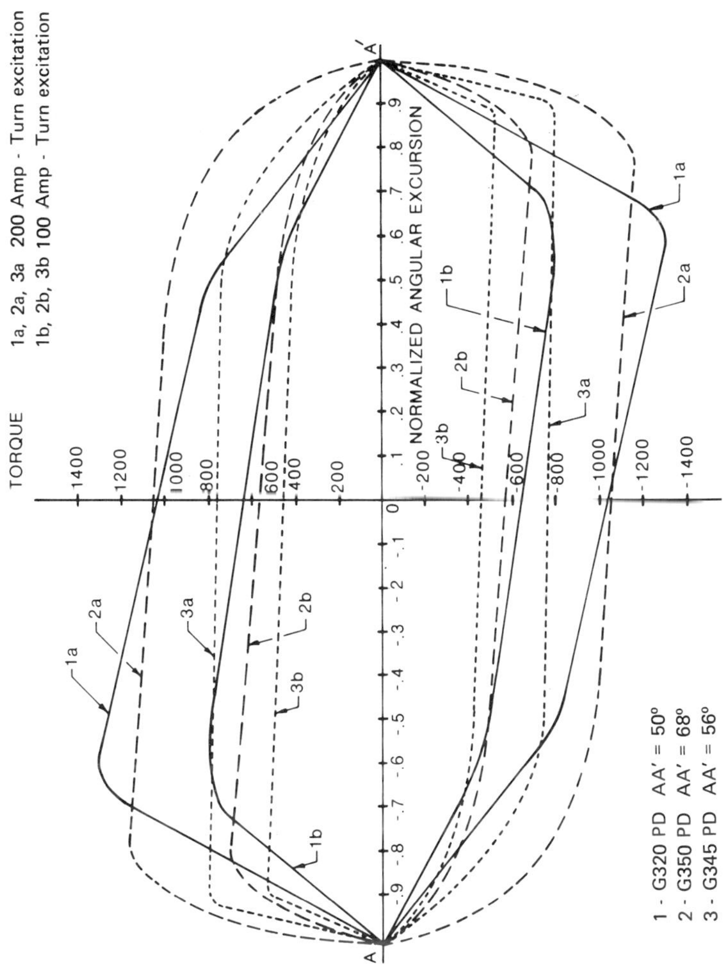


Figure 30 Comparative linearization curves.

Optical techniques are limited in their resolving ability by the wavelength of light and by the temperature characteristics of available detectors. When used, they are normally integrated with the system. Both interferometric and grating designs are common.

Magnetic position transducers such as the "metresite" or the "microsyn" of Mueller (1949) as well as solid-state devices interact with the magnetic circuits of the driving stage and are not used with high-performing scanners.

Capacitive sensors are most commonly used as integral elements of LIS. They are compact, easily integrated with the geometry of the magnetic drive section, and capable of repeatable performances such as those listed in Table 9.

These position sensors measure extremely small capacitive imbalances between rotor and stator in a circuit, analogous to that of Rohr (1949) shown in Figure 29. Such sensors have an essentially infinite lifetime, a linearity often exceeding 0.1%, no hysteresis, a wide bandwidth (up to 50 kHz), a high signal-to-noise ratio, rigid mechanical coupling to the rotor, and essentially no electrical cross-talk with the drive coil circuits.

## 5.2. Resonant Low Inertia Scanner Design

The desirable features of resonant scanners have long been recognized and used in laboratory environments; however, before resonant scanners were accepted in the industrial or office environment, mirror perturbations caused by the scanner and mount had to be reduced to a few arcseconds.

1. Mirror flatness. The mirror characteristics are the responsibility of the scanner supplier, as the mirror is an integral part of the dynamic system. For graphic equipment, mirrors must remain flat to a half wave over a reasonable excursion and temperature range. Mirror thermal and dynamic requirement design as well as anchoring methods are analyzed in Section 4.3.

2. Cross-axis stability. As high-Q systems, resonant LIS amplify driver-induced as well as external perturbations. The isogonic suspension (Montagu, 1976a) solved the problem of driver-induced perturbations. The IDS design (Montagu, 1983) solved both problems, driver-induced and external perturbations.

3. Amplitude control. The built-in tachometer is used to define the mirror position and trigger data transfer. Its signal should have a high level of stability for both phase and amplitude information over the applicable temperature range. Tachometer designs have been proposed by Dostal (1971), Selverstone (1976), and Montagu (1976a, 1980).

4. Availability. Supplies must be available in quantity and at a quality level suitable for volume production at an acceptable cost. The complexity of drive electronics must be included in these considerations.

### 5.2.1. IDS Resonant Scanner

One of the most novel resonant LIS to have been introduced is the balanced torsion bar design with an induction torque driver. The device is illustrated in Figure 31. The mirror is suspended by torsion bars T1 and T2 in a fully symmetrical arrangement designed to be mass balanced, so that accelerations along the three principal axes of translation will not cause torsional excitation of the mirror.

The drive coil and the armature are magnetically linked in a transformer-like fashion by a soft iron core. The armature is a single-turn, rectangular drive loop with an edge colinear with the torsion current in the drive loop. This current interacts with the return path of the flux of a permanent magnet to create the drive torque:

$$T = \frac{\mu ANIBlr}{LR}$$

where $\mu$ is the permeability of the iron core, A is the cross section of the iron core, N is the number of turns of the drive coils, I is the driver coil current amplitude, B is the field of the permanent magnet, l is the length of the drive loop in the magnetic field, r is the drive loop acting radius, L is the length of the iron core path, and R is the resistance of the drive loop. In this manner, a moving coil driver is obtained, without having to provide leads or brush contacts to the moving armature. The motion of the armature induces a voltage and current which are sensed by a pick-off coil. The actual voltage measured is the sum of this induced voltage, which is proportional to velocity, and of the portion of the drive coil voltage induced by transformer couplings. It is expressed as

$$E = \mu \frac{AIN^2}{L} \omega \cos \omega t + NBLr\frac{\partial \theta}{\partial t}$$

where $\omega$ is the resonant frequency. The transformer coupling component is easily subtracted electronically. This velocity sensor

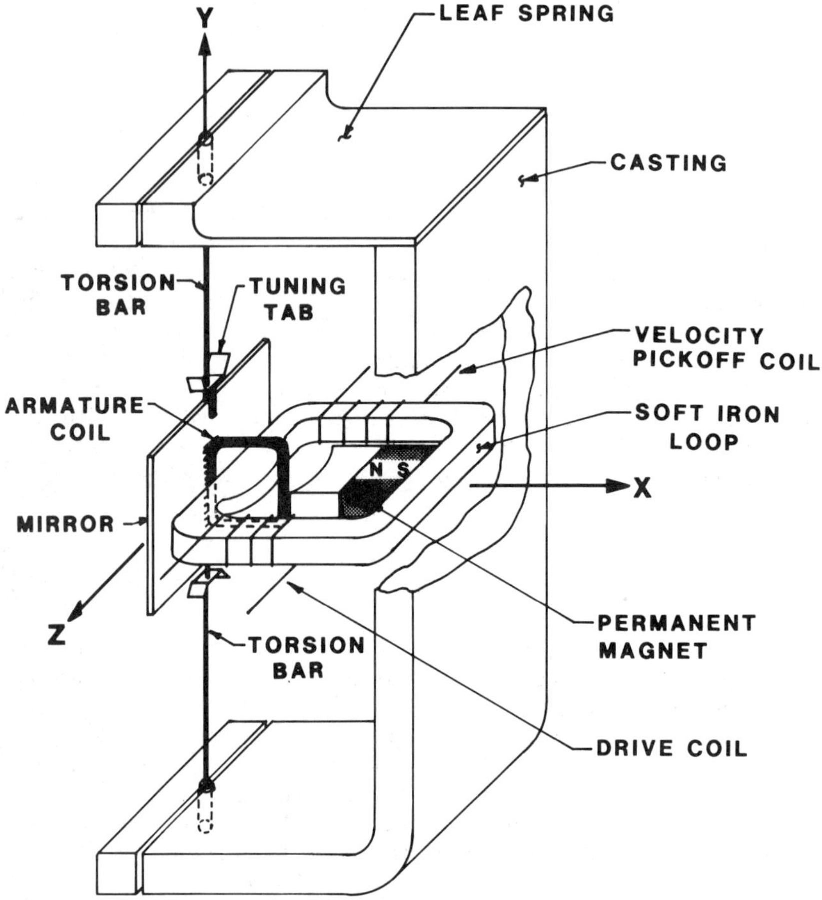


Figure 31 Resonant LIS.

permits extremely good, simple, and external amplitude control with drift below 100 ppm/°C of the peak-to-peak excursion. The resonant frequency is stable to -160 ppm/°C. A theoretical derivation of the pertinent parameter is presented in the Appendix (Sec. 9.3).

## 6. DESIGN OF ELECTRONICS FOR LOW INERTIA SCANNERS

### 6.1 Electronics for Galvanometric Low Inertia Scanners

The general considerations for the design of closed-loop driver amplifiers to drive scanning galvanometers are similar to those of other precision servo designs. Engineering considerations start with defining the necessary power supply stability to minimize induced drifts. The use of low-noise, wide-bandwidth, low-drift operational amplifiers throughout the amplifier is needed when the application is demanding in the area of speed or stability. Careful attention must be given to isolation of the position signal from the output driver to eliminate spurious oscillations due to coupling modes. Unwanted ground paths must be studiously avoided, particularly in the presence of case-grounded position feedback sensors. "Zeroing" potentiometers should be avoided owing to their excessive drift characteristics. The use of temperature-stable components throughout the design, including resistors and capacitors, is highly advisable.

As all moving iron galvanometers present a significant inductive load to the driver amplifier, particular attention needs to be given to the output stage. Typically the configuration of the output amplifier looks like a transconductance amplifier, so as to remove the unwanted pole caused by the load inductance and give a second-order response. Since the maximum rate of rise of current in the load is limited by the ratio of drive voltage to inductance, significant increases in servo response can only be achieved by raising the driver voltage substantially. Under these conditions it is necessary to check that safe operating area restrictions on the bipolar output transistors are met at all times.

Provisions to limit the rotational excursion should be incorporated in the design if damage to either the mirror or the galvanometer may occur when inadvertently overdriven. Such overdriving can occur during the turn-on/turn-off transient when power supplies are not properly sequenced. Likewise, as current rise times are improved and servo bandwidths extended, care must be taken to guarantee that $I^2R$ losses in the galvanometer do not exceed the steady-state limit. It may be desirable to provide thermal shutdown protection of the load under the control of a thermistor or thermal switch mounted in the galvanometer.

A galvanometer can be simply modeled by a second-order differential equation such as

$$KI = K_s \theta + D \frac{\partial \theta}{\partial t} + J \frac{\partial^2 \theta}{\partial t^2}$$

where I is the drive current, $K_s$ is the spring constant, D is the damping constant, J is the rotor and load inertia, $\theta$ is the shaft rotation, and K is the torque constant.

When driven by an amplifier with position error and velocity feedback as shown in Figure 32, the transfer function is

$$\frac{\theta}{e_{in}} = \frac{KA}{JS^2 + (D + KK_D K_\theta)S + (K_s + KK_\theta A)}$$

where $K_D$ is the scanner damping constant and $K_\theta$ is the transducer position constant. There is a finite error in the response to any nonzero excitation. Magnetic hysteresis and frictional losses will also result in position errors. The effect of the closed-loop amplifier is to extend the bandwidth of the galvanometer with the addition of servo stiffness and to improve its static linearity to that of the position detector. The dynamic performance of the system is second order only as long as there are no additional perturbations. In actual devices perturbations exist as a result of the nonlinearity of the torque versus angle and of the torque versus current. Additionally, the inductance presented to the amplifier is angle dependent, and many of the characteristics are thermally sensitive.

The most frequent limit to the achievable servo stiffness is torsional resonance of the rotor and load inertias. Errors in position are determined by the error of the position sensor *and* by the hysteresis and torque nonlinearities of the galvanometer. The transient response is also angle dependent owing to the nonlinear inductance of the device. Such limitations occur in all practical devices, and the effect of nonlinearities must be analyzed on a case-by-case basis.

For those applications requiring the utmost in position accuracy, the simple servo can be modified to include error integration. Then, given enough time, all errors associated with the internal spring, the hysteresis losses, the magnetic torque variations, and the frictional losses are removed, and the nonlinearity of the position sensor is the sole remaining contributor to position error. Unfortunately, dynamic behavior is slowed by the addition of error integration, and no improvement in dynamic errors results, as the achieved servo stiffness is no greater than without error integration. One benefit associated with the introduction of error integration is tracking a ramp input with no error after decay of the initial velocity transient.

## 6.2. Electronics for Resonant Low Inertia Scanners

The need in applications such as nonimpact printing for highly accurate and repeatable scanning motions which can be generated at affordable prices has renewed interest in resonant scanner technology. The availability of laser diodes has simplified the process of amplitude modulation of the source, which, in turn, simplifies dealing with the sinusoidal scanner velocity. The need remains, however, to synchronize the scanner to external drive electronics and to generate a pixel clock to strobe data either in or out of user hardware. Low-cost microprocessors and high-speed complex logic elements make this easier and more affordable. Although producing such a clock for applications requiring high throughputs, accuracy, and precision continues to be complex, the necessary components are now available.

There are many methods by which the need for accurate, repeatable, and affordable scanning may be achieved. Four such techniques are described here. All of the techniques have the usual engineering trade-offs of ease of implementation, cost, and system performance. The major issues are addressed as the alternative schemes are explored. The first of the four schemes assumes the scanner is used as either an input or output mechanism, as, for example, in a laser printer. General Scanning chose the last method for implementation as a practical, versatile test bed for the evaluation of scanners in specific applications.

### 6.2.1. Theory of Operation

A scanner operating at mechanical resonance executes a motion described by the following equation, where $\theta$ is the scanner shaft angle, $\theta_s$ is the maximum angle of excursion, t is time, and $T_s$ is the period of the scanner:

$$\theta = \theta_s \sin\left(\frac{2\pi t}{T_s}\right)$$

It therefore follows that

$$t = \frac{T_s}{2\pi} \sin^{-1}\left(\frac{\theta}{\theta_s}\right)$$

and that

$$\frac{d\theta}{dt} = \frac{2\pi}{T_s}\theta_s \sqrt{1 - \left(\frac{\theta}{\theta_s}\right)^2}$$

Normalizing the velocity of the scan to the maximum at $\theta_s = 0$ yields

$$\frac{d\theta/dt}{2\pi\theta/T_s} = \sqrt{1 - \left(\frac{\theta}{\theta_s}\right)^2}$$

It can be seen that this function has a maximum value of unity at $\theta = 0$ and a minimum of zero at $\theta = \theta_s$. Taking a desired scan angle efficiency of 95% of the rated excursion as an arbitrary example yields a velocity ratio of 0.312. Thus the maximum velocity is 3.2 times the minimum. The clocks need to have periods in this ratio, and the light source or automatic gain control systems need to compensate their amplitudes by the same amount. Clearly, the velocity ratio reduces dramatically with reduced usable scanner angle and makes the clocking requirements easier to meet.

When mapping to or from a flat image plane, an additional nonlinearity is encountered due to the tangent error. The equations below describe the motion in the flat image plane; X is the displacement in the image plane and S is the distance from the scanner mirror to the image plane:

$$X = S\tan(2\theta)$$

$$X = \frac{\tan\theta}{1 - \tan^2\theta}\, 2S$$

$$\frac{dx}{dt} = 2S(1 + \tan^2 2\theta)\,\frac{d\theta}{dt}$$

$$\frac{dS}{dt} \equiv 0$$

It is thus possible to modify the scanner $d\theta$ in order to yield a constant displacement in the image plane.

Note that in applications additional constraints are placed upon the clock generation method. Compensation for amplitude and frequency shifts with time and temperature of the scanning element need to be implemented. In printer applications the time inaccuracies of the pixel clock result in errors of positioning of the beam. Unfortunately, the human eye is particularly sensitive to positioning errors if they are coherent line to line. This problem is usually addressed by requiring a high degree of pixel accuracy, typically better than 10% of the pixel spacing (Bestenreiner et al., 1976). Likewise, auxiliary driving components and sensors require stabilities and time resolutions proportionally higher due to the need for greater accuracy. As the system performance demands increase, truly impressive clock rates and time stability can be required. Figure 32 relates for a typical printing application the number of pixels per line to the scanner frequency and the resulting equivalent clock frequencies needed to yield the 10% pixel placement accuracy at the 95% angle utilization of a resonant scanner.

### 6.2.2. Analog Clocks

Analog clocks represent conceptually and physically the simplest implementation of a scanner linearization scheme. Unfortunately, they also suffer from relatively poor performance characteristics

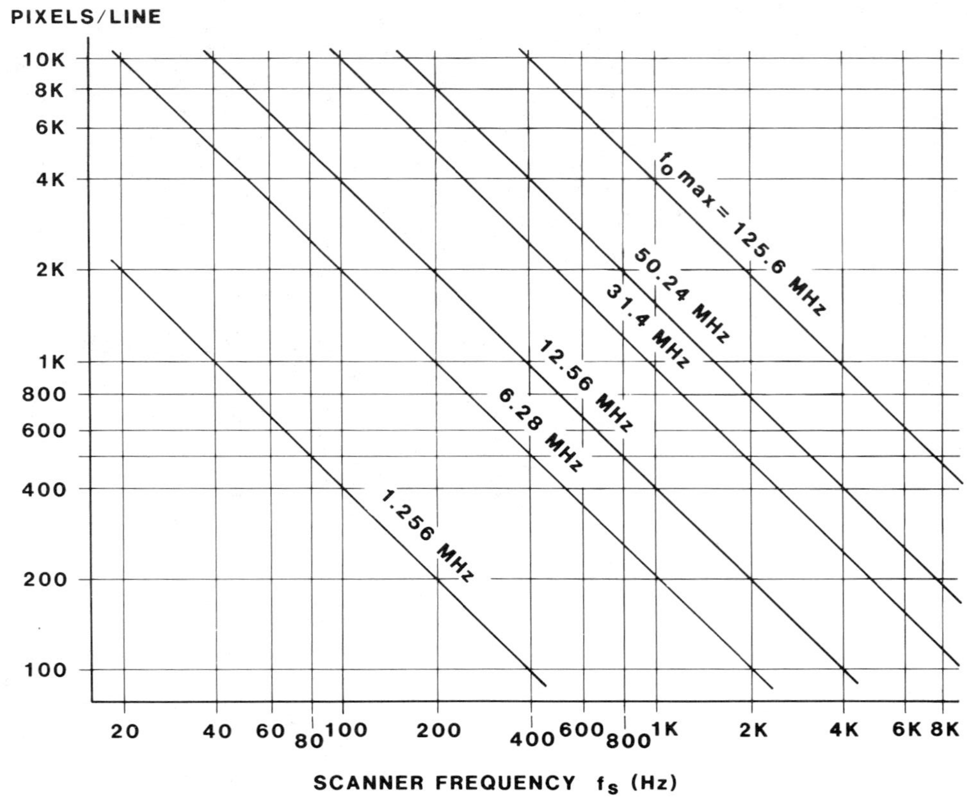


Figure 32 Pixel density of function of scanner and clock frequency as applied to resonant scanner based image generating system.

either from thermal drifts, resolution limits, or maximum pixel and scanner frequencies.

One analog scheme consists of controlling a voltage-controlled oscillator from a full-wave-rectified signal representing the scanner mirror position (Burrer, 1976). Another technique beats a pair of high-frequency oscillators against each other in order to generate a wide range of pixel clock rates from an oscillator whose frequency can deviate linearly by only a small percentage. The method described here uses some analog and digital techniques in a cost-effective manner for a low-performance system.

A relatively simple analog controller (Fig. 33), such as the GSI AX730, permits the scanner to oscillate in a sinusoidal motion. Any amplifier which provides the means to cause the scanner to oscillate at its natural frequency at an adjustable stable amplitude is sufficient. Either external feedback sensors or the signal supplied by the scanner may be used to sustain self-resonance. A sinusoidal output signal proportional to the scanner amplitude is derived from the feedback mechanism and used by the clock generating method.

The clock generator compares a digitally generated linear signal to the feedback signal representing the scanner mirror location. When the two signals are equal, an analog comparator causes the digital signal generator to increase (or decrease) by a fixed amount and wait until equality has again been achieved. In this manner, the clock pulse increasing the digital signal generator represents equal angles of motion of the scanner mirror and produces time-displaced pulses at a frequency proportional to the velocity of scanner mirror motion.

A buffered digital-to-analog converter, whose input is derived from an up/down counter, forms the digitally generated signal. A programmable switch register sets the upper and lower bounds of the counter. The digital magnitude comparator compares the output of the switch register, or its 1's complement, with the current contents of the counter register. When either the upper or lower bound is reached, the control logic terminates output clocks until the scanner passes through the same limit in the opposite direction. If a function other than the arcsine is desired, a read-only memory (ROM) with the corresponding angular mapping can be inserted between the counter and the digital-to-analog converter. In this manner the sinusoidal mirror motion can be linearized and other geometric errors removed. Additional circuitry gates signals so that the clock pulses are generated during the desired fraction of scanner rotation. A switch inhibits scanning in both directions. A phase shift network eliminates any

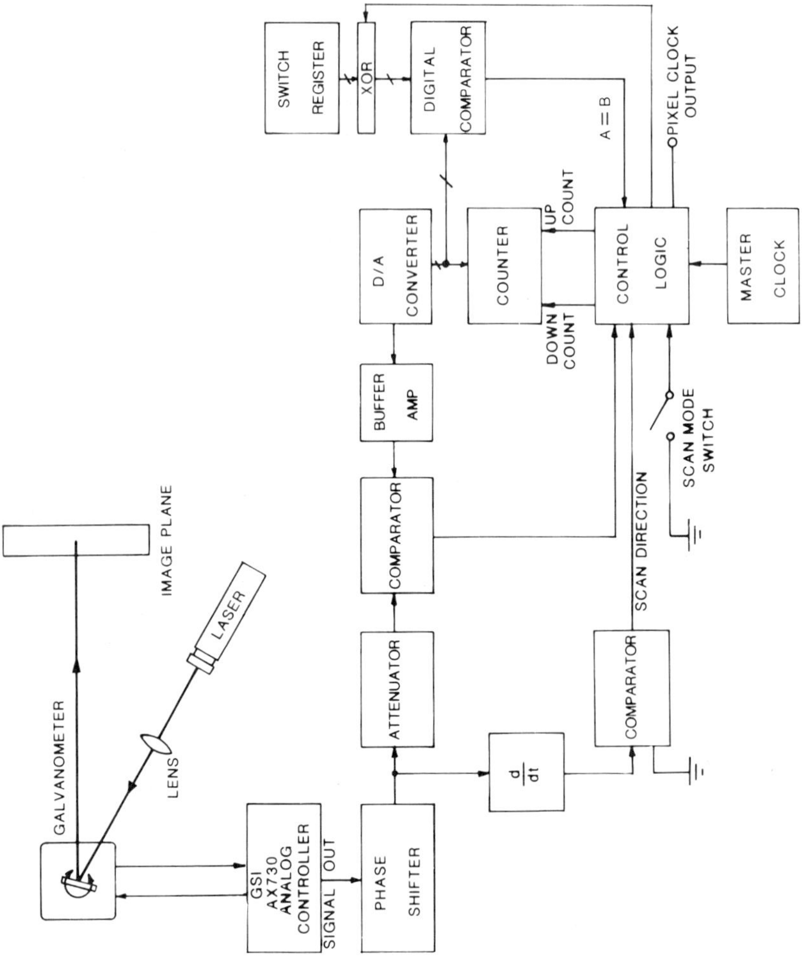


Figure 33 Analog clock.

phase shifts that may occur between the actual scanner mirror location and the analog feedback signal. A differentiation network determines the direction of scan and may also be used to derive a signal proportional to the scan velocity.

While quite straightforward, this structure has several inherent limitations. The number of clock pulses output is set by the resolution of the counter and the digital-to-analog converter. Also, the total number of clock pulses per scan is set by the number stored in the switch (or storage) register. The precision of the linearization is limited by the combined errors of the digital-to-analog converter, the distortion of the analog feedback signal, and the assumption of sinusoidal motion of the scanner. Frequency drifts of the scanner from thermal changes cause errors in the starting and ending of the scan due to the resulting phase shifts. Errors in amplitude of the feedback mechanism result in changes of the percentage of scanner angle used. As the master clock is not locked to the scanner, random pixel clock phase errors occur unless an "on demand" master clock is employed. A practical upper bound of resolution elements and scanner frequency is imposed by the need for settling time for the digital-to-analog converter and comparator. A practical typical bound is about 1-μsec clock rates for a 10-bit converter. Thus the linearization scheme is limited to a maximum scanner frequency of 125 Hz if 80% of the scanner amplitude and all of the available digital codes are used. To make this scheme more attractive, the number of bits in the conversion needs to be reduced. Inexpensive 8-bit converters are capable of running at rates up to 10 MHz, allowing scanner operation at up to 5 kHz. Clearly there are significant speed—resolution trade-offs to be considered with this approach. Bidirectional operation can be achieved with the scheme; however, to expect pixel placement going in opposite directions to overlap each other is unrealistic owing to the effects of phase delays and their associated drift rates.

### 6.2.3. Grating Method

Conceptually, the grating technique for linearizing a resonant system is the easiest to understand (Fig. 1). Systems employing gratings are used to generate clocks for the control of rotating polygons as well as resonant scanners (Morrill, Mason, and Green, private communication; Ohnishi et al., 1980; Grant, 1979). A simple analog driver at its resonant frequency sets the scanner into sinusoidal motion (Fig. 34). The feedback source is used to maintain scanner amplitude. The linearization system consists of a light source

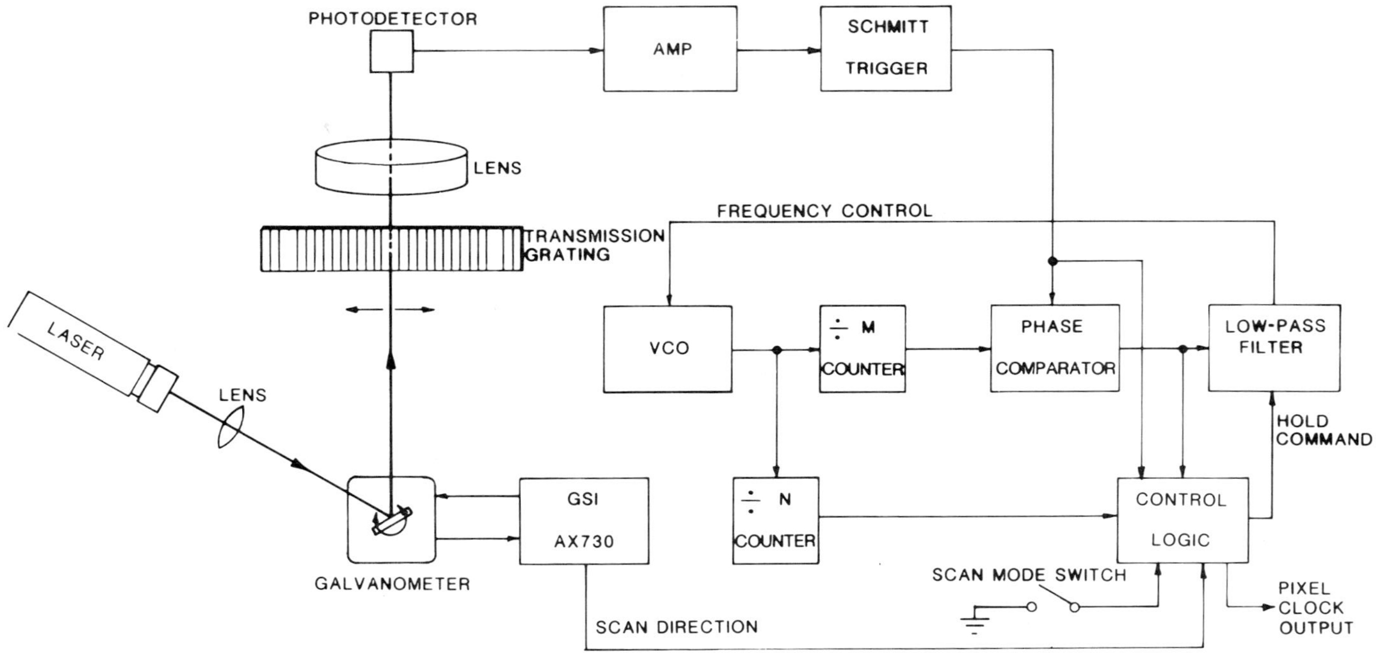


OUTPUT PULSES = # OF GRATING LINES X (M/N)

Figure 34 Grating clock generator.

deflected by the scanner mirror and imaged onto an accurately ruled grating. The transmitted or reflected light is collected by imaging optics and focused onto a simple photodetector. The photodetector output is amplified and passed onto a Schmitt trigger in order to generate nominal square waves at the spatial frequency of the grating. The grating incorporates clear areas near either end consisting of several missing lines for the purpose of defining the start and end of scan limits. An additional light source is deflected by the scanner mirror and used for the output of data onto the desired image plane.

A phase-locked loop is then constructed locking the master oscillator frequency divided by a counter to the grating pulse stream. The same master oscillator is divided down by another counter chain to yield the basic pixel clock. In this manner the pixel clocking rate can be synthesized from the grating without any severe restrictions; that is, the pixel clocks are related to the number of grating lines in the active scan path by the following equation:

$$\text{Pixel clock} = \text{number of grating lines} \times M/N$$

where M and N are integer numbers set by counters.

The control logic generates suitable gating signals to produce pixel clocks for either or both scan directions and also provides the means to do both the start and end of scan detection derived from the clear areas of the grating. The additional lines on either side of the clear areas of the grating provide the time for the phase-locked loop to acquire and lock onto the grating pulse stream. The availability of data-stripping integrated circuits (ICs) used in applications such as Winchester drives makes this phase-locked loop implementation easier than ever before.

This linearization technique has much to offer the user. It represents the minimal configuration of electronic hardware needed to achieve high accuracy and stability. It simultaneously handles both the sinusoidal scanner speed variation and the tangent correction in flat-field scanning applications. Scanner drifts in frequency and amplitude are compensated for by the clock. The spatial frequency of a relatively simple grating can be easily altered to meet the system demands; however, all of this is not achieved without difficulty and expense.

The drawbacks include the need to multiplex a second imaging system onto the scanner mirror with all the inherent optics and attendant costs. The grating accuracy and differences in the two optical systems determine, to a large extent, the system linearity and pixel placement accuracy. While a combined optical path might be implementable, in practice it leads to difficult packaging restrictions. Also, since the grating must be imaged by a high-resolution spot, local imperfections yield significant phase jitter in the phase-locked loop, which needs both fast acquisition and good tracking characteristics. It is also clear that phase delays in both the clocking mechanism and in the image modulation system result in pixel placement errors when an attempt is made to use the system in the bidirectional scanning mode. While these concerns are real, this approach may yield the best overall system configuration when the requirements call for high pixel counts at high scan rates with a high degree of pixel placement accuracy and repeatability and with virtual immunity to scanner and optical component drifts.

### 6.2.4. Digital Linearizer: Method 1

The two digital systems about to be described are intended to eliminate the second optical path needed in the grating scheme while still addressing the need for high accuracy, high repeatability, and the large pixel count. Both digital systems rely upon the inherently repeatable scanner rotation characteristics from scan to scan. Both techniques allow flexibility in the number of pixels to the line, the pixel density, and the scan angle efficiency. As before, the scanner is driven by a simple analog driver to achieve sinusoidal motion at a preset amplitude. In this case a second entry point is included to slave the scanner's amplitude to the linearization controller with the automatic gain control (AGC).

Method 1 contains a start-of-scan sensor placed in the image plane outside the active user area but within the minimum scanner deflection amplitude. Circuitry is provided to phase-lock a high-frequency oscillator to the scanner. A second amplitude control loop is used to adjust the scanner amplitude and hold it precisely at a fixed amplitude. The remaining logic is used to count the number of master clocks needed between pixel elements in order to correct to the desired accuracy the sinusoidal scanner velocity variation and, when desired, to remove the tangent error or any other describable, repeatable system error.

The start-of-scan detector (SOS) consists of a split-cell photodetector whose outputs are amplified and compared both to a fixed level and to each other (Fig. 35). With such a configuration it is possible to determine the exact time when the source beam has passed the sensor and in which direction the beam was traveling.

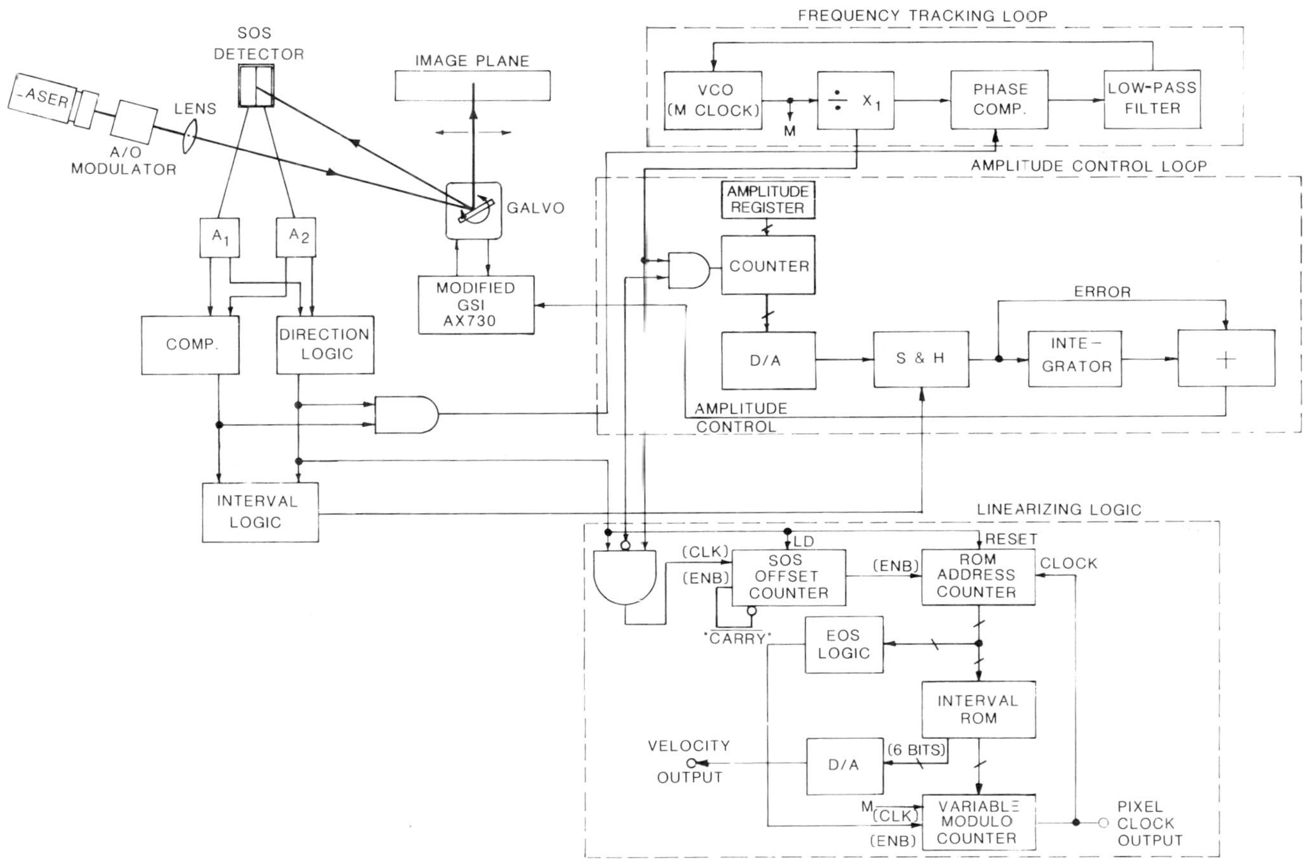


Figure 35 Digital linearizer: method 1.

Suitable logic makes this determination and provides signals that represent the time interval between beam passages over the sensor in both directions and also the time of occurrence of the passage. The SOS logic and sensor is a crucial element in achieving a high-accuracy system, and it needs to be repeatable and stable with regard to its phase delay, rise and fall times, and light sensitivity.

The output of the SOS locks the master oscillator to the scanner by a simple analog phase-locked loop circuit. The master oscillator runs at a frequency of X times the scanner oscillation frequency, where X is determined by the desired system parameters, including usable scanner amplitude, total number of pixels, and desired pixel placement accuracy. It should be noted that this clock runs at very high frequencies, often upward of 150 MHz, in a high-resolution, high-speed scanning application requiring good pixel placement accuracy. For example, an application calling for 20,000 pixels placed with an accuracy of 1 part in 200,000 using 80% of the scan angle with the scanner running at 200 Hz needs a master clock frequency of approximately 136 MHz. This clock is divided down by counters which are tapped at points suitable for transistor-transistor logic (TTL) counters used for determining the scanner amplitude and for the pixel clock generation.

The scanner amplitude is determined by comparing the number of master clocks that have occurred between consecutive passages of light beam over the SOS with a value stored in either a switch or storage register. The output of this magnitude comparison is converted into a voltage by a digital-to-analog converter whose output is retained by a sample-and-hold circuit. That value represents the amplitude error of the scanner and is summed with the integral of the amplitude error and used to drive the AGC input of the controller. In this fashion the amplitude of the scanner is held to a fraction of a pixel interval. Note that the scanner operating angle is always held constant by this technique, since the master oscillator is phase and frequency locked to the scanner.

The scanner pixel linearization is accomplished by altering the number of master clocks between each pixel clock through the action of an ROM and a counter. This, in effect, results in controlling the pixel period on a pixel-by-pixel basis. In many applications the pixel period update interval can extend over many pixels if the tolerable absolute noncumulative pixel placement error does not exceed 10% of the pixel-to-pixel spacing. The divided-down master clock is gated by the SOS detector logic and used to drive the clock input to the SOS offset counter. This counter is used to set the interval between the SOS detector and the beginning of the image area. After this counter has overflowed, the ROM address counter is enabled allowing the pixel period information to be loaded into the variable modulo counter. This counter is then counted down to zero at the master clock rate. It then generates a pixel clock output and also increase by one clock increment the pixel interval ROM address counter. This process continues until the address counter reaches a predetermined limit corresponding to the end of the active image area at which point the end-of-line logic inhibits pixel clock generation. The interval ROM also is programmed to contain bits to generate an analog signal proportional to the velocity of the scan for use in programming the image intensity modulation.

The advantages of this technique include a high degree of repeatability coupled with independence of performance from scanner drifts. Also, only a single optical path is used. Known optical corrections can be incorporated into the ROM by modifying the individual pixel periods. The accuracy of the system can be calculated in advance, since the total number of master clocks in the scan interval is known ahead of time. The disadvantages include the need to implement two high-accuracy phase-locked loops and to deal with clock frequencies in the hundreds of megahertz. Likewise, the absolute resolution of the system is set by the period of one master clock interval, and all pixel intervals are an integer multiple of that period. Figure 36 is a photograph of the linearizing electronics board. The method described next addresses some of these issues.

### 6.2.5. Digital Correction: Method Two

This method implements a digital linearization scheme based upon the computing capability of an inexpensive microprocessor. The structure allows a large degree of flexibility to adapt to various usable angles of excursion, optical corrections, and scanner frequencies. As before, a start-of-scan sensor with appropriate amplification and decoding logic is incorporated. Additionally, an end-of-scan (EOS) detector is needed. As always, the scanner is set into motion by an analog controller at approximately the correct amplitude.

A master voltage-controlled oscillator (VCO) running at the pixel clock rate implements the linearization scheme (Fig. 37). Thus in practical systems the pixel clocking rate is unlikely to exceed 50 MHz, reasonably within the reach of TTL logic. Clock generation proceeds with the SOS logic allowing the ROM address counter to start counting at the rate set by the master oscillator.

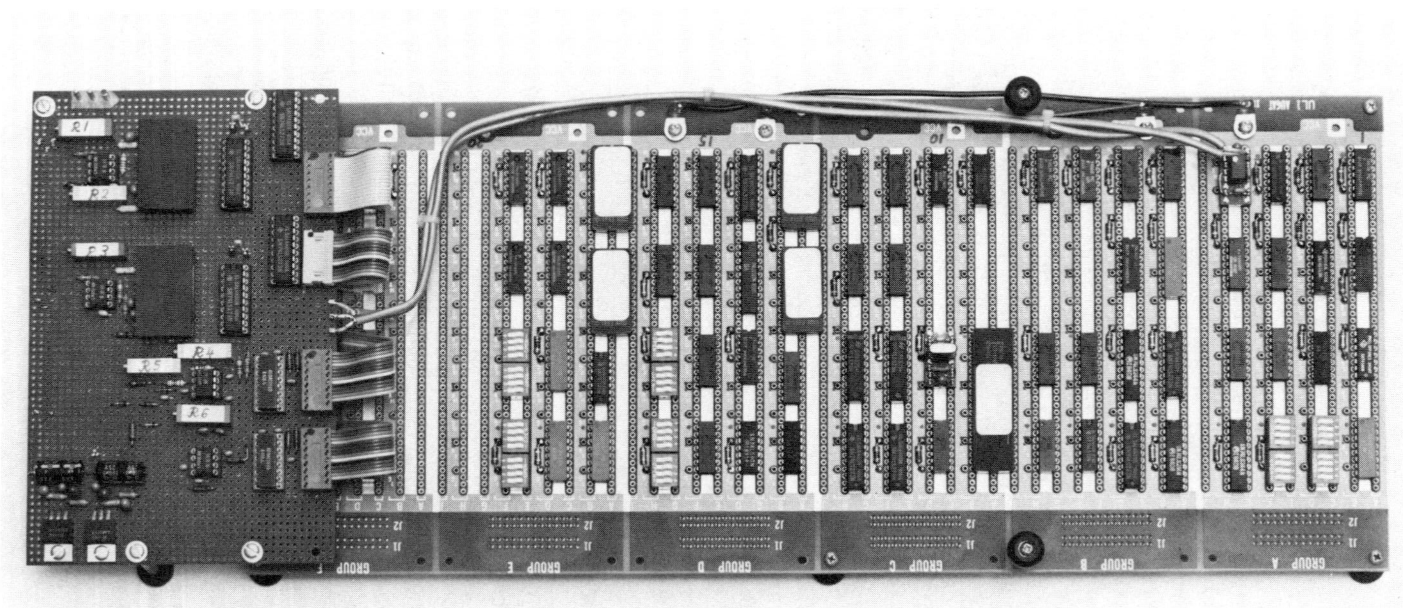

Figure 36 Linearizing electronics board.

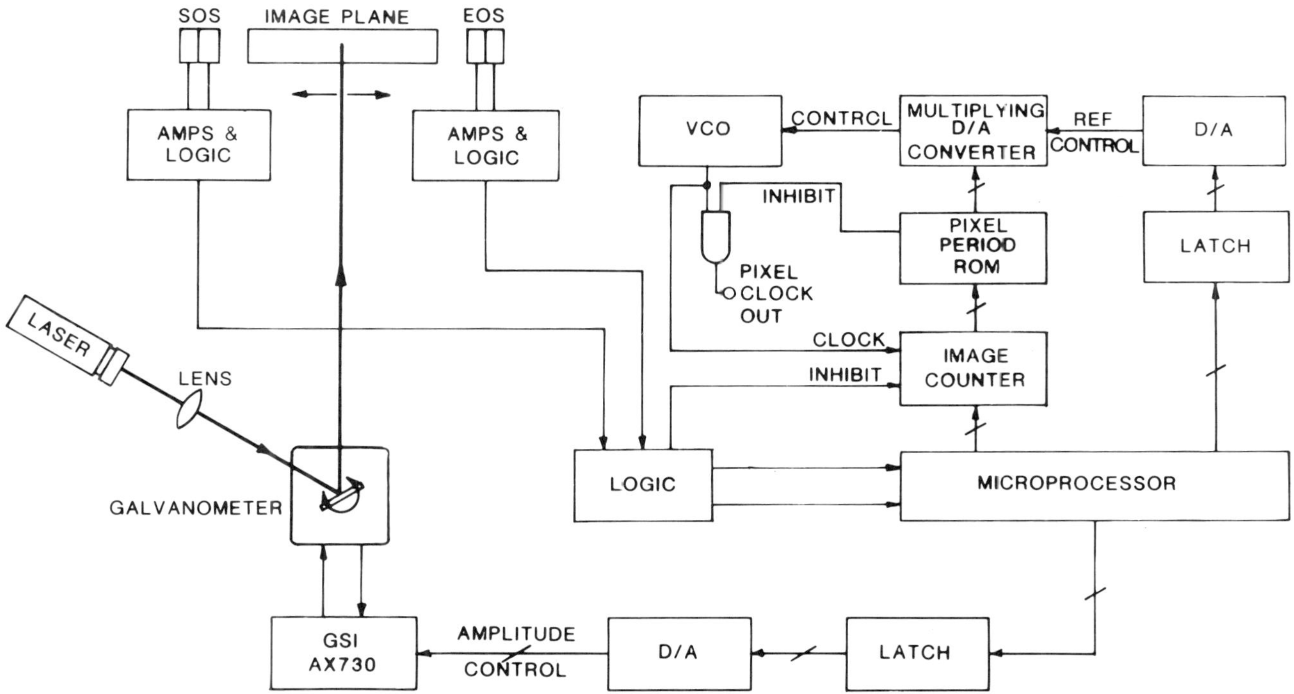


Figure 37 Digital linearizer: method 2.

This counter counts continuously until the light beam crosses the end-of-scan detector, at which time further counting is inhibited. During the counting interval the counter output is addressing an ROM containing the next desired master clock frequency (or period). An additional bit gates the master clock output in order to eliminate clocks during the out-of-image area interval. It also contains bits to allow amplitude modulation of the imaging source.

The ROM address counter can be loaded both from the microprocessor as well as read back into the microprocessor. The initial count loaded into the ROM counter sets the spacing between the SOS detector and the beginning of the image area. The final count in the ROM counter at the time of passage of the beam over the end-of-scan detector (EOS) is read back into the microprocessor as input information related to the amplitude and frequency drift of the scanner during the previous line scan interval. The microprocessor uses this information to output data to latches and a digital-to-analog converter to provide a "fine" tuning adjustment to the VCO master oscillator. The microprocessor can also be interrupted periodically by the ROM address counter to enable corrections to be inserted during the scan interval. The microprocessor also receives the outputs of the SOS and EOS detector logic to determine the direction of rotation of the scanner and to provide for the resetting, loading, and reading of the ROM counter. As before, the ROM may contain programming to correct for known geometric distortions from the optics path and may even correct for phase delays when the scanner is used bidirectionally.

Stability to ±1 pixel count is achievable with this technique, as is bidirectional operation. Amplitude and frequency drifts in the scanner are corrected by the microprocessor, eliminating the need for sophisticated analog phase-locked control loops. Another major advantage is that the fastest element in the hardware is running at the pixel clock rate, which can be substantially lower than in the previous method. The drawbacks to this system are the implied stability and repeatability required of the analog VCO and the associated driving digital-to-analog converters. Also, an EOS detector needs to be implemented; however, this technique is likely to yield the most cost-effective solution for systems requiring high pixel rates and densities with moderate to high pixel placement accuracy.

## 7. SYSTEM DESIGN FOR LOW INERTIA SCANNERS

Products which incorporate low inertia scanners, such as laser printers, phototypesetters, or forward-looking infrared sensors, also include digital logic electronic controls. The digital controls regulate the internal sequences of events, as well as the process and protocols needed to interface with incoming and outgoing data. As far as the issues affecting system performances are concerned, logic controllers are critical. Digital control issues frequently cause the components' erratic behavior. The digital control design requirements as encountered by other scanning systems can be found in Urbach et al. (1982) and Pelsue (1983). Some of the subjects to be addressed are the following:

1. Pixel position accuracy and master clock frequency
2. Digital sources of timing errors, pulse length, and pulse references
3. Analog sources of timing errors
4. Stability of lasers
5. Mechanical perturbations

## 7.1. Pixel Position Accuracy and Master Clock Frequency

The method used to synchronize data streams to the physical motion of the scanner, as well as the method used to command the scanner motion, strongly influences pixel placement accuracy. The scanning process combines both digital and analog functions. Because controls with digital electronics have a high degree of predictability and noise immunity, they are preferable to schemes which are inherently analog; however, the bandwidth limitations of currently realizable digital systems mean that trade-offs are still necessary.

The ability to locate a pixel accurately in a resonant sinusoidal scanning application is highly dependent on the ability to quantize time. This ultimately results in the need to generate high system clock rates as the need for increased throughput and accuracy is demanded. Most schemes require the master clock rate be a minimum of 120 MHz in order to locate pixels to at least 10% of the pixel spacing during any portion of a sinusoidal scan using 95% of the available scanner angle at a rate of 1000 scans per second and a total number of pixels of 2400. Although this master clock rate is not extremely high, it requires relatively sophisticated design techniques and care. Clearly, increasing the system performance requires computing at higher rates.

Protecting the master oscillator against line frequency and random noise pickup is mandatory, since they cause short-term phase jitter which is detectable as pixel location errors on a one-to-one basis. Long-term frequency drifts cause scan amplitude variation.

In galvanometric systems the scanner amplitude is not independent of the driving frequency, even well below the system's natural frequency. Frequency modulation (FM) of the system clock appears as pixel placement error and can be particularly troublesome to track down. It frequently appears to be synchronous with the AC mains, and therefore is attributed to grounding problems, power supply ripple, or mechanical vibration in any part of the system.

### 7.2. Digital Timing Errors

Additional sources of pixel placement errors can occur because of other timing errors in digital systems. The attempt to synchronize an analog event to the system clock results in time jitter equivalent to the clock period. A frequent mistake is to choose a relatively low-frequency digital clock to be the synchronizing source, causing large errors. These errors appear as time jitter in the pixel data stream and can be easily mistaken for mechanical motion instability in the scanner. Likewise, propagation delays through long logic chains, such as ripple counters, appear as sources of time jitter owing to the variability of the propagation delay and the time uncertainty of the clock pulse transition. Finite rise and fall times of digital logic levels are the cause of those uncertainties. Clearly, timing devices such as monostables are ruled out in high-performance applications owing to their inherent instability.

### 7.3. Analog Sources of Timing Errors

Systems incorporating a resonant scanner generally use its resonant frequency for synchronization. It ensures line-to-line repeatability. The slightest frequency change of the source or the resonance of the scanner, as well as any changes in Q, strongly affect the data and phase synchronization of the external clock-driven systems.

To generate rasters, galvanometric low inertia scanners are often operated in a periodic pattern such as an over-scanned full sawtooth (Fig. 27a), or a truncated sawtooth where a delay is entered prior to the forward scan period. In order to achieve a desired angular accuracy, either the scanner must have stabilized to that same level of accuracy prior to the useful time segment of the scan or the data stream clock must be synchronized with the flyback initiation pulse.

## 7.4. Stability of Lasers

In a gas laser photons of a pumped gas bouncing between the two end mirrors create light. Structural, thermal, electrical, and gas variances affect the aim, divergence, spot size, shape, and intensity of the beam stability. They are sources of errors or instability and must be recognized when the system performances are being defined. Obviously, errors or instability have different consequences on a laser TV projector than with a flatbed scanner printing an art book.

Solid-state lasers exhibit random source motion of approximately 50 μm, wavelength instabilities, and strongly astigmatic spot shapes. When solid-state lasers are used in a printer, the print quality degradation which can result may mistakenly be attributed to other elements of the system. Analyzing their behavior is complex, as solid-state lasers usually emit in the infrared.

There are suitable lasers, power supplies, and optical designs available for demanding applications, but often efficiency, cost, and bulk are the price to pay.

## 7.5. Mechanical Perturbations

Imperfect $f\theta$ correction lenses can cause apparent linear position and velocity distortions similar to those on a nonlinear scanner.

Inertia and the instrument chassis design are important considerations, specifically in using resonant low inertia scanners. Section 3.2 briefly reviews the parameters: unpredictable Q, variable power consumption, and cross-axis motions. Typical solutions are either a soft scanner interface or a high inertia and a lossy chassis.

Raster-type instruments of high precision, such as flatbed scanners, phototypesetters, and laser printers, translate or rotate the photoreceptor across the scanned area. The photoreceptor is usually motor driven at comparatively slow speeds via a worm screw, gears, or belt. Although speed variations have numerous sources, they often are confused with scanner wobble. Typical sources are load variation, out-of-round drums, supply voltage, gear train irregularities, shaft winding, or stepper motor tooth irregularities. At advances of one increment per scan, the perceptible "periodic artifacts" generated are readily visible (see Sec. 2.1).

## 8. CONCLUSIONS

Assembling in a concise and logical presentation material which indirectly affects the daily life of individuals in modern society has been a most gratifying experience. Optical scanners are only one of many techniques which facilitate the interface between man, computer systems, and ultimately machines; however, more than 50% of human information is obtained visually, and optical scanners can be made to closely parallel the human method of acquiring data.

In our age robotics will supplant automation, and it is to be expected that humanlike vision will replace the fixed television camera. Just as a robotic arm is similar to a human arm (with four to six degrees of freedom), so computer vision will duplicate human vision.

The human vision system has an array of light sensors, the retina, and muscles which aim the eye and focus the image. In artificial vision, charge couple devices (CCD) or vidicon cameras are the equivalent of the retina, and two-dimensional scanners with focusing capabilities fill the other requirements necessary for humanlike vision.

Camera-based electronic vision systems are presently available from General Electric, Reticon, Cognex, Octek, Machine Intelligence, Object Recognition Systems, Automatix and Unimation, and others. The scene is set for the next application of optical scanners.

## 9. APPENDIX

### 9.1. Torque Calculation for Moving Iron Galvanometers

Moving iron galvanometric scanners of the type shown in Fig. 38, derive their mechanical torque from the interaction of an electromagnetic circuit and the magnetic field of a permanent magnet. In the calculations below, it is assumed that soft iron pole pieces never reach saturation and that they have infinite permeability when compared to the permeability of the air gap.

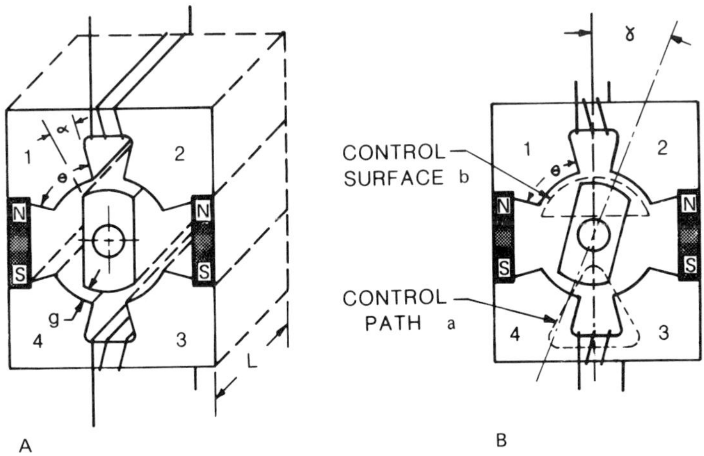


**Figure 38** Moving iron galvanometer.

The steady-state torque T is calculated by taking the derivative of the total energy U in the air gap with respect to the angular rotation $\gamma$. Two types of energy are present (Feynman, 1964). The first is the mechanical energy $U_m$. The second, the electrical energy $U_e$, is related to the work on the conduction electrons that sustain a constant current in the conductor. The galvanometer under power cannot be considered an independent control volume because it exchanges energy with the power source. Variations of electrical energy within the galvanometer can be measured either as the change of inductive energy stored or as the quantity of energy exchanged between the device and the electrical power source. Under constant current I conditions, electrical energy is exchanged through a change in drive voltage V.

The total energy in the air gap is the mechanical energy $U_m$ plus the electrical energy as defined above. It is expressed as:

$$U = U_m + U_e \tag{1}$$

If the electrical energy is forced to remain constant, the torque is the derivative with respect to the angle of the magnetic energy in the air gap. According to the first Maxwell equation for magnetic matter, the total electric potential V at the terminal of an iron-cored coil is related to the electric potential E along the coil winding and the magnetic field created $B_c$. It is expressed as:

$$V = \int E \cdot dl = \frac{d}{dt}\int B_c \cdot da = \frac{d(\phi_c)}{dt} \tag{2}$$

where $B_c = \mu H_c$ and $\phi_c$ is the magnetic flux derived from the coil current. Consequently:

$$\phi_c = \int V\, dt \tag{3}$$

but as $U_e = \int VIdt$ where the coil resistance is assumed negligible. Thus:

$$\phi_c = \frac{U_e}{I} \tag{4}$$

The constraint to have $U_e$ constant forces $\phi_c$ to be a constant as I also has constant value for a steady-state condition. Therefore the torque T can be expressed by the two equations:

$$T = \frac{dU_m}{d\gamma} = \frac{dU}{d\gamma} \tag{5}$$

$$\frac{d\phi_c}{dt} = \frac{d\phi_c}{d\gamma} = 0 \tag{6}$$

Symmetry simplifies the torque calculations for the device in Fig. 38. Gaps 1 and 3 have identical energy conditions, as do gaps 2 and 4. Since permanent magnets have permeability comparable to air, the assumption is that poles 1 and 2 are disassociated from poles 3 and 4. Thus the effects in gaps 1 and 2 are assessed and multiplied by 2.

Because the electromagnetic fields are linear, the superposition method is used to calculate the energy in the gaps as the sum of the values of all independent fields. Each air gap supports the sum of the fields generated by the permanent magnets plus the field induced by the drive coil. Each can be derived independently as the two fields do not interact.

When the drive coils are not powered, B is the component of the magnetic field in the air gaps generated by the permanent magnets. The value of B is independent of rotor angular displacement $\gamma$ from position of symmetry as long as the following two conditions for absolute angular values are met:

1. $$|\gamma| + |\alpha| < |\theta| < |2\alpha| \tag{7}$$

   where $\alpha$ is the angular overlap of one pole and rotor at position of symmetry and $\theta$ is the maximum arc of each stator pole piece.

2. The pole pieces of the stator and of the rotor are perfectly concentric segments of cylinders. Therefore the air gap g is constant.

The assumption is made that B is constant across each pole face and equal for all air gaps as well as independent of rotor angular position. This can be proven from either Ampere's law or Gauss's law. They are the line integral of the field along control path a and the surface integral along control surface b of Fig. 38.

$$\int B \cdot dP = 0 \tag{8}$$

$$\int B \cdot da = 0 \tag{9}$$

The conclusion is that B is constant.

Using Ampere's law for magnetic fields derived from current source, fields $B_1$ and $B_2$, generated by current I in a coil of N turns under pole pieces 1 and 2, is expressed:

$$\int B \cdot dl = \mu NI \tag{10}$$

As all air gaps have equal value g, the magnetic fields are uniform in each air gap. The above equation simplified is:

$$B_1 + B_2 = \frac{\mu NI}{g} \tag{11}$$

By applying Gauss's law to the control volume b indicated in dotted lines on Figure 39B and by naming $A_1$ and $A_2$ the areas of the magnetic field under poles 1 and 2, respectively:

$$B_1 \cdot A_1 = B_2 \cdot A_2 \tag{12}$$

The next two equations use Eqs. (11) and (12) to solve for $B_1$ and $B_2$:

$$B_1 = \frac{\mu NI}{g} \cdot \frac{A_2}{A_1 + A_2} \tag{13}$$

$$B_2 = \frac{\mu NI}{g} \cdot \frac{A_1}{A_1 + A_2} \tag{14}$$

The energy U/2 in gaps 1 and 2, which represents only half the total energy, is:

$$\frac{U}{2} = \frac{1}{2\mu}\int_1 (B + B_1)^2 \, dV + \frac{1}{2\mu}\int (B - B_2)^2 \, dV \tag{15}$$

As the fields in each gap are uniform, Eq. (15) simplifies to:

$$U = \frac{g}{\mu}(B + B_1)^2 A_1 + \frac{g}{\mu}(B - B_2)^2 A_2 \tag{16}$$

It can be expressed as a function of the flux in each pole, which is:

$$\phi_1 = (B + B_1)A_1 \tag{17}$$

$$\phi_2 = (B - B_2)A_2 \tag{18}$$

Therefore Eq. (16) becomes:

$$U = \frac{g}{\mu}\left[\frac{(\phi_1)^2}{A_1} + \frac{(\phi_2)^2}{A_2}\right] \tag{19}$$

Taking the derivative with respect to $\gamma$ within the constraints of Equation 6, $d\phi/d\gamma = 0$ yields the torque of the device.

$$T = \frac{dU}{d\gamma} = \frac{g}{\mu}\left[(\phi_1)^2 \cdot \frac{d(A_1)^{-1}}{d\gamma} + (\phi_2)^2 \cdot \frac{d(A_2)^{-1}}{d\gamma}\right] \tag{20}$$

$$T = -\frac{g}{\mu}\left[\left(\frac{\phi_1}{A_1}\right)^2 \cdot \frac{dA_1}{d\gamma} + \left(\frac{\phi_2}{A_2}\right)^2 \cdot \frac{dA_2}{d\gamma}\right] \tag{21}$$

By using Eqs. (17) and (18), it simplifies to:

$$T = \frac{g}{\mu}\left[(B + B_1)^2 \frac{dA_1}{d\gamma} + (B - B_2)^2 \frac{dA_2}{d\gamma}\right] \tag{22}$$

Referring to Fig. 39B:

$$A_1 = rL\ (\alpha - \gamma) \tag{23}$$

$$A_2 = rL\ (\alpha + \gamma) \tag{24}$$

With respect to $\gamma$, their derivatives are:

$$\frac{dA_1}{d\gamma} = -rL \tag{25}$$

$$\frac{dA_2}{d\gamma} = rL \tag{26}$$

Therefore the torque of Eq. (22) simplifies to:

$$T = \frac{grL}{\mu} [(B + B_1)^2 - (B - B_2)^2] \tag{27}$$

and with Eqs. (13) and (14), becomes:

$$T = BLNID - \mu DL(NI)^2 \cdot \frac{\gamma}{2g\alpha} \tag{28}$$

where D = 2r.

Note that as long as $\mu NI\gamma / 2g\alpha << B$, Eq. (28) approximates the familiar Lorentz equation. If 10% of best straight line nonlinearity is tolerable at 75% of maximum peak-to-peak excursion, then for a galvanometer where the air gap is $10^{-4}$ meter and B = 1 telsa, the maximum value of NI is limited to approximately 400 Ampere turns. These are typical values for most commercial scanners, galvanometers, and pen motors of this design. Much larger torques can be obtained without reaching magnetic saturation if very large torque nonlinearities can be tolerated. Section 9.2 below analyzes alternate magnetic designs that correct these deficiencies.

## 9.2 Linearization Techniques for Moving Iron-Limited Rotation Devices

From Eq. (28), the torque of a simple galvanometer can be assumed proportional to current only at low current level and/or when limited to the central region of the total possible excursion. The penalties for operating beyond these constraints are nonlinearities of torque versus current and current polarity as well as angular position.

Techniques have been devised to extend the useful capabilities of this class of devices. The application with its required speed, accuracy, and cost leads to the selection of the type of scanner and linearization. The two major modes of operation are the open loop design, and the position servo or closed loop design. In the open loop design, positioning or tracking is achieved by a linear mechanical spring that is deflected in torsion to equal the torque of the magnetic driver. In the position servo or closed loop design, the desired position or motion is measured by a separate transducer and used to energize the magnetic torque drive. Each has different requirements.

### 9.2.1. Open Loop Operation

Two designs for scanners with a linear relationship between angular shaft positions and currents, where the magnetic torque is opposed by a linear mechanical spring, are presented by Montagu (1965, 1971). Both designs create a magnetic spring that weakens the mechanical spring at the correct angle in order to compensate for the torque short fall expressed by Eq. (28). Their dynamic behavior is nonlinear; that is, the damping ratio $\xi$ for such a second order system is expressed as:

$$\xi = 0.5C\ (KJ)^{-0.5} \tag{29}$$

where C is the electronically generated back emf damping, K is the combined mechanical and magnetic spring stiffness, and J the inertia of the rotor and load.

In order to have the same overshoot at all angular positions, C must track the nonlinearities of K. Techniques to obtain a signal that generates a controlled damping compensating exactly for the nonlinearities of the spring constant are explained by Montagu (1977, 1980).

Scanners constructed according to the above designs yield an approximately constant damping ratio at all angular positions but a greater bandwidth in the mid range of the useful excursion than at the extremes. This is a successful solution when the magnetic torque, in order to reach a proper pointing position, is opposed by a mechanical spring. However, errors due to current polarity changes at given angles are not addressed, and only limited help is offered to expanding the bandwidth of a servo system. So long as the minimum bandwidth is attained at all angular positions, this is satisfactory. These systems are used in optical instruments where speed and positional accuracy are not critical.

### 9.2.2. Closed Loop Operation

Linearizing the torque/current relationship is the most desirable method to optimize the closed loop performances of a second-order system. Referring to Fig. 39B, the nonlinear term in Eq. (28) comes from the increase of the reluctance in the gap of pole piece 1. The coil-generated magnetic fields, $B_1$ and $B_2$ in gaps 1 and 2, respectively, are in series. The cross section of gap 1 decreases. Values for the fields $B_1$ and $B_2$ can be derived from Eqs. 13, 14, 23, and 24.

$$B_1 = \frac{\mu NI}{g} \cdot \frac{\alpha + \gamma}{2\alpha} \tag{30}$$

$$B_2 = \frac{\mu NI}{g} \cdot \frac{\alpha - \gamma}{2\gamma} \tag{31}$$

Equation (31) shows that the value of $B_2$ converges toward zero as pole piece 2 becomes totally uncovered. Simultaneously $B_1$ goes to zero as the same flux threads both poles.

The torque T expression of Eq. (27) can be expanded to:

$$T = \frac{grL}{\mu} [B(B_1 + B_2) + (B_1^2 - B_2^2)] \tag{32}$$

If at all angles $B_1 = B_2 = \mu NI/g$ the torque in Eq. (32) becomes only the first linear term of Eq. (28).

A rotor design with four possible paths for the coil-generated magnetic field can approach this requirement. Referring to Fig. 39a, gaps 1b and 2b are added in parallel to gaps 1a and 2a. Gaps 1a and 2a are equivalent to gaps 1 and 2 of Fig. 38. Gaps 1b and 2b are generated by two concentric cylindrical surfaces; stator pole pieces 1, 2, 3, and 4 forming one boundary and cylindrical region "b" of the rotor forming another boundary. As this region has a lesser radius than the major pole piece regions "a," a larger air gap, G, is created.

In this design, when the armature fully rotates as shown in Fig. 39B, the reluctance of pole piece 2 converges toward a finite value and magnetic fields $B_1$ and $B_2$ never go to zero. The limit value is expressed as:

$$B_1 = \frac{\mu NI}{g + G} \tag{33}$$

As gap G approaches value g, $B_1$ is no longer a function of angle; neither is the energy in the air gaps. Therefore it is not possible to generate a torque. A compromise value for G has been obtained experimentally as the exact theoretical solution is quite complex.

It is evident that the smaller diameter "cylindrical" need not have cylindrical shape and that gap G can be a function of angle so as to obtain virtually any desirable torque versus angular position function.

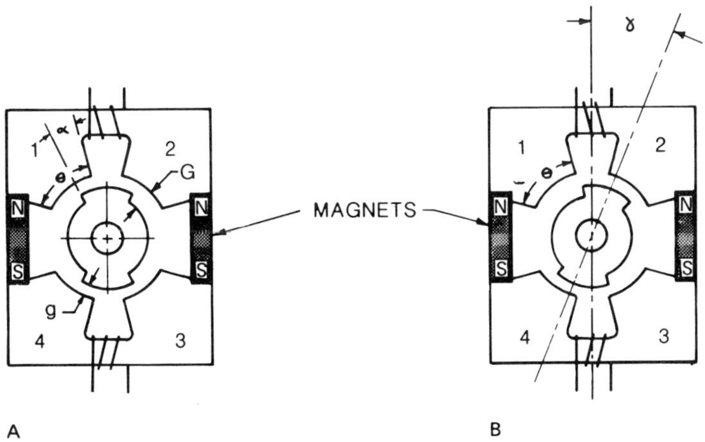


Figure 39 Moving iron galvanometer.

Note that the presence of two additional reluctance paths calls for enlarged permanent magnets and that flux paths with larger sections are needed. Also, the null peak torque will be less for this new design, and the inductance will be higher. In conclusion, linear performances are not achieved without penalties.

Experimental results indicate that best compromise is obtained when secondary air gaps b are four to ten times the dimension of primary gaps a. For example, Fig. 30 shows torque, current, and angular position for stators with 1/2 in. diameter bore.

### 9.3. Induction Driven Moving Coil Resonant Scanners

The type of scanners addressed here belong in the category of moving coil devices where a current-carrying conductor reacts to the field of a permanent magnet. They are commonly known as voice coils or meter movements, and have a drive coil energized through flexible leads. The devices of interest here are different; their drive coil is energized by induction. Flexible leads are not practical for resonant scanner drives because they cause energy losses and secondary resonances, and are frequently affected by fatigue failure. Also, anchor points add penalizing inertia.

The induced moving coil equivalent of a d'Arsonval resonant scanner is presented in Fig. 31. Its torsion bar suspension and mass balanced aluminum mirror armature combination is such that acceleration along the three translation axes does not cause unwanted angular mirror motions. The drive coil and the armature loop are magnetically linked in a transformerlike manner by a soft iron core. The iron core material selected must have high permeability $\mu$ at the operating frequency. The permeability is assumed constant through the iron core. The armature is a single-turn secondary coil of the transformer and is hinged at radius r on a pair of torsion bars. The primary of the transformer is the drive coil with N turns and carries an alternating current $I = I_o \sin \omega t$ of frequency $\omega$. This causes a magnetic field H to be induced in the iron core of length L which according to Ampere's law is expressed as:

$$HL = NI_o \sin \omega t \tag{34}$$

In the armature an electric potential e is induced and is expressed according to Faraday's law where A is the section of the iron core threading the loop. It is:

$$e = -\mu A \frac{dH}{dt} \tag{35}$$

Calling the armature resistance R, the current $I_a$ in the loop is derived from Ohm's law:

$$I_a = \frac{e}{R} = -\frac{\mu ANI_o\omega}{RL} \cdot \cos \omega t \tag{36}$$

This current interacts along pole piece length l with the field B of the permanent magnet to generate the drive torque T according to Lorentz's law:

$$T = -\frac{\mu ANI_o lrB\omega}{RL} \cdot \cos \omega t \tag{37}$$

As the armature is free to move with angular velocity $d\theta/dt$, a potential E is induced according to Faraday's law and can be calculated from

$$E = Blr \cdot \frac{d\theta}{dt} \tag{38}$$

This potential generates a current $I_d$:

$$I_d = \frac{E}{R} = \frac{Blr}{R} \cdot \frac{d\theta}{dt} \tag{39}$$

and a torque $T_d$ according to Lorentz' law:

$$T_d = \frac{(Blr)^2}{R} \cdot \frac{d\theta}{dt} \tag{40}$$

The motion of the armature is that of a second order system with inertia J and torsional spring stiffness K as well as mechanical damping C. When subjected to the torques of Eqs. (40) and (37), its angular motion is described as:

$$J \frac{d^2\theta}{dt^2} + \left[\frac{(Blr)^2}{R} + C\right] \frac{d\theta}{dt} + K\theta = \mu \frac{ANI_o}{RL} Blr\omega \cos \omega t \tag{41}$$

When the mechanical resonant frequency $\omega = \sqrt{K/J}$ is equal to the excitation frequency, Eq. (41) yields solutions for angular position $\theta$ and loss factor Q:

$$\theta = \frac{\mu ANI_o Blr}{L[(Blr)^2 + CR]} \sin \omega t \tag{42}$$

$$Q = \frac{R\omega J}{(Blr)^2 + CR} \tag{43}$$

As expected, both the excursion and the Q are a function of the mechanical damping C that must be minimized.

A second coil, identical to the drive coil, senses current of Eq. (39) that is generated by the motion of the armature. It is also linked to the drive coil as a transformer and the potential $E_p$ at its terminals is the sum of both induced signals.

$$E_p = \mu \frac{A}{L} N^2 I_o \omega \cos \omega t + NBlr \cdot \frac{\mu A}{RL} \frac{d\theta}{dt} \tag{44}$$

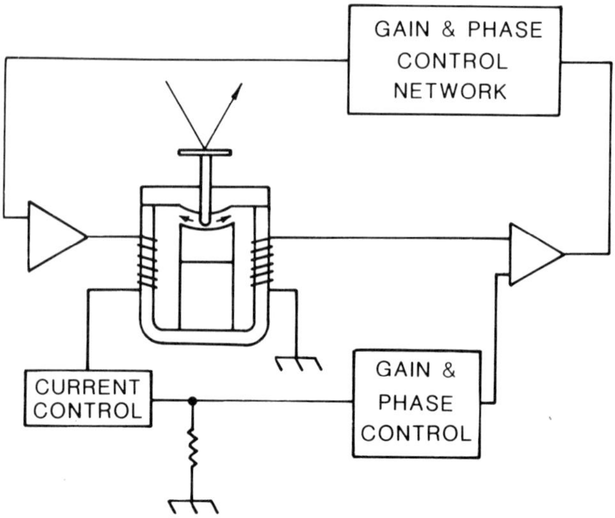


Figure 40 Resonant scanner driving circuit.

The value of $d\theta/dt$ can be derived from Eq. 42 and is expressed as:

$$\frac{d\theta}{dt} = \frac{\mu A N I_o B l r}{L[(Blr)^2 + CR]} \omega \cdot \cos \omega t \qquad (45)$$

and therefore Eq. (44) simplifies to:

$$E_p = \mu \frac{A}{L} N^2 I_o \omega \left[ 1 + \frac{(BPr)^2 \mu A}{[(B\ell r)^2 + CR]RL} \right] \cos(\omega t) \qquad (46)$$

The velocity term is a function of $\mu^2$ and can be much larger than the transformer coupling term while both have the same phase. At high operating frequency when $\mu$ is small, it is possible to subtract from the pickoff signal a voltage proportional to the drive current and have again a clean velocity signal. This velocity signal is then used as positive feedback to induce self oscillation of the scanner. The amplitude of motion is controlled by limiting the available drive current. Figure 40 shows such a circuit design.

## REFERENCES


Allen, D., Montagu, J., et al. (1978). Low inertia scanner performance and selection. In *Electro-Optical System Design*, November.

Bestenreiner, F., Greis, U., Helmberger, J., and Stradler, K. (1976). In *Journal of Applied Photo Engineering*, Vol. 2, pp. 86–92.

Bieser, L. (1974). *Laser Applications*, Academic, New York.

Blitzer, S. D. (1948). The design of a torque motor for hydraulic servo applications. Master of science thesis, Department of Electrical Engineer, Massachusetts Institute of Technology, Boston, Massachusetts.

Brosens, P. J. (1972). Dynamic mirror distortions in optical scanning. In *Applied Optics*, Vol. 2, No. 12.

Brosens, P. (1976). Scanning accuracy of the moving-iron galvanometer scanner. In *Optical Engineering*.

Burrer, G. J. (1976). Patent #3978,281.

Dostal, F. (1971). *Resonant Torsional Oscillators*. U.S. Patent #3,609,485.

Feynman, P., Leighton, R., and Sands, M. (1964). *Lectures on Physics*, Vol. 2, Addison Wesley, Reading, § 15–2.

Grant, D. E. (1979). Optical–mechanical design of the IBM 6670 laser printhead. In *SPIE Proceedings*, Vol. 200, pp. 195–199.

Laws, P. E. (1952). An electro-mechanical transducer with permanent magnet polarization. *Technical Note No. G.W. 202.* Royal Aircraft Establishment, Farnborough, England.

Montagu, J. I. (1965). *Electric Motor for Limited Rotation.* U.S. Patent #3,177,385.

Montagu, J. I. (1971). *Actuator.* U.S. Patent #3,624,574.

Montagu, J. I. (1976a). *Oscillation Motor.* U.S. Patent #3,959,673.

Montagu, J. I. (1976b). A practical approach to low inertia scanner selection. *SPIE*, Vol. 84.

Montagu, J. I. (1976c). *Limited Rotation Motor with Velocity Sensing*. U.S. Patent #3,970,979.

Montagu, J. I. (1980). *Limited Rotation Moving Iron Motor with Novel Velocity Sensing and Feedback*. U.S. Patent #4,186,332.

Montagu, J. I. (1983). Laser scanners for office automation. In *Electro Optics*, May.

Mueller, R. (1949). *Dynamo Transformer*. U.S. Patent #2,488,734.

Ohnishi et al. (1980). Patent #4,212,018.

Paul G. (1983). Image data synchronization techniques in rotating polygon scanners. In *SPIE Proceedings*, Vol. 396.

Pelsue, K. (1983). Post-objective flatfield laser imaging system. *Proceedings of Laser in Graphics 4th Annual Conference.*

Rohr, B. (1949). *Position Detector.* U.S. Patent #4,185,215.

Selverstone, P. (1976). *Electrically Damped Oscillating Motor.* U.S. Patent #4,090,112.

Timoshenko, S., and Goodier. (1951). *Theory of Elasticity*, McGraw-Hill, New York.

Urbach, J., et al. (1982). Laser scanner for electronic printing. In *Proceedings of the IEEE*, June.

Zook, J. David. (1974). Light beam deflector performance: A comparative analysis. In *Applied Optics*, Vol. 13, p. 875.

# 6

# Gaussian Laser Beam Diameters

Gerald F. Marshall
*Energy Conversion Devices, Inc., Troy, Michigan*

## I. INTRODUCTION

A laser is notably an intense monochromatic source of radiation. The source radiates at wavelengths in the electromagnetic spectrum according to its design. Lasers emit a very narrow cone of radiation referred to as a *laser beam*. The lowest order transverse electromagnetic (TEM) mode output of a laser is a beam with a spherical gaussian wavefront (Young, 1972a). The characteristics of this beam are typically radially symmetric about its axis, the principal direction of propagation. At any perpendicular cross section of the beam its radiance, that is, its brightness as perceived by the eye in the visible region of the spectrum, is also gaussian; it is at a maximum on the axis and falls off with distance from the axis. For a helium-neon laser, the beam of light at a wavelength of 632.8 nm may be characterized to a first order by a nominal divergence of about 1 mrad for a nominal diameter of about 1 mm. It is the definition and clarification of the "diameter" of a laser beam that I address.

There are several terms in vogue used to express the diameter of a laser beam. Some are misnomers, some are ambiguous which, like some road signs, are misleading unless we know the territory. Interrelated with this confusion in the diameter of a laser beam is the misuse of the more accepted radiometric terminology (Nicodemus, 1978; Nicodemus et al., 1977). This terminology itself seems all too frequently to be under revision and, not suprisingly, in a "state-of-flux." I shall review the relevant Commission Internationale de l'Eclairage (CIE) terms that are most widely accepted internationally and use them appropriately to discuss, clarify, and define the diameter of a laser beam.

## 2. RADIOMETRIC TERMINOLOGY

Radiometry is the technology surrounding the measurement of electromagnetic radiant power (Watts) and energy (Joules) across the entire spectrum of wavelengths. Photometry, which involves the sensitivity of the eye and has its own units, is that part of radiometry that deals only with the visible portion of the electromagnetic spectrum ranging approximately 400–720 nm (Jenkins and White, 1951a). In this wavelength band, because of the spectral response characteristic of the eye, the conversion factor between photometric and radiometric units is a nonlinear relationship with wavelength (Young, 1977b). As radiometric terms cover the whole spectrum I shall discuss the subject in these terms and, where appropriate, write the photometric terms in parentheses.

### 2.1. Radiometric Terms

The important CIE terms to review are those associated with: (1) radiation from a source, (2) radiation traversing a medium, and (3) radiation incident at a surface.

| | | |
|---|---|---|
| Radiant flux (luminous flux): | $\Phi$ | Radiant power [Watts] |
| Radiant intensity (luminous intensity): | I | Radiant power emitted per unit solid angle in a specified direction from a source. [Watts/steradian] |
| Radiance (luminance): | L | Radiant power emitted per unit solid angle in a specified direction per unit projected area on the source, receiver, or intermediate intersecting reference surface in the specified direction. [Watts/square meter/steradian] |

| | | |
|---|---|---|
| Irradiance (illuminance): | E | The radiant power per unit area either traversing a plane or incident at a surface. [Watts/ square meter] |

## 2.2. Radiant Flux $\Phi$

The term *radiant flux* represents the power in a bundle of rays, and is often simply referred to as flux, provided there is no ambiguity with luminous flux.

## 2.3. Radiant Intensity I

The terms *radiant intensity* and *radiance* both relate to the radiant properties of a source and relate to one another by

$$\text{Radiant intensity}_{\theta} = \int_s \int \text{Radiance} \cdot \cos \theta \cdot ds$$

where ds is an elementary area of a finite source of area s and $\theta$ represents the angular direction from the normal (Fig. 1).

## 2.4. Radiance L

*Radiance* and its matching photometric term *luminance* also corresponds to the subjective term we observe and sense with the eye and refer to as *brightness*. The surface of each elementary area ds of an isotropically diffuse radiator in the form of a uniformly radiating luminous disk will look equally bright from any angle of view; however, the integrated intensity observed at a distance in a specified direction from the normal to the disk falls off according to a cosine function. Radiators whose intensity distribution approximate closely to this function are said to be lambertian (Fig. 1).

Radiance L also appears to be an invariant parameter associated with the conservation of radiant flux $\Phi$ within a bundle of light rays that emanates from a small source object of finite area as it traverses an optical system of nonabsorbing media (Jenkins and White, 1951b).

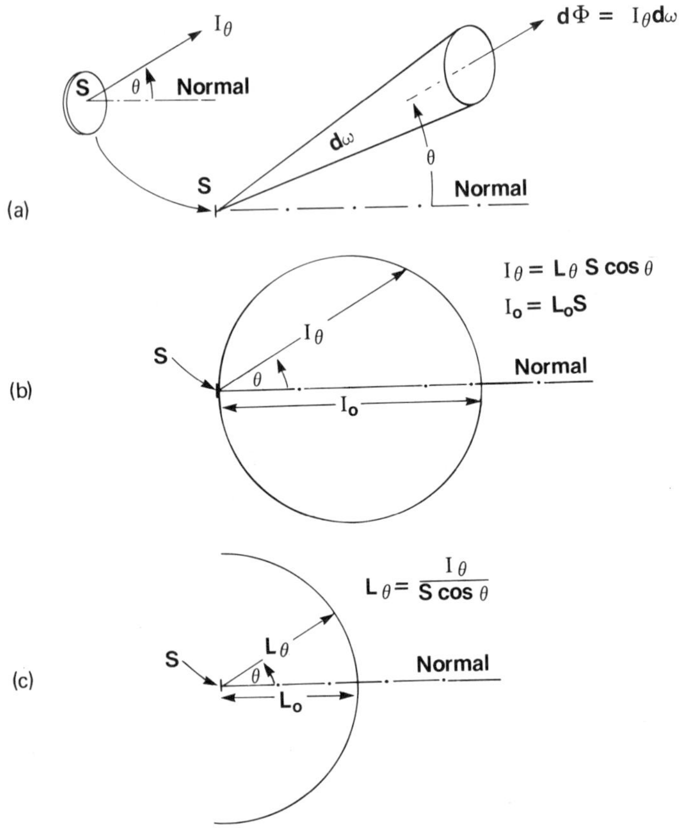


Figure 1 Lambertian radiator in the form of a small disk which has a surface area S. (a) Elementary cone of radiant flux $d\Phi$ within a solid angle $d\omega$. (b) Polar distribution of the radiance $L_\theta$ versus the angular direction $\theta$ to the normal. For a lambertian radiator $L_\theta = L_o$, a constant, and the polar surface characteristic is hemispherical with a radius $L_o$. (c) Polar distribution of the radiant intensity $I_\theta$ versus the angular direction $\theta$ to the normal. The polar surface characteristic is spherical with a diameter $L_o/S$.

### 2.5. Irradiance E

*Irradiance* corresponds to the passage of radiant power per unit as a bundle of light rays traverses a plane or is incident on a plane.

## 3. GAUSSIAN BEAM

With respect to cylindrical coordinates let the direction of propagation of a laser beam be in the z direction and let the laser beam axis be colinear with the z axis with the origin O at the beam waist (Fig. 2). Let r be the radial coordinate.

Consider any perpendicular cross section of the beam through the axial point Q(0,z) (Fig. 2). The distribution of radiant flux density traversing this plane is gaussian with a standard deviation of $\sigma(z)$. $\sigma(z)$ is a function of z because the beam is either diverging or converging; $\sigma$ equals $\sigma(z)$.

Figure 2 shows a laser beam envelope for which $r/\sigma = k$ and which is a chosen constant. Thus, the beam envelope radius $r_k$ and the beam envelope diameter $2r_k$ at any section may be expressed in terms of the parameter k, that is, in multiples of the gaussian standard deviation $\sigma(z)$ namely

$$2\,r_k = 2\,k\,\sigma(z) \tag{1}$$

where

$$\sigma(z) = \sigma_0 \left[ 1 + \left( \frac{\lambda z}{4\pi(\sigma_0)^2} \right)^2 \right]^{1/2} \tag{2}$$

and where $\lambda$ represents the wavelength of light. In other words, k represents the relative or normalized beam radius with respect to the standard deviation $\sigma(z)$ at any cross-section. It will be our choice in the value of k that will define the diameter of a laser beam and to which I shall return.

### 3.1. Beam Irradiance

Using a radiometric detector, which has a minute active area A compared with the cross-sectional area of the laser beam, we can

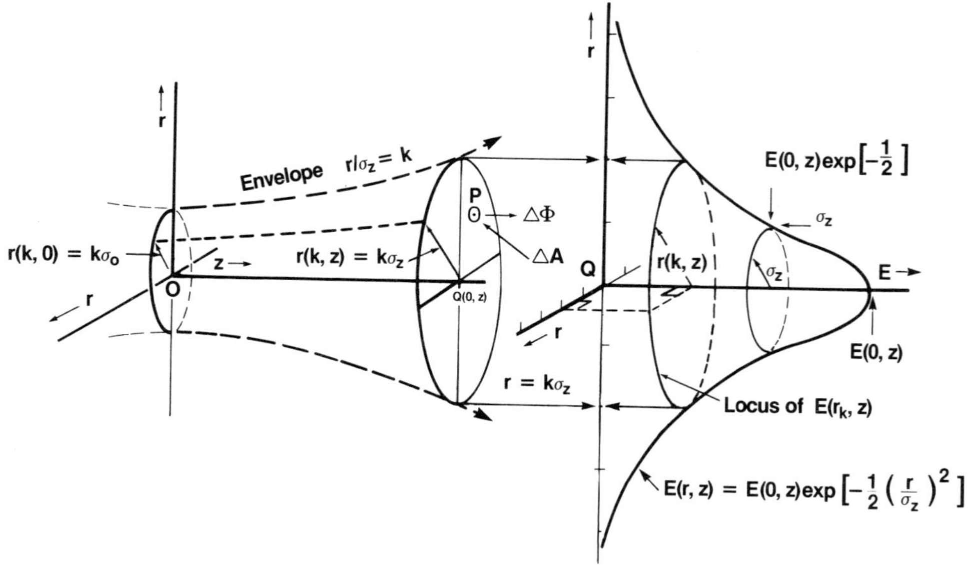


**Figure 2** Beam envelope for $r_k = k\sigma_z$ at which radius the irradiance $E(r_k, z)$ of the beam is a constant fraction $\varepsilon(r_k)$ of the peak irradiance $E(0, z)$ at the axis. More importantly, in a nonabsorbing medium the envelope contains a constant beam power $\Phi_k$ which is a constant fraction $f_r'$ of the total radiant beam power. For a gaussian beam the fraction $f_k$ of excluded beam power which is outside the beam envelope also corresponds to $\varepsilon(r_k)$.

intercept and explore the radiant flux $\Phi$ in the beam, for example, at the point P(r,z). The quotient

$$\frac{\Delta\Phi}{\Delta A} \cong \frac{d\Phi}{dA}$$

is, by definition, the irradiance E(r,z) incident on the detector surface and, therefore, the irradiance of the beam at the point P(r,z). The three-dimensional radially symmetric gaussian distribution of E(r,z) in the perpendicular plane through Q(0,z) is depicted to the right of Figure 2 and resembles a Mexican sombrero. Mathematically it is expressed by

$$E(r,z) = E(0,z) = \exp\left[-\frac{1}{2}\left(\frac{r}{\sigma_z}\right)^2\right] \tag{3}$$

where E(0, z) represents the peak irradiance on the beam axis. From Eq. (3) the normalized irradiance, shown in Fig. 3, is

$$\varepsilon(r,z) = \frac{E(r,z)}{E(0,z)} = \exp\left[-\frac{1}{2}\left(\frac{r}{\sigma_z}\right)^2\right] \tag{4}$$

Frequently the term *intensity* is heard in loose conversation and the term is also seen, with its symbol I, in the literature, including my own [2], instead of the term irradiance and its symbol E. Happily, I have seen the term *irradiance* used appropriately with regard to a laser beam but regrettably, yet understandably, with the symbol I for irradiance instead of the appropriate symbol E.

### 3.2. Beam Diameter Criteria

For the diameter of a laser beam we may choose any value of k. The important criteria in selecting this value of k for the beam envelope diameter $2k\sigma(z)$ are:

1. The resulting fraction $f_k'$ of encircled radiant flux $\Phi_k$ with respect to the total radiant flux $\Phi_\infty$, $k = \infty$; namely

$$f_k' = \frac{\Phi_k}{\Phi_\infty} = \frac{\text{Encircled beam power}}{\text{Total beam power}} \tag{5}$$

See Figure 4.

2. The complementary fraction $f_k$ of the excluded radiant flux $(\Phi_\infty - \Phi_k)$ with respect to the total radiant flux $\Phi_\infty$, $k = \infty$; namely

$$f_k = \frac{\Phi_\infty - \Phi_k}{\Phi_\infty} = \left[1 - \frac{\Phi_k}{\Phi_\infty}\right] = \frac{\text{Excluded beam power}}{\text{Total beam power}} \tag{6}$$

I will go on to show that the normalized irradiance $\varepsilon(r_k, z)$ given in Eq. (4) corresponds also to $f_k$ such that

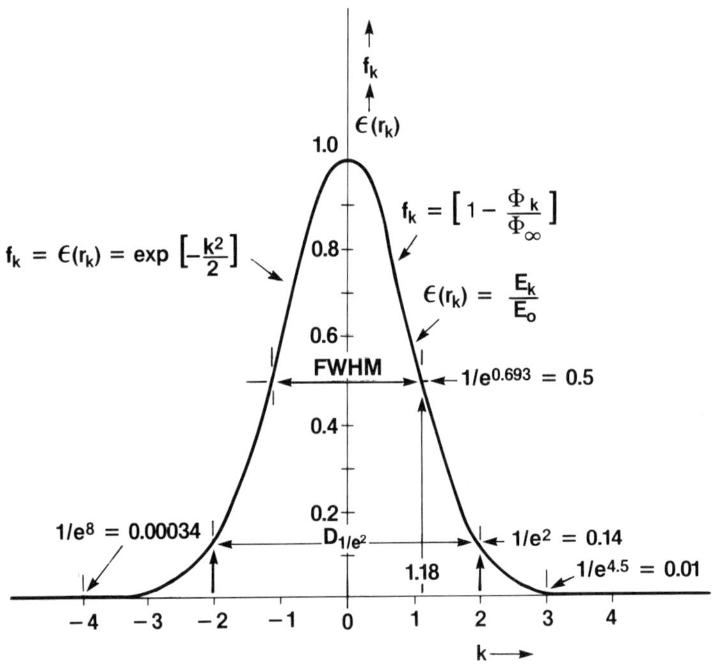


Figure 3 Gaussian distribution of normalized irradiance $\varepsilon(r_k)$ and excluded fraction $f_k$ of beam power against relative beam radius k.

$$f_k = \left[ 1 - \frac{\Phi_k}{\Phi_\infty} \right] = \varepsilon(r_k, z) = \frac{E(r_k, z)}{E(0, z)} \tag{7}$$

As the beam diverges from or converges toward the beam waist the beam diameter increases or decreases respectively but the criteria for defining the beam diameter remain the same.

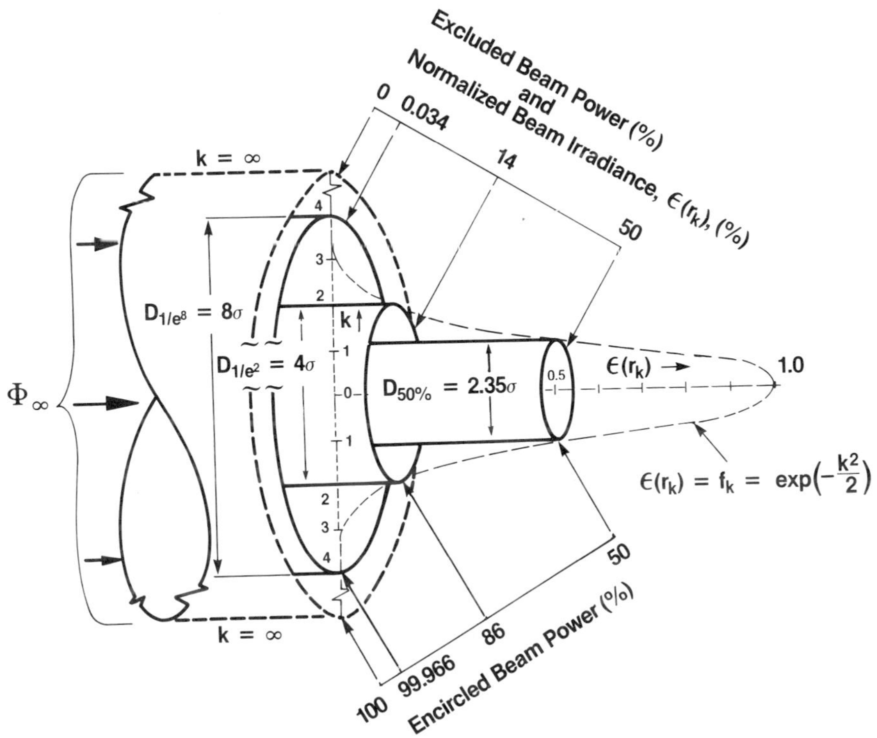


Figure 4 Laser beam envelopes encircling the three most widely used fractions of the total beam power $\phi_\infty$; the beam envelope diameters and the fraction $f_k$ of excluded beam power. Also depicted is a gaussian profile of the normalized irradiance $\varepsilon(r_k)$ with respect to radius.

### 3.3. Encircled Beam Power

Consider the radiant flux encircled $d[\Phi(r)]$ between two envelopes of radii r and (r + dr) at any plane section such as that through the axial point Q(0,z)

$$d[\Phi(r)] = 2\pi r dr . E(r,z)$$

$$\text{since } 2rdr = d(r^2)$$

$$\text{then } d[\Phi(r)] = \pi E(r,z) d(r^2) \tag{8}$$

Substituting for E(r,z) from Eq. (3) gives

$$d[\Phi(r)] = \pi E(0,z) \exp\left[-\frac{1}{2}\left(\frac{r}{\sigma_z}\right)^2\right] d(r^2) \tag{9}$$

Integrating r = o to r = $k\sigma_z$ gives

$$\Phi(k\sigma_z) = \pi E(0,z) \int_0^{(k\sigma_z)} \exp\left[-\frac{1}{2}\left(\frac{r}{\sigma_z}\right)^2\right] d(r^2) \tag{10}$$

$$\Phi_k = 2\pi(\sigma_z)^2 E(0,z)\left[1 - \exp\left(-\frac{k^2}{2}\right)\right] \tag{11}$$

where $\Phi_k$ in Eq. (11) represents the encircled radiant flux for a beam envelope of radius $k\sigma_z$.

The total radiant flux, that is, the total beam power, is obtained by substituting k = ∞ into Eq. (11). This leads to

$$\Phi_\infty = 2\pi(\sigma_z)^2 E(0,z) \tag{12}$$

Dividing Eqs. (11) and (12) gives the fraction of encircled beam power.

$$f_k' = \frac{\Phi_k}{\Phi_\infty} = \left[1 - \exp\left(-\frac{k^2}{2}\right)\right] \tag{13}$$

For a fixed value of k Eq. (13) is constant, independant of z and satisfies all sections of the beam envelope.

### 3.4. Excluded Beam Power

It follows from Eqs. (5) and (6) that the fraction of excluded radiant flux $(\Phi_\infty - \Phi_k)$ from the beam envelope is given by

$$f_k = (1 - f_k') = \exp\left(-\frac{k^2}{2}\right) \tag{14}$$

Equation (4) is:

$$\varepsilon(r,z) = \frac{E(r,z)}{E(0,z)} = \exp\left[-\frac{1}{2}\left(\frac{r}{\sigma_z}\right)^2\right]$$

Substituting $r = r_k = k\sigma_z$ into the above expression gives:

$$\varepsilon(r_k, z) = \frac{E(r_k,z)}{E(0,z)} = \exp\left(-\frac{k^2}{2}\right) \tag{15}$$

Inspection and comparison of Eqs. (14) and (15) leads to

$$f_k = \varepsilon(r_k) = \exp\left(-\frac{k^2}{2}\right) \tag{16}$$

The function in the expression of Eq. (16) is a gaussian distribution independent of z. Hence the fraction of the excluded beam power and the normalized irradiance fall-off in an identical manner as we choose increasing values for the relative beam radius k. This convenient corresponding relationship is entirely due to the unique mathematical properties of the gaussian function.

Table 1 gives $f_k'$ and $f_k$ for selected values of k. Inspection of Table 1 shows that for all practical purposes the effective envelope of the total beam power may be limited to a relative radius of k - 4, at which the beam envelope encircles 99.966% of the total beam power.

Table 1 Radiant Flux Versus Beam Radius

| Relative beam radius $k$ | Fraction of encircled beam power $f'_k = \frac{\Phi_k}{\Phi_\infty}$ [%] | Fraction of excluded beam power $f_k = \left(1 - \frac{\Phi_k}{\Phi_\infty}\right)$ [%] | Normalized irradiance $\varepsilon(r_k) = \frac{E(r_k)}{E(0)}$ [%] | $f_k = \varepsilon(r_k) = \exp -\left(\frac{k^2}{2}\right)$ |
|---|---|---|---|---|
| 0 | 0 | 100 | | 1 |
| 1 | 39 | 61 | | $1/e^{0.5}$ |
| 1.1744 | 50 | 50 | | $1/e^{0.963}$ |
| $\sqrt{2}$ | 63 | 37 | | $1/e$ |
| 2 | 86 | 14 | | $1/e^2$ |
| 3 | 98.9 | 1.1 | | $1/e^{4.5}$ |
| 4 | 99.966 | 0.034 | | $1/e^8$ |
| $\infty$ | 100 | 0 | | 0 |

## 4. BEAM DIAMETER DEFINITIONS

From Table 1 we can selectively choose several radii to define the diameter of a gaussian laser beam. The three most widely used definitions are:

### 4.1. Half-Power Beam Diameter, $D_{50\%}$

The half-power beam diameter is that diameter of the beam core which contains 50% of the total radiant beam power, $k = 1.1744$. At the same time the irradiance at a section normal to the beam axis and at the beam envelope is 50% of the axial irradiance. This beam diameter is sometimes called the "full width at half maximum," which in turn, is abbreviated to FWHM.

### 4.2. $1/e^2$ Excluded Power Beam Diameter, $D_{1/e^2}$

The $1/e^2$ excluded power beam diameter is that diameter of the laser beam core which contains all but $1/e^2$ (14%) of the total radiant beam power; $k = 2$.

Alternatively this diameter corresponds to the

### 4.3. $1/e^2$ Irradiance Level Beam Diameter, $D_{1/e^2}$

The $1/e^2$ irradiance level beam diameter is the diameter of the beam envelope at which the irradiance at a section normal to the beam axis and at the beam envelope is $1/e^2$ (14%) of the axial irradiance.

Unless otherwise stated $D_{1/e^2}$ is the diameter of a laser beam most generally understood when referring to the diameter of a laser beam. However, it is frequently contracted and loosely referred to as the "$1/e^2$ diameter" which, in a literal sense, is ambiguous.

### 4.4 Full-Power Beam Diameter, $D_{1/e^8}$

The full-power beam diameter is that diameter of the laser beam core which contains all but $1/e^8$ (less than 0.04%) of the total beam power; k = 4.

It is also the diameter of the beam envelope at which the irradiance at a section normal to the beam axis and at the beam envelope is $1/e^8$ (<0.04%) of the axial irradiance. This diameter is twice the $1/e^2$ excluded power beam diameter, or twice the $1/e^2$ irradiance level beam diameter; $D_{1/e^8} = 2 \times D_{1/e^2}$.

## REFERENCES

Jenkins, F. A., and White, H. E. (1951). *Fundamentals of Optics*, 2nd ed., McGraw-Hill Publishing Co. Ltd., New York, (a) p. 189; (b) Section 7.12, pp. 104, 105.

Marshall, G. F. (1980). Scanning Devices and Systems, in Kingslake, R. and Thompson, B. J., eds., *Applied Optics and Optical Engineering*, Vol. 6, Academic Press, New York, p. 218.

Nicodemus, F. E. ed. (1978). *Self-Study Manual on Optical Radiation Measurements: Part 1 - Concepts, Chapters 4 and 5*, National Bureau of Standards, Washington, D.C., p. 3.

Nicodemus, F. E., Richmond, J. C., Hsia, J. J. et. al. (1977). *Geometrical Considerations and Nomenclature for Reflectance*, National Bureau of Standards, Washington, D.C., p. 35.

Young, M. (1977). *Optics and Lasers*, Springer-Verlag, Berlin-Heidelberg-New York, (a) pp. 156–158; (b) pp. 45–46.

# 7

# Optics for Data Storage

## Optical Disk Technology

Paul Kuttner
*Optische Werke G. Rodenstock, Munich, Federal Republic of Germany*

## 1. PRINCIPLE OF OPTICAL DATA STORAGE

Digital storage and retrieval of information on optical disks offer widespread applications in several versions, such as laser vision, compact disk, and data storage. The advantages of this technique are its high storage capacity, rapid access, and favorable manufacturing costs. The technique generally consists of burning small holes called pits in a thin film deposited on a rotating optical disk. Tellurium deposited on a carrier is an especially suitable material. Holes less than 1 μm in diameter can be burned into a layer 0.06 μm thick. This results in a storage density of about $10^8$ bits/cm$^2$. Storage capacities of $5 \times 10^{10}$ bits or even more are quite feasible.

A laser is controlled by a light modulator and focused onto the film by a high-numerical-aperture objective lens. The holes are formed in the film with an illuminance of over $10^6$ W/cm$^2$, which heats up the material, partly melts the metal, and opens the hole. The required illuminance is attained with lasers of 10-mW output. The irradiation time used for burning holes is typically 0.5 μsec. Thus the information is stored as a progressing sequence of pits arranged along a spiral in the optical disk.

To read the stored information, the disk is rotated and an individual pit is illuminated by a laser beam focused onto the disk by a high-numerical-aperture objective lens. Means are provided for automatically focusing the objective lens relative to the disk and for radial tracking of the spiral-shaped information trace. The light source may be either a gas laser or a semiconductor laser, and therefore they have to be either expanded or collimated, respectively. Using semiconductor lasers, anamorphic imaging components are added to compensate for the elliptical cross section of the beam and cylindrical lenses to compensate, and thereby eliminate, the astigmatism.

The beam reflected at the optical disk contains the read information. After a double pass through a quarter-wave plate, the plane of polarization of the returning beam is at right angles to that of the illuminating beam. Therefore, it may be separated at a polarizing beam splitter.

After reflection the beam is focused onto a detector unit. The detector unit converts the optical signal into an electrical signal and also provides the signal to control the servo loop for automatic focusing and tracking. A schematic representation of the optical layout of an optical data storage device is shown in Figure 1. The light source depicted is a semiconductor laser.

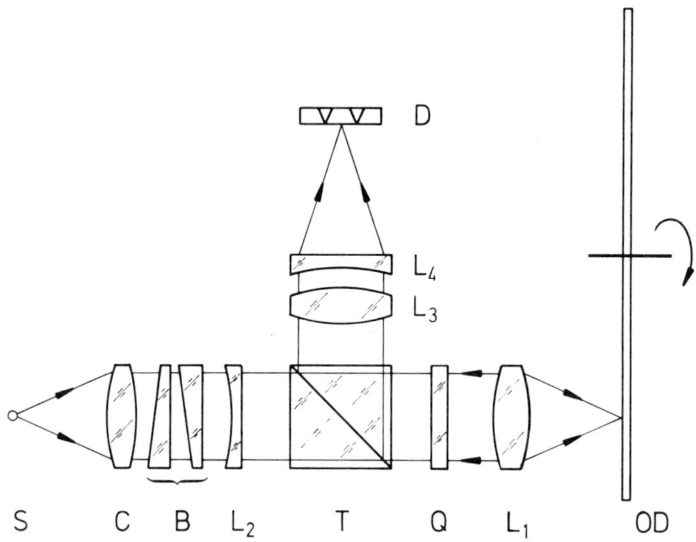


**Figure 1** Schematic optical layout of an optical data storage system. S, laser; C, collimator; B, anamorphic beam expander; $L_2$, cylindrical lens; T, beam splitter; Q, quarter-wave plate; $L_1$, objective lens; OD, optical disk; $L_3$, collimator; $L_4$, cylindrical lens for the automatic astigmatic focusing; D, detector.

Therefore, a collimator C is used to obtain a parallel light beam. Added are an anamorphic beam expander B and a cylindrical lens $L_2$. The cubic beam splitter T is based on a polarizing layer stack. Following the collimator $L_3$ that operates in reverse, is a cylindrical lens $L_4$, which is used for automatic astigmatic focusing.

## 2. THEORY OF OPTICAL IMAGE FORMATION

A point source radiates spherical waves. Surfaces orthogonal to the emitted light beams are called wavefronts. A wavefront is characterized as a surface of equal phase in the emitted wave. A light source at an infinite distance results in a plane wavefront. Passing through an objective lens, a divergent light bundle of rays is transformed into a convergent bundle of rays that for an ideal lens will intersect at a single point. For a real lens free of geometrical optical aberrations, the intensity distribution in the point image follows diffraction theory. For optical data storage the size of the point image is important. For diffraction-limited lenses with uniform illuminance over the pupil, the intensity distribution I(x) through the image depends on both the numerical aperture (NA) and the wavelength $\lambda$, and is given by Airy's formula (1835).

The normalized expression can be written as

$$\frac{I(x)}{I(0)} = \left(\frac{2J_1(x)}{x}\right)^2 \tag{1}$$

where

$$x = \frac{2\pi}{\lambda}(NA)r \tag{2}$$

$J_1$ is the Bessel function of the first kind and first order, and r is the radial coordinate in the image plane. This diffraction pattern consists of a central core with concentric diffraction rings. The diameter $d_A$ of the central core is determined by the first zero of the function given in Eq. (1). Thus

$$d_A = 1.22\,\frac{\lambda}{NA} \tag{3}$$

**Table 1** Diameter of the Airy Disk as a Function of NA and the Wavelength of Light

| Numerical aperture, NA | Diameter of the Airy disk, $d_A$ (μm) | |
|---|---|---|
| | For λ = 632.8 nm | For λ = 820 nm |
| 0.4 | 1.92 | 2.50 |
| 0.5 | 1.54 | 2.00 |
| 0.6 | 1.28 | 1.66 |

Approximately 84% of the total incident energy is concentrated into the central Airy disk of the Airy pattern. For 632.8 and 820 nm, the wavelengths frequently used in optical data storage technology, the diameter of the Airy disk is given in Table 1 for objective lenses with different NA values.

## 2.1. Apodization and Image Quality

Laser beams have a gaussian intensity distribution in sections perpendicular to the direction of light propagation. This attribute of lasers is described by Kogelnik (1965) and especially for semiconductor lasers by Finck et al. (1980–81). Using a laser as a light source, allowance has to be made for the gaussian intensity distribution. This gaussian characteristic remains invariant even after expanding the beam diameter and is given by

$$\varepsilon(\rho) = \varepsilon_0 e^{-h^2\rho^2} \tag{4}$$

where $\varepsilon(\rho)$ is the intensity of the beam as a function of the radial coordinate, $\varepsilon_0$ is the intensity at the center of the beam, $\rho$ the radial coordinate, and h a parameter. Expansion of the beam changes the value of the parameter h. The more the beam is expanded, the smaller will be the value of h.

$2\rho_1$ represents the diameter of the entrance pupil, while the energy $\phi$ collected by the lens is given by

$$\phi(\rho_1) = \int_{\psi=0}^{2\pi} \int_{\rho=0}^{\rho_1} e^{-h^2\rho^2} \rho \, d\rho \, d\psi = 0 \tag{5}$$

where $\varphi$ represents the azimuth angle.
The fraction of the gaussian beam energy collected by a circular pupil is given by

$$\frac{\phi(\rho_1)}{\phi_{total}} = 1 - \exp\left[-h^2\rho_1^2\right] \tag{6}$$

Assuming the definition of the diameter for a laser beam to be at the $1/e^2$ intensity level, this means that at the diameter where the normalized intensity is 0.14, about 86% of the total energy is encircled within the diameter. Figure 2 shows gaussian laser beams with different degrees of truncation.

The diameter at which the normalized intensity reaches the value of 0.5 will, by Eq. (6), encircle 50% of the total laser energy. ergy. As shown, for example, by Born and Wolf (1959, p. 391) and Goodman (1968), the diffraction theory of optical image formation is based on the principle that the amplitude distribution in the image point is given by the Fourier transform of the pupil function. The square of this amplitude function is the intensity distribution across the image. The pupil function G of a rotationally symmetrical lens is given by

$$G(\rho) = e^{(2\pi i/\lambda)W(\rho)} |G| \tag{7}$$

$W(\rho)$ is the wave aberration, that is, the deviation of the real wavefront from the gaussian reference sphere. The wave aberration reduces to zero for an ideal lens. $\rho$ is the radial coordinate in the pupil of the lens and $|G(\rho)|$ is the amplitude distribution of the incident wave. Generally, the wave aberration and the amplitude distribution are given for the exit pupil of the lens. For an ideal lens, the wave aberration equals zero and when there is a constant-amplitude distribution across the pupil, the Fourier transformation of the pupil function provides the amplitude distribution in the image point. Furthermore, by squaring this distribution, we obtain the Airy formula as given in Eq. (1).

A nonuniform amplitude distribution of the image-forming wavefront is called apodization and is described, for example, by

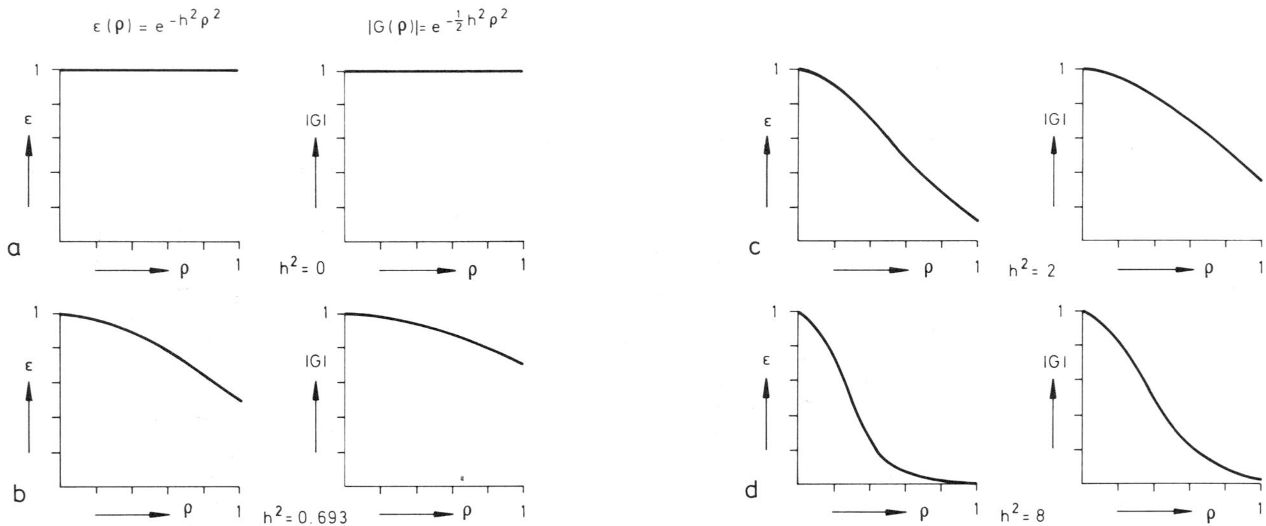


| Type of truncation | Laser beam diameter defined at: |
|---|---|
| a | Uniform illumination |
| b | 0.5 value of gaussian intensity distribution |
| c | $1/e^2$ intensity level of gaussian intensity distribution, called $1/e^2$ diameter |
| d | Twice the $1/e^2$ diameter |

**Figure 2** Normalized amplitude and intensity distributions across a truncated gaussian laser beam. $\rho$ = radial coordinate, h = truncation parameter, $\varepsilon(\rho)$ = intensity distribution, $|G(\rho)|$ = amplitude distribution.

Marechal et Francon (1960). A laser beam with a gaussian intensity distribution according to Eq. (4) can therefore be interpreted as an apodization with an amplitude distribution across the pupil described by

$$|G| \sim e^{-(1/2)h^2\rho^2} \tag{8}$$

For an ideal lens with apodization, Kirchhoff's diffraction integral for the amplitude distribution in the image point is

$$U(p,q) = \int_A\int |G(\xi,\eta)| e^{-(2\pi i/\lambda)(p\xi + q\eta)} d\xi \, d\eta \tag{9}$$

The integration has to be extended over the whole area of the pupil. $\xi$ and $\eta$ are cartesian coordinates of any point in the pupil. For a circular aperture in polar coordinates these may be written as

$$\xi = \rho \cos \theta \quad \text{and} \quad \eta = \rho \sin \theta \tag{10}$$

Let $(\omega, \psi)$ be coordinates of a point in the diffraction pattern and let s be the normal to the wavefront with the components p and q in the planes $(\xi, \zeta)$ and $(\eta, \zeta)$, where $\zeta$ is the direction normal to $\xi$ and $\eta$. Let $\omega$ be the sine of the angle for the direction (p,q) with the $\zeta$ axis; it follows that

$$p = \omega \cos \psi \quad \text{and} \quad q = \omega \sin \psi \tag{11}$$

and $\omega = \sqrt{p^2 + q^2}$. Applying polar coordinates to Eq. (9) and integrating over the angular coordinate, one obtains the amplitude distribution across the image given by

$$U(x) = 2\pi \int_0^a e^{-(1/2)h^2\rho^2} J_0 \frac{2\pi}{\lambda} \rho\omega \; \rho \, d\rho \tag{12}$$

where $x = (2\pi/\lambda)(NA)r$, $J_0$ is the Bessel function of order zero, and a is the radius of the circular aperture. To derive this relationship, the integral representation of Bessel functions was applied [see, e.g., Jahnke and Emde (1945)].

When h = 0 there is uniform illumination across the pupil and the evaluation of Eq. (12) results in the amplitude distribution across the image, the square of which is Airy's formula. Then the recurrence formula for Bessel functions

$$\frac{d}{dx} xJ_1(x) = xJ_0(x) \tag{13}$$

is applied [see, e.g., Jahnke and Emde (1945)]. Given the aperture's diameter, the value of the parameter h determines the amount of the apodization in the beam diameter used by the lens.

Figure 2 shows the normalized gaussian intensity distributions of a laser beam for different degrees of truncation as well as the associated amplitude distributions. The radius of the aperture is also normalized to the value 1. These distributions are shown for those laser beam diameters as they are defined in Table 2. The fraction of the total incident energy contained within these differently defined laser beam diameters is also given in Table 2.

**Table 2** Definition of Laser Beam Diameter

| Apodization case | Laser beam diameter defined at: | Contained fraction of total energy (%) |
|---|---|---|
| a | Uniform illumination | 0 |
| b | 0.5 level of the gaussian intensity distribution | 50 |
| c | $1/e^2$ intensity level of the gaussian intensity distribution, called the $1/e^2$ diameter | 86 |
| d | Twice the $1/e^2$ diameter | 99.96 |

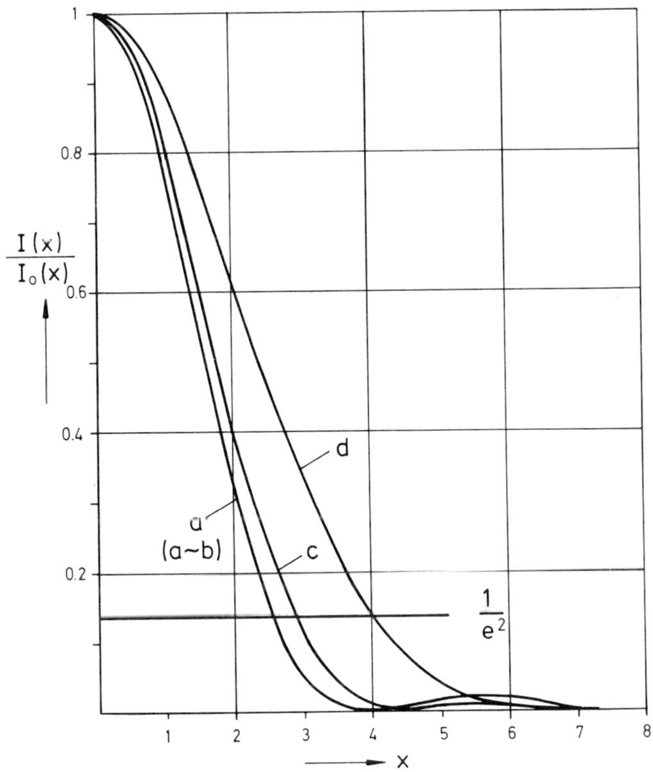


Figure 3 Intensity distribution across a point image spot for an incident gaussian laser beam with different degrees of truncation. For apodization cases a–d, the laser beam diameters are defined as in Figure 2. $I(x)/I(0)$ = normalized intensity distribution, $x = (2\pi/\lambda)(NA)r$, r = radial coordinate in the image plane.

If the beam diameter is expanded to such a degree that it results in an almost uniform illumination of the aperture, we will utilize only a small fraction of the total incident energy. Frequently, the beam diameter is defined at the $1/e^2$ intensity level of the guassian intensity distribution. Then 86% of the total energy is encircled. With twice the $1/e^2$ intensity level diameter, almost the total energy is encircled. Figure 3 shows the intensity distributions across the image point for truncated gaussian laser beams, as illustrated in Figure 2.

For a given apodization the amplitude distribution across the image can be obtained from Eq. (12). After normalization the intensity distribution across the image point is obtained by squaring. The intensity distributions are given as functions of

$$x = \frac{2\pi}{\lambda}(NA)r$$

Hence for given $\lambda$ and NA, the size of the point image can be determined. The smallest possible point image is achieved by uniform illumination of the aperture, but at the expense of energy lost by beam expansion and the consequent truncation at the aperture. Increasing the apodization, that is, decreasing the degree of truncation of the gaussian beam distribution, causes the central core in the point image to increase in size and causes a shift of the diffraction rings toward larger diameters and a deterioration of the energy contained within these rings [case (c)]. Case (b), small apodization: shows only an insignificantly wider intensity distribution in the point image. With negligible truncation the image intensity distribution becomes gaussian [case (d)]. The radii x of the first two diffraction rings and the intensity $I(x)/I(0)$ are given in Table 3.

Generally, if laser light is used, the point image diameter is defined at the $1/e^2$ intensity level of the intensity distribution in the spot. As Figure 2, case (d), shows, the $1/e^2$ intensity-level diameter of the incident gaussian beam equals half the diameter of the aperture of the lens. The intensity level across the point image caused by this apodization reduces to the value $1/e^2$ at x = 4. This is also illustrated in Figure 3. The image spot diameter $d_{1/e^2}$ at the $1/e^2$ intensity level in the point image can be derived from Eq. (2) when d = 2r and is given by

$$d_{1/e^2} = \frac{4}{\pi}\frac{\lambda}{(NA)} \tag{14}$$

where NA stands for the numerical aperture of the lens used. This relationship can also be written as

$$d_{1/e^2} = \frac{4}{\pi}\frac{\lambda f}{D_b} \tag{15}$$

Table 3 Diffraction Rings in the Point Image for a Truncated Gaussian Laser Beam

| Apodization case | Diffraction ring x coordinate | | | I(x)/I(0) of the first diffraction ring maximum |
|---|---|---|---|---|
| | 1st, min. | 1st, max. | 2nd, min. | |
| a | 3.8 | 5.1 | 7.0 | 0.017 |
| b | 4.0 | 5.3 | 7.1 | 0.012 |
| c | 4.5 | 5.6 | 7.3 | 0.005 |
| d | No diffraction rings | | | |

where $D_b$ is the $1/e^2$-intensity-level diameter of the laser beam [see, e.g., Zock and Lee (1972) and Hopkins and Buzawa (1976)].

### 2.2. Encircled Energy

The encircled energy L states what fraction of the total incident energy is contained within a circle of the diameter $2r_0$. This fraction is given by

$$L(r_0) = \frac{1}{\phi} \int_0^{2\pi} \int_0^{r_0} I(r)r \, dr \, d\psi \tag{16}$$

where I is the intensity distribution in the image point and $\phi$ is the total luminous flux passing the pupil of the lens. For an ideal lens with apodization it is given by

$$\phi = \int_0^{2\pi} \int_0^{a} G^2 (\rho)\rho \, d\rho \, d\theta \tag{17}$$

The encircled energies as a function of $x = (2\pi/\lambda)(NA)r$ are shown in Figure 4 for the point image mentioned in Figure 3. Comparing the encircled energy functions of less pronounced apodization with functions without apodization, rather small effects can be observed in the central region. Beyond a diameter corresponding to the first dark diffraction ring in the Airy pattern, an increase of apodization, which means a less pronounced truncated incident gaussian beam, will reduce the fraction of total energy contained in the outer region of the point image. The characteristics of cases (a), (b), and (c) show this behavior. Illumination of the lens with a gaussian beam causes a reduction of the fraction of energy contained in the central region of the point image. For values of $x > 6$, almost the total energy is encircled, whereas in the case of uniform illumination, the encircled energy is only about 90%. This behavior is shown in curves (a) and (d).

When reading out information stored as small pits in optical disks, it is important to know what fraction of the point image energy is encircled within a diameter $d_1 = 2r_1$. This diameter has to cope with the diameter of a pit or can be chosen somewhat larger as long as there remains enough margin of safety from the next pit. The energy fraction lying beyond this region is then given by $L(\infty) - L(r_1)$. This energy fraction produces background illumination and may be the cause of interfering cross-talking effects. The ratio Q, the energy fraction contained within a circle of the diameter $d_1$ to the total energy in the point image, can be used for quality assessment of optical data storage objective lenses.

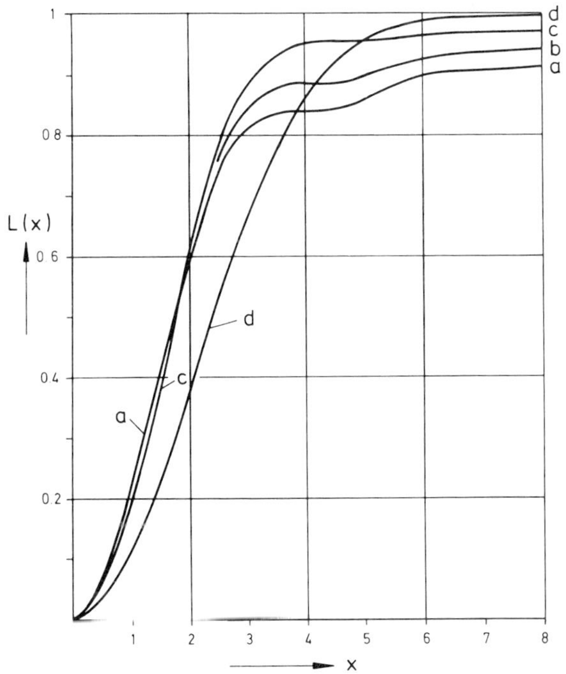


Figure 4 Encircled energies as function of $x = (2\pi/\lambda)(NA)r$ of the image spot truncated gaussian laser beams. For apodization cases a→d, the laser beam diameters are defined as in Figure 2. $L(x)$ = encircled energy function, $r$ = radial coordinate in the image plane.

This criterion is written as

$$Q = \frac{L(r_1)}{L(\infty)} \tag{18}$$

The image quality of real lenses can represent only an approximation to the diffraction-limited goal of an ideal lens. Within the small image field needed, optical data storage lenses come close to the limit given by diffraction theory. In the calculation process of an objective lens there are always residual geometric optical aberrations to which residual errors, inevitable in the production process, have to be added. Both cause a reduction of the energy fraction contained in the central core of the point image and an appropriate increase in the outer regions. In the quality assurance of real objective lenses, uniform illumination of the pupil eliminates the problem of aligning the gaussian laser beam to the objective lens. A quality assurance criterion, which is based on practical experience, is obtained if the central core is defined by the diameter of the first dark ring in the Airy formula. The total image spot energy is approximately the encircled energy within a diameter $d_2$ which is large enough to contain the broadening of the point image caused by the aberrations. Taking the third dark Airy ring for the diameter $d_2$, the quality criterion Q may be written as

$$Q \cong \frac{L(r_1)}{1.06L(r_2)} \tag{19}$$

Thus allowance is made for the 6% of the total spot energy that lies beyond the third dark Airy ring with radius x = 10.2. A statement on the quality of an optical data storage objective lens can be made either by measuring the intensity distribution in a point source image spot or by determining the ratio of the energy contained within the central core to the total energy in the image spot.

### 2.3. Depth of Focus

Due to the increase in spot size either side of the best focus position, each lens has an associated depth of focus. For the requirements of optical data storage technology, defocusing up to $\Delta z$ can be tolerated where the intensity on the axis has deteriorated to 95% of the intensity at the best focal setting, z = 0. For an ideal lens, uniformly illuminated, the intensity along the optical axis is given by

$$I(u) = I(0)\left[\frac{\sin(u/4)}{u/4}\right]^2 \tag{20}$$

where u stands for

$$u = \frac{2\pi}{\lambda}(NA)^2 z \tag{21}$$

[see, e.g., Born and Wolf (1959, p. 440)]. For u = 1.6 the function I(u) is 0.95I(0). Hence it follows from Eq. (21) that for tolerable defocusing

$$\Delta z \approx \pm \frac{\lambda}{4(NA)^2} \tag{22}$$

and consequently, the depth of focus is given by

$$2\ \Delta z \approx \frac{\lambda}{2(NA)^2} \tag{23}$$

Values for this depth of focus 2 $\Delta z$ are given in Table 4 for objective lenses with different NA values. The less the truncation of the incident laser beam, the greater is the energy concentrated in the pupil's central region and thus the smaller is the effective numerical aperture of the lens and laser beam combination, resulting in a larger depth of focus. The intensity along the optical axis for an ideal lens is given in Figure 5 for different degrees of apodization. The curves are shown for NA = 0.5 at a wavelength of 800 nm. The depth of focus for a gaussian beam is 50% larger than that for a beam with a uniform intensity profile. This is evident by comparing curves (a) and (d). With gaussian illumination of an objective lens of 0.5 NA and 800 nm wavelength the intensity on the axis deteriorates from 100% for the best focus setting to 95% for defocusing by the amount of ±1.2 μm. With uniform illumination, however, the 95% point calls for ±0.8 μm. During the design stage and in the general layout of an optical data storage system, allowance has to be made for this difference. A gaussian laser beam not truncated by the pupil yields a larger

**Table 4** Values for the Depth of Focus for an Ideal Lens with Uniform Illumination of the Entrance Pupil as a Function of NA and the Wavelength of the Light

| Numerical aperture, NA | Values for the depth of focus, 2$\Delta$z (μm) | |
|---|---|---|
| | For λ = 632.8 nm | For λ = 800 nm |
| 0.4 | 2.0 | 2.5 |
| 0.5 | 1.2 | 1.6 |
| 0.6 | 0.9 | 1.1 |

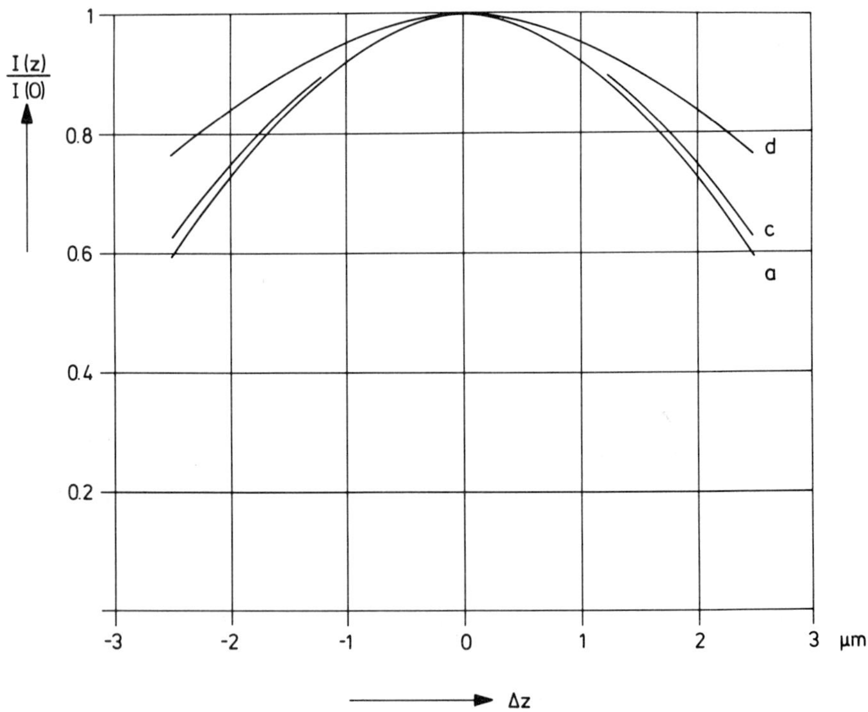


Figure 5 Intensity along the optical axis for an incident gaussian laser beam with different degrees of truncation for an ideal lens, NA = 0.5, λ = 800 nm. For apodization cases a, c, and d, the laser beam diameters are defined as in Figure 2. I(z)/I(0) = normalized intensity along the optical axis, z = optical axis of the lens.

point image such that the attained storage capacity is smaller, but the tolerable range for focusing the optical disk becomes larger. However, uniform illumination of the pupil produces a substantial energy loss due to beam truncation, while the central core of the point image becomes as small as possible, and therefore a high storage capacity is attainable. Nevertheless, because of the small depth of focus, the required precision in focusing increases.

As shown in Table 4, an increase in the numerical aperture reduces the depth of focus. When laying out the optical system, allowance must be made for the increased precision in focusing when higher-numerical-aperture objective lenses are used.

Allowances are required, in particular, for the mechanics and for the servo loop for automatic focusing. Moreover, objective lenses with higher-numerical-apertures call for tighter tolerances of the cover glass used to protect the information stored in the optical disk.

## 2.4. Wave Aberration

In accordance with Eq. (7), the pupil function is described by the wave aberration W and the type of apodization $|G|$. For a lens with geometric optical aberrations, the emergent wavefront is no longer spherical and therefore the wave aberration is not zero. A reference sphere may be associated with such a wavefront.

The mean square difference between the wavefront and the reference sphere is described as the variance, that is, the square of the standard deviation and consequently given as

$$(\Delta W)^2 = \overline{W^2} - (\overline{W})^2 \tag{24}$$

A reference sphere for which there is a balance between the positive and the negative wavefront deviations from the reference sphere is called a gaussian reference sphere. Thus the mean value of the wave aberration reduces to zero, $\overline{W} = 0$. Hence expression (24) simplifies to

$$(\Delta W)^2 = \overline{W^2} \tag{25}$$

The center of curvature of the gaussian reference sphere is the gaussian image point P. This is shown in Figure 6. The wavefront can be calculated numerically by determining equal optical path lengths

$$\sum_i n_i s_i = \text{const.} \tag{26}$$

for several rays emerging from an object point and evenly distributed over the pupil. The connecting surface through the endpoints represents the wavefront. In Eq. (26), $n_i$ is the refractive index and $s_i$ the geometrical path length in the medium i. The deviation of this surface from the reference sphere is the wave

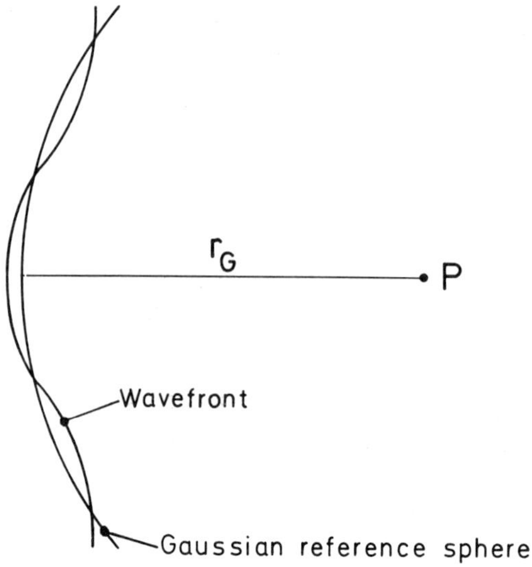


**Figure 6** Wavefront and a gaussian reference sphere. $r_G$ = radius of the gaussian reference sphere.

aberration. The wave aberration can be measured interferometrically. Therefore, it is usual and expedient to present the wave aberration in units of the wavelength used. Hence the amount of the wave aberration is given for each point of the reference sphere as

$$W = \frac{\lambda}{N} \tag{27}$$

where N stands for a positive number. As the wavefront is the orthogonal surface to the emergent light rays, when a wave aberration exists the light rays will therefore no longer converge to a single point in the gaussian image plane.

## 2.5. Strehl's Definition

For a wave with a wave aberration the point image spot is enlarged compared with that of a diffraction-limited ideal lens. The intensity at the center of the image point is reduced because of the broadening of the energy distribution. Thus, as Strehl has shown (1902), the ratio of the peak intensity in the point image to the peak intensity of an appropriate ideal lens is a measure of the image quality. This ratio i(P) at the center P of the gaussian reference sphere is called Strehl's definition. For small aberrations Strehl's definition may be written as

$$i(P) \approx 1 - \left(\frac{2\pi}{\lambda}\right)^2 (\Delta W)^2 \tag{28}$$

with the mean square wave aberration $(\Delta W)^2$ given by Eq. (24) [see, e.g., Born and Wolf (1959, p. 463)]. This formula implies that for small aberrations, the reduction of the normalized peak intensity at the center of the point image is independent of the nature of the aberration. In Table 5 are values of Strehl's definition and the corresponding root mean square (rms) of the wave aberration with respect to the gaussian reference sphere and expressed by

$$\Delta W = \sqrt{\overline{W^2}}$$

The results of Eq. (28) are uncertain if $\Delta W \gtrsim \lambda/10$.

Table 5 Values of Strehl's Definition of Image Quality and the Corresponding RMS Value of the Wave Aberration

| Strehl's definition, i | RMS value of the wave aberration, $\sqrt{\overline{W^2}}$ |
|---|---|
| 0.61 | $\lambda/10$ |
| 0.83 | $\lambda/15$ |
| 0.90 | $\lambda/20$ |
| 0.94 | $\lambda/25$ |
| 0.96 | $\lambda/30$ |
| 0.985 | $\lambda/50$ |

### 2.6 Modulation Transfer Function

It is state of the art to assess the image quality of optical systems with the aid of the modulation transfer function (MTF). Because of geometric optical aberrations and diffraction, if a luminous object point is imaged into the image plane, an image spot of finite size with an intensity distribution called a point spread function $B(x,y)$ is built up.

From an extended intensity distribution $I_0(x,y)$ in the object plane a linear imaging process forms an intensity distribution $I_1(x,y)$ in the image plane. This operation may be described by the convolution of the intensity distribution in the object plane with the point spread function of the lens. Thus

$$I_1(x,y) - \int_{-\infty}^{+\infty} \int_{-\infty}^{+\infty} I_0(x',y')B(x - x',\ y - y')\, dx'\, dy' \qquad (29)$$

where x,y and x',y' are cartesian coordinates in the image and object fields, respectively. This convolution means that the intensities from different elements of the object plane are additive. Since the imaging process is linear, the different elements in the object plane are assumed to emit incoherent radiation; otherwise, unwanted interference effects could occur in the image.

The extent of these interference effects depends on the degree of coherence. In describing an optical imaging process by Eq. (29), the point spread function may not vary around the image point under consideration. Such a region is called isoplanatic. Without loss on the range of validity the two-dimensional representation may be replaced by a one-dimensional representation that can be accomplished metrologically, provided that the image analysis is carried out separately for the x and y directions. The frequency analysis of any intensity distribution function determines the component sinusoidal spatial frequencies of the function. A frequency analysis may be described mathematically by the Fourier transformation of the intensity distribution. The characteristic of the Fourier transform indicates the amplitudes associated with the respective spatial frequencies. Under the given assumptions the one-dimensional optical transfer function (OTF) may be written as

$$\mathrm{OTF} = \frac{\int_{-\infty}^{+\infty} I_1(x)e^{-2\pi iRx}\, dx}{\int_{-\infty}^{+\infty} I_0(x)e^{-2\pi iRx}\, dx} \qquad (30)$$

where R is the spatial frequency, measured in cycles/mm (c/mm). The OTF is composed of a modulus and a phase.

In optical data storage technology only lenses with almost rotationally symmetrical point images should be used. For those point images that are almost symmetrical, only the modulus is significant. With only the modulus component the OTF is called the modulation transfer function (MTF).

In accordance with Eq. (30), the MTF is the ratio of the amplitude of a sinusoidal function in the image plane to the amplitude of the corresponding sinusoidal function in the object plane for each spatial frequency. For a sinusoidal intensity $I_0$ in the object plane and $I_1$ in the image plane,

$$I_0(x) = \frac{1}{2}\left[1 + A_0 \sin(2\pi Rx)\right] \tag{31}$$

and

$$I_1(x) = \frac{1}{2}\left[1 + A_1 \sin(2\pi Rx)\right] \tag{32}$$

where $A_0$ is the amplitude in the object plane and $A_1$ in the image plane and where $A_1 < A_0$. If $A_0$ is normalized to 1, the MTF value for this spatial frequency is given by

$$T(R) = A_1 \tag{33}$$

This behavior is shown in Figure 7. For a sinusoidal intensity distribution the MTF value is determined by the ratio

$$T(R) = \frac{I_{max} - I_{min}}{I_{max} + I_{min}} \tag{34}$$

where $I_{max}$ and $I_{min}$ are the maximum and minimum of the intensity distribution across the image. Increasing the spatial frequency, the modulation becomes smaller and the imaged structures show less and less contrast. For a diffraction-limited lens the limit of the spatial frequencies transferred is attained for

$$R_l = \frac{2NA}{\lambda} \tag{35}$$

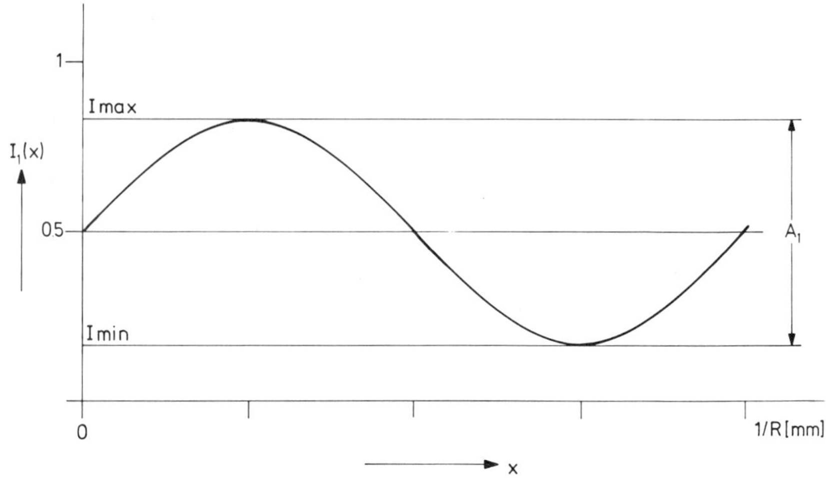


Figure 7 Sinusoidal image intensity distribution across the image, R = spatial frequency, x = coordinate in the image plane, $A_1$ = amplitude of the intensity distribution.

For this limiting frequency $R_l$ the MTF goes to zero. Equation (35) indicates that the spatial frequencies transferred are higher and thus the transferable structures are finer the higher the numerical aperture and the shorter the wavelength used. Table 6 gives the limiting spatial frequencies for diffraction-limited objective lenses.

For a diffraction-limited lens of 0.5 NA and 800 nm wavelength, Figure 8 gives the MTF curves for various pronounced

Table 6 Limiting Spatial Frequencies for Diffraction Limited Objective Lenses as a Function of NA and the Wavelength of Light

| Numerical aperture, NA | Limiting spatial frequency, $R_l$ (c/mm) | |
|---|---|---|
| | For $\lambda$ = 632.8 nm | For $\lambda$ = 800 nm |
| 0.4 | 1264 | 1000 |
| 0.5 | 1580 | 1250 |
| 0.6 | 1896 | 1500 |

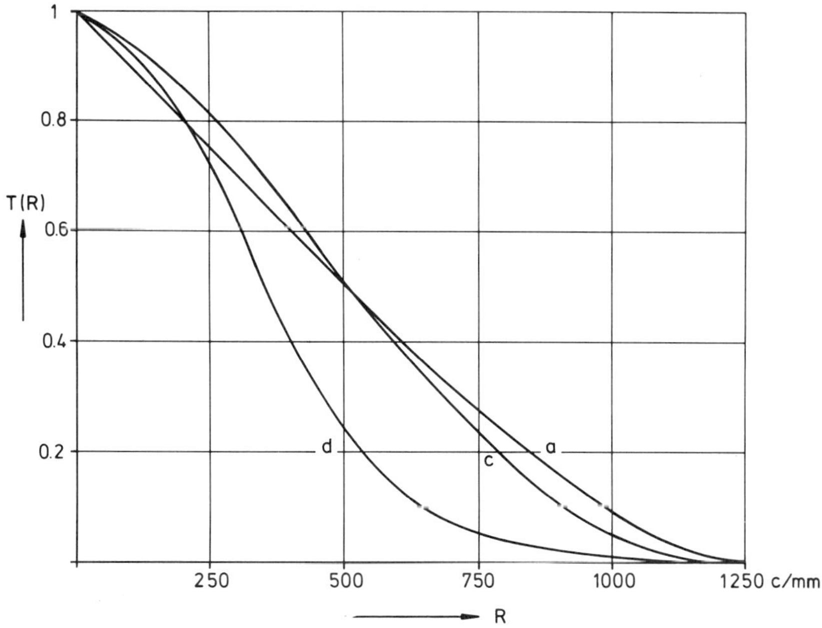


**Figure 8** MTF of a diffraction limited lens with NA = 0.5, $\lambda$ = 800 nm for different degrees of apodization. For cases a, c, and d, the laser beam diameters are defined as in Figure 2.

truncations of the incident gaussian beam, that is, for different degrees of apodization. Apart from the lower frequency range, the best MTF is attained with uniform illumination of the lens. A nontruncated gaussian beam significantly reduces the modulation at higher frequencies. Small aberrations in nearly diffraction limited lenses reduce the MTF values and are most pronounced in the middle-frequency range, whereas there is a smaller reduction at the higher frequencies. This effect depends somewhat on the nature of the aberration. An approximately valid relationship between the reduction of the MTF at half the limiting frequency and the rms of the wave aberration is given in Table 7.

In optical data storage technology the information is stored in the storage medium as a progressing sequence of nearly circular or elliptical pits. A square-wave intensity distribution is composed of the fundamental frequency $R_0$ and the harmonics with the frequencies $3R_0$, $5R_0$, . . . . Thus

Table 7 Reduction of MTF at Half the Limiting Spatial Frequency and the RMS of Wave Aberration for Almost-Diffraction-Limited Lenses

| Reduction in MTF response, $-\Delta T$ (for $R_l/2$)[a] | RMS value of the wave aberration, $\sqrt{\overline{W^2}}$ |
|---|---|
| 0.15 | $\sim\lambda/10$ |
| 0.05 | $\sim\lambda/20$ |
| 0.02 | $\sim\lambda/30$ |
| 0.01 | $\sim\lambda/50$ |

[a] $R_l$ = limiting spatial frequency.

$$I_0(x) = \frac{1}{2} + \frac{2}{\pi}\left(\sin\left(2\pi R_0 x\right) + \frac{1}{3}\sin\left(6\pi R_0 x\right) + \frac{1}{5}\sin\left(10\pi R_0 x\right) + \cdots\right) \quad (36)$$

with $R_0 = 1/\ell$, where $\ell$ is the length of the cycle. In accordance with Fourier analysis the frequencies contained are sharply defined for long sequences, but they are rather broadened for frequencies for short sequences.

Typical center-to-center spacings of 1.5–2 μm as used in data storage technology give rise to fundamental frequencies of 660–500 c/mm, respectively. As shown in Figure 8, the harmonics of these fundamental frequencies are no longer transferred by typical objectives used in optical data storage technology. If, as shown in Table 8, the length of the bright portions of the object intensity distribution are no longer equal to the length of the dark portions, the additional harmonics $2R_0$, $4R_0$, . . . occur. The MTF value for frequencies of 1000 c/mm and higher is very small, so that their contribution to the image formation is minute. Unequal lengths of the bright and the dark positions of the object structure also affect the amplitude of the fundamental frequency. If $l$ is the length of the cycle and $ml$ is the length of the bright portion, $l(1 - m)$ expresses the length of the dark portion.

For the fundamental frequency $R_0$, Table 8 gives several ratios of the modulation $A_0(m)$ to the modulation $A_0(m = \frac{1}{2})$. For

**Table 8** Ratio of Modulation $A_0(m)/A_0$ (m = 1/2) for the Fundamental Frequency $R_0$

| Object structure | Duty cycle, m | Relative modulation, $A_0(m)/A_0(m = 1/2)$ |
|---|---|---|
| 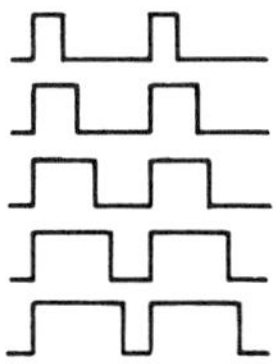 | 1/4 | 0.70 |
| | 3/8 | 0.92 |
| | 1/2 | 1 |
| | 5/8 | 0.92 |
| | 3/4 | 0.70 |

m = Ratio of the bright part of the cycle to the length of the cycle.

$m = \frac{1}{2}$, that is, for equal lengths of the bright and dark portions, maximum modulation is attained.

Even if the object intensity distribution, presented by a progressing sequence of pits is not accurately described by a square-wave, the attainable signal modulation of two consecutive pits may be expressed by

$$A_1 \approx A_0(R_0)T(R_0) \tag{37}$$

If $l$ is the center-to-center spacing of two consecutive pits, the fundamental spatial frequency is given by

$$R_0 \approx \frac{1}{\ell} \tag{38}$$

## 3. OBJECTIVE LENSES FOR OPTICAL DATA STORAGE

The technology of digital storage and retrieval of information on optical disks has found widespread application in the following three domains:

1. Laser vision (LV), formerly video long play (VLP)
2. Compact disk (CD) audio systems
3. Data storage (DS) or digital optical recording (DOR)

### 3.1. Characteristic Parameters of Optical Data Storage Systems

Typical light sources and recording media parameters associated with optical data storage technology are compiled in Table 9.

### 3.2. Spectral Region of Lasers

The information stored as digital pits in the storage medium may be read optically. The effected modulation of the reflected beam is converted into electrical signals by a photodiode. As shown by Heemskerk (1978), the light source has to have a radiance of at least 2.5 kW/cm$^2$-sr to obtain a detector output with a sufficiently high signal-to-noise ratio. This demand can be met only by lasers.

In optical data storage technology, gas lasers are used, but they are now being replaced more and more by semiconductor lasers, commonly called laser diodes. Gas lasers emit monochromatic radiation. Generally, semiconductor gain guided AlGaAs lasers are used in optical data storage technology. These are lasers with a constricted double heterojunction (CDH) structure. Their advantage is their compact form and their low cost in production. The active layer of AlGaAs semiconductor lasers is approximately 0.2–1.5 μm thick, some 5 μm wide, with a length between the reflecting surfaces of 250 μm. The amount of dopant in the effective layer determines the central emitted wavelength. These spectral lines are longitudinal modes of the resonator and fulfill the resonance condition

$$l_R = \frac{m\lambda}{2n} \qquad (39)$$

where $l_R$ is the resonator's length, n the refractive index of the resonator material, and m an integer corresponding to the number of half-waves in the resonator. Thus when $l_R$ = 250 μm, λ = 780 nm, and n = 3.59, one obtains m = 2301. The separation Δλ of two neighboring lines is determined by Δm = ±1 and consequently Δλ = ±0.3 nm, as Fink et al. (1980–81) have shown. Their

Table 9 Typical Light Sources and Recording Media Parameters for Optical Data Storage Systems

| | LV | CD | DS Write | DS Read |
|---|---|---|---|---|
| (a) The light source | | | | |
| Radiation source | Gas laser: HeNe; semiconductor laser | Semiconductor laser | Gas laser: Ar, HeCd, HeNe; semiconductor laser | Gas laser: HeNe; semiconductor laser |
| Nominal wave-length (nm) | 632.8, 780 | 780 | 441.6, 488.0, 632.8, 820, 870 | 632.8, 820 |
| Spectral region (nm) | 632.8, 760–800 | 780–820 | 780–820, 800–840 | 632.8, 780–860, 800–840 |

| | LV | CD | DS |
|---|---|---|---|
| (b) The storage medium | | | |
| Trace interval (μm) | 1.6 | 1.6 | 1.6–2.0 |
| Trace width (μm) | 0.4, 0.6 | 0.6 | 0.6–0.8 |
| Trace depth (μm) | 0.1 | 0.1 | 0.06–0.1 |
| Disk material | Plastic | Plastic | Glass, plastic, aluminum |

amplitudes decrease compared with the main central mode. The envelope covers a band of some 5 nm. A multimode spectral emission has a short coherence length. At a constant operating temperature of 30°C the band of wavelengths may vary in the range ±10 nm, due to tolerances in production. This variation is shown schematically in Figure 9. For an operational temperature range from 0 to 60°C the total wavelength band spread may be ±20 nm.

## 3.3. Design Characteristics of Objective Lenses for Optical Data Storage

### 3.3.1 Color Correction

No chromatic correction is required for objective lenses using a given monochromatic wavelength. Lenses designed without color correction are used with gas lasers, and are generally used for LV systems and some development systems in the DS field. In Figure 10 the allowed longitudinal chromatic aberration of LV-type objective lenses used with HeNe lasers is shown as a function of wavelength. If semiconductor lasers are used, the objective lenses require correction of longitudinal chromatic aberration over

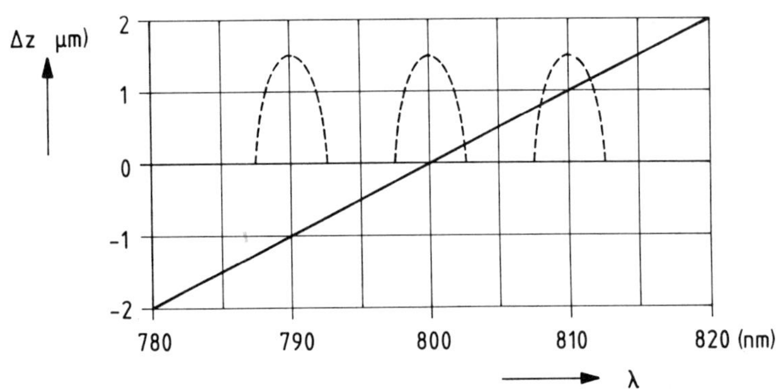


Figure 9 Longitudinal chromatic aberration $\Delta z$ as a function of wavelength $\lambda$ for CD-type objective lenses used with a semiconductor laser; NA = 0.5, f = 5 mm. Dashed line: the spectral band of a laser diode radiation and the variation in range.

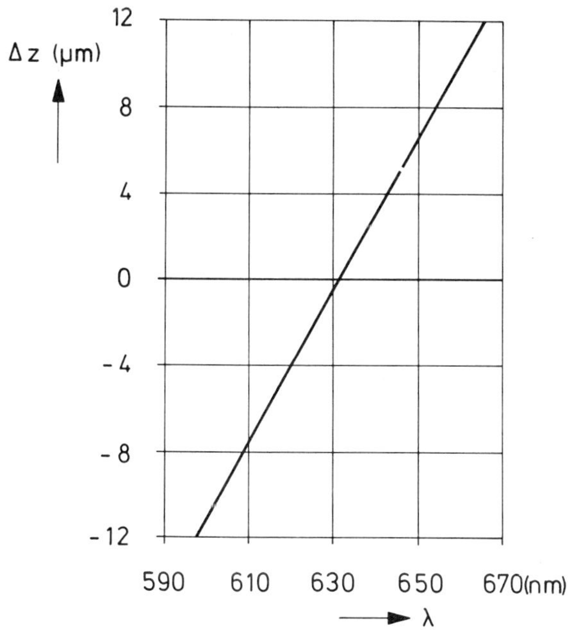


**Figure 10** Longitudinal chromatic aberration $\Delta z$ as a function of wavelength $\lambda$ for LV-type objective lenses with a HeNe laser; NA = 0.4, f = 8.3 mm.

the bandwidth of 5 nm. To avoid unwanted degradation of the image quality, the admissible longitudinal chromatic aberration at the border of the emitted bandwidth should be less than the depth of focus given in Table 4. If the objective lens is used for reading only, a reduction in the allowable longitudinal chromatic aberration to about one-third of the value of that for a non-color-corrected lens is necessary. Therefore, a longitudinal chromatic aberration of some ±2 μm is admissible for the total spectral range of ±20 nm. For a rise in temperature up to 60°C the emitted spectral band may be shifted by 10 nm. The resultant defocusing of about 1 μm is compensated for by the automatic focusing actuator. CD objective lenses and some DS lenses belong in this category. Figure 9 shows the typical longitudinal chromatic aberration versus wavelength of CD-type objective lenses used with semiconductor lasers.

If writing and reading are simultaneous and at different wavelengths, the objective lens has to be achromatic for the two wavelengths. In this case the lateral chromatic aberration also has to be reduced. For trace widths of 0.4 or 0.6 μm, the lateral chromatic aberration in the whole image field has to be reduced to 0.05—0.1 μm to ensure the very same image position for both wavelengths. DS objective lenses especially belong in this category.

Figure 11 shows the typical longitudinal chromatic aberration of an achromatic lens between the wavelengths 632.8 and 820 nm. LV systems now use semiconductor lasers more and more. In changing from HeNe to semiconductor lasers, allowance has to be made for the longer wavelengths of the semiconductor laser. To obtain the same image resolution, the numerical aperture of the objective lens has to be increased in proportion.

### 3.2.2. Image Field and Field Curvature

Reading the information stored in the spiral-shaped trace of LV systems, a mirror M in front of the objective lens has to be tilted more and more to follow the trace. This mirror, in combination with an LV objective lens, is illustrated in Figure 12a. This technique requires an image field of 2y = 0.8 mm. Only when the reading beam reaches the border of the image field is the objective lens reset in the radial direction of the optical disk. Taking advantage

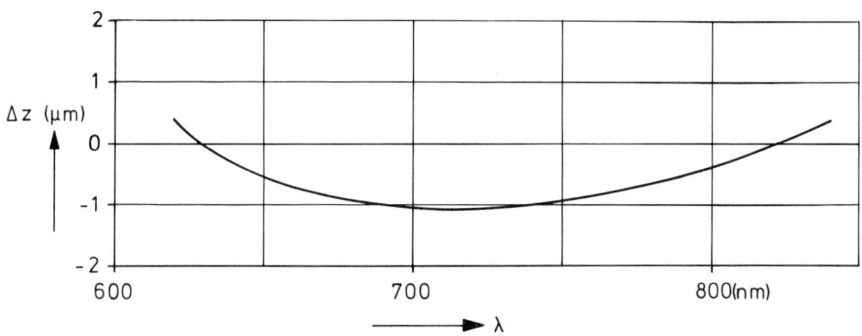


**Figure 11** Longitudinal chromatic aberration Δz as a function of wavelength λ for DS-type objective lenses with achromatic correction for the wavelengths 632.8 and 820 nm; NA = 0.6, f = 3.5 mm.

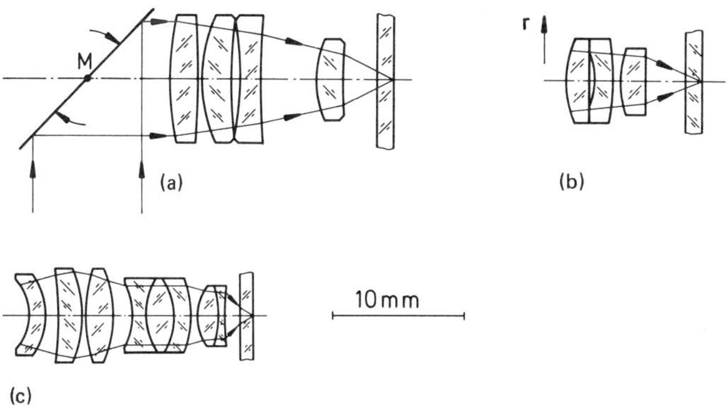


Figure 12 Schematic of objective lenses for optical data storage technology. (a) LV objective lens with mirror M in front of the objective lens for radial tracking. (b) CD objective lens with radial reset for radial tracking in direction r. (c) DS objective lens with full color correction.

of the automatic focusing of the objective lens under operating conditions, a field curvature up to

$$r \sim \frac{f}{2} \tag{40}$$

is admissible. Here r stands for the radius of the image field and f for the focal length of the objective lens. The sag $\Delta z$ of the field curvature at a height y from the lens axis is

$$\Delta z = \frac{y^2}{2r} \tag{41}$$

Table 10 shows values for the defocusing caused by field curvature sag as a function of image height y for an objective lens with a focal length of 8.3 mm. Thus the permissible radius of field curvature is 4.6 mm in the image field. The acceptance of this permissible field curvature allows for the correction of astigmatism over the whole image field and a much reduced coma for an objective lens that meets the specification for volume, weight, and size.

Table 10 Permissible Values for Field Curvature Sag for an Objective Lens f = 8.3 mm

| Image height, y (mm) | Sag of field curvature, $\Delta z$ (μm) |
|---|---|
| 0.1 | 0.5 |
| 0.2 | 2 |
| 0.3 | 5 |

In CD and DS systems the objective lens is reset continuously for radial tracking. This is shown schematically in Figure 12b. CD- and DS-type objective lenses need only a small image field, up to 2y = 0.2 mm, depending on the system. This field size is required to meet the mechanical manufacturing tolerances associated with the assembly and the alignment of the objective lens and the collimator. With this small image field the specification for CD systems is met with a three-element lens. Hence smaller and lighter lenses may be realized. For short focal lengths, for example, 4.5 mm, the weight of these lenses is reduced to 0.6 g. However, a corrected lens for two wavelengths, including a lateral chromatic aberration of less than 0.1 μm, requires an eight-element lens as shown schematically in Figure 12c. Such an objective lens is required for some DS applications.

### 3.3.3. Types of Objective Lenses

A schematic representation of the types of objective lenses for optical data storage technology is shown in Figure 12. The element illustrated on the right is the cover glass and is part of the image-forming system because the information is stored on its back surface. Table 11 gives typical objective lens data. The full specifications for objective lenses used in LV, CD, and DS systems are displayed in Table 12.

The birefringence seen in objective lenses is caused by both the birefringence of the glass types used and the mechanical strain resulting from the mounting of the elements. A wavefront degradation of less than $\lambda/60$ during a single pass through a three- or four-element objective lens requires glass types with a birefringence

Table 11 Data Associated with Objective Lenses

| Objective lens type | Application | NA | f (mm) | λ (nm) | Color correction | Image field (mm) | Weight (g) |
|---|---|---|---|---|---|---|---|
| a | LV | 0.4 | 8.3 | 632.8 | Monochromatic | 0.8 | 3.6 |
| b | CD, DS | 0.5 | 5 | 780–820 | 5-nm spectral band | 0.2 | 0.6 |
| c | DS | 0.6 | 3.5 | 632.8, 800–840 | Achromatic for 3.32 and 820 nm; correction of lateral aberration | 0.1 | 4.3 |

Table 12 Full Specifications for Objective Lenses Used in Optical Data Storage Technology

| | LV | CD | DS |
|---|---|---|---|
| NA | 0.40–0.65 | 0.40–0.45 | 0.45–0.60 |
| Focal length (mm) | 4.5–9 | 4.5, 5 | 3.5–6 |
| Clear diameter (mm) | 5.4–7.2 | 4.0, 4.1 | 4.2–5.4 |
| Field diameter (mm) | 0.2, 0.4, 0.8 | 0.08, 0.2 | 0.04, 0.2, 0.4 |
| Cover glass thickness (mm) | 1.3 | 1.2 | 0.2, 1.2–1.5 |
| Image scale | 1:20; Infinite conjugate | Infinite conjugate | Infinite conjugate |
| Free working distance (mm) | 0.8–3.5 | ∿2.3 | 0.75–2 |
| Range of longitudinal chromatic correction (nm) | None | 5 | 5; Achromatic for other wavelengths; some lenses require additional lateral color correction |
| Vignetting (%) | 0 | 0 | 0 |
| Weight (g) | 1.6–3.7 | 0.6–1.6 | 0.6, 3.5, 4.3 |

of less than 15 nm/cm, and for an eight-element lens less than 8 nm/cm. The same amount of birefringence is reserved for the assembly.

### 3.3.4. RMS of Wave Aberration

In accordance with Eqs. (25) and (27), the image quality of objective lenses may be described by $\Delta W$, the RMS wave aberration, in the forms

$$\sqrt{\overline{W^2}} = \frac{\lambda}{N} \tag{42}$$

and

$$\overline{W^2} = \frac{\lambda^2}{N^2} \tag{43}$$

both of which are in common use.

Figure 13 shows the computed rms wave aberration as a function of the half-field diameter y for the three objective lenses depicted in Figure 12. The LV objective lens (a) has been computed for monochromatic light. This lens has an rms wave aberration of less than $\lambda/100$ on-axis, which drops to $\lambda/12$ at y = 0.4 mm, the half-field diameter.

The objective lens (b) is color corrected for a spectral range of 5 nm. The rms wave aberration is given for the central wavelength, $\lambda_c$ = 800 nm, and for $\lambda = \lambda_c \pm 2.5$ nm. When focused for the central wavelength $\lambda_c$ the objective lens has an rms wave aberration of $\lambda/70$ on axis, which drops to $\lambda/50$ at $\lambda = \lambda_c \pm 2.5$ nm. At the border of the field rms values of approximately $\lambda/25$ are attained. The objective lens (c) is achromatic for the two wavelengths 632.8 nm and 820 nm. The rms wave aberration is approximately $\lambda/50$ across the whole field. In the same focal plane the image quality for $\lambda$ = 632.8 nm is somewhat better on-axis and somewhat less across the field.

Objective lenses for optical data storage are manufactured with the highest precision. A symmetrical point image is essential. Root-mean-square wave aberration values of $\lambda/20$ through $\lambda/25$ are realized on-axis and across a large portion of the field.

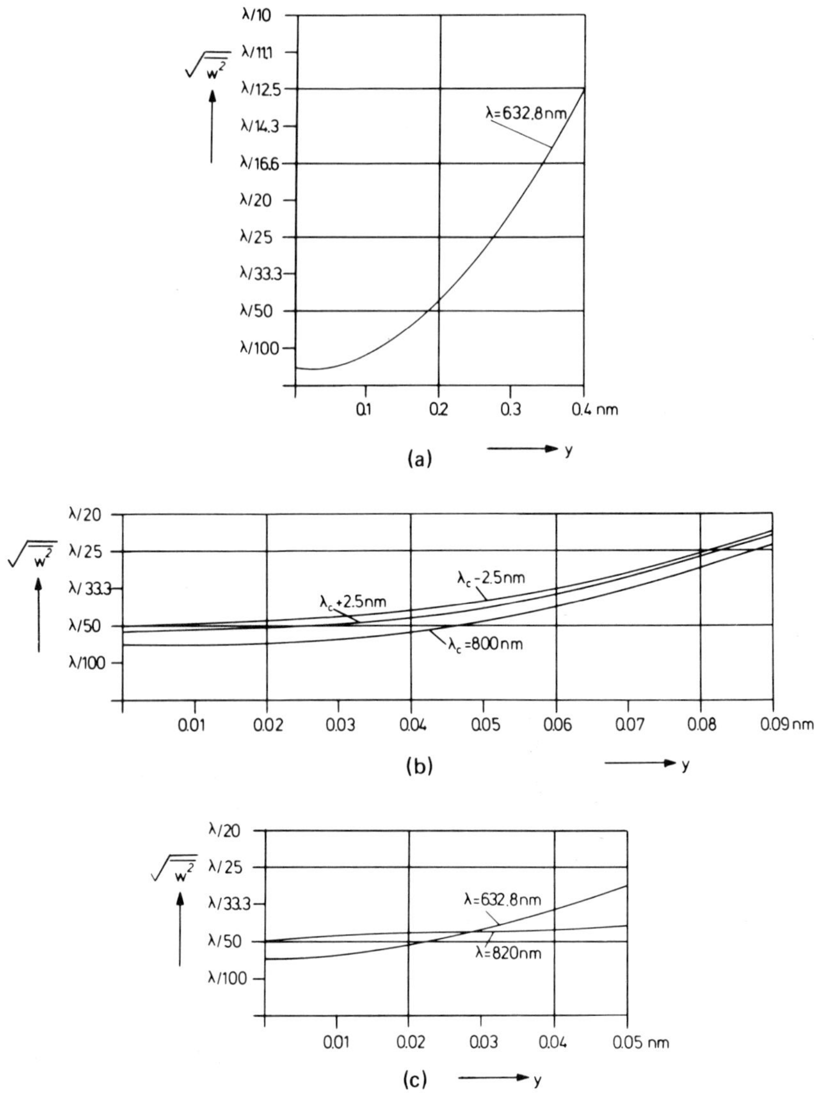


**Figure 13** Calculated values for the rms wave aberration $\sqrt{\overline{W^2}}$ of objective lenses as a function of the image height y for optical data storage technology. (a) LV objective lens. (b) CD, DS objective lens with a reduced image field. (c) DS objective lens with full color correction.

### 3.3.5. Cover Glass

The digitalized information stored in optical disks is protected by its cover glass. Very often the thickness of this cover glass ranges between 1.2 and 1.5 mm. Normally, polymethyl methacrylate (PMMA) or polycarbonate with a refractive index of 1.49 or 1.55, respectively, is used for the cover glass. The planoparallel cover glass is part of the image-forming system and allowance has to be made in the lens design for its refractive index and thickness. Variations in thickness, refractive index, and tilt influence the attainable image quality.

The effects of variations in thickness and in refractive index are of the same nature. A refractive index variation of $\Delta n$ changes the optical path length by the same amount as a thickness variation of $\Delta h$ and is related by

$$\Delta h = \frac{h}{n} \Delta n \tag{44}$$

The influence of thickness variation on image quality may be expressed as the difference between the rms wave aberration for a thickness $h + \Delta h$ and for the nominal thickness $h$. Thus

$$\Delta \sqrt{\overline{W^2}} = \sqrt{\overline{W^2}}_{h+\Delta h} - \sqrt{\overline{W^2}}_{h} \tag{45}$$

An automatic focusing system is required to compensate for cover glass thickness variations. Typical values for image-quality degradation caused by thickness variations are given in Table 13 for a 1.2-mm-thick cover glass associated with the corrected lens types shown in Figure 13.

The values given for image degradation are also valid for other cover glass thicknesses. Because the image quality across the field is uniform, the values of Table 13 are valid for the entire image field. Table 13 illustrates that the image degradation of high-aperture objective lenses is much more sensitive to thickness variations of the cover glass. Similarly, the image degradation caused by a tilted cover glass can be expressed as the difference between the rms wave aberration for a tilted cover glass and that for one without a tilt. Thus

**Table 13** Image-Quality Degradation Caused by a Thickness Variation of the Cover Glass

| Thickness variation, $\Delta h$ (mm) | Image degradation, $\Delta\sqrt{\overline{W^2}}$ | | |
|---|---|---|---|
| | For NA = 0.4 | For NA = 0.5 | For NA = 0.6 |
| 0.04 | Negligible | $\lambda/160$ | $\lambda/50$ |
| 0.08 | Negligible | $\lambda/55$ | $\lambda/20$ |
| 0.12 | $\lambda/330$ | $\lambda/30$ | $\lambda/12$ |
| 0.20 | $\lambda/70$ | $\lambda/18$ | $\lambda/7$ |

$$\Delta\sqrt{\overline{W^2}} = \left(\sqrt{\overline{W^2}}\right)_{\delta} - \left(\sqrt{\overline{W^2}}\right)_{\delta=0°} \tag{46}$$

where $\delta$ is the tilt angle.

For the objective lenses mentioned, typical values for the image degradation caused by a tilted cover glass 1.2 mm thick are given in Table 14. To avoid unwanted image-quality degradation, the cover glass of optical disks should have a local surface figure of about $\lambda/25$.

**Table 14** Image-Quality Degradation Caused by a Tilted Cover Glass 1.2 mm Thick

| Tilt angle, $\delta$ (deg) | Image degradation, $\Delta\sqrt{\overline{W^2}}$ | | |
|---|---|---|---|
| | For NA = 0.4 | For NA = 0.5 | For NA = 0.6 |
| 0.5 | Negligible | $\lambda/160$ | $\lambda/50$ |
| 1 | $\lambda/140$ | $\lambda/40$ | $\lambda/16$ |
| 1.5 | $\lambda/70$ | $\lambda/20$ | $\lambda/10$ |

## 4. OPTICAL COMPONENTS FOLLOWING THE LIGHT SOURCE

The optical components immediately following the light source convert the emitted radiation into a light beam with a plane wavefront. The angular radiation distribution of the emitted laser radiation determines the required optical components.

### 4.1. Beam Expander

Gas lasers of the appropriate type emit parallel beams with beam diameters ranging from 0.5 to 0.75 mm and divergence angles 2u of about 3.5 arcmin. The laser beam is expanded by a beam expander to fill the entrance pupil of the imaging objective lens. The amount of expansion of the gaussian laser beam determines the degree of beam truncation at the pupil of the objective lens. According to Figure 3, a desirable beam expansion is where the $1/e^2$ intensity level matches with the rim of the pupil in order to obtain a small point image for any given NA value. For an optical image forming system the Lagrange aperture, given by

$$y_i \sin u_i = \text{const.} \tag{47}$$

remains invariant. Here $2y_i$ is the beam diameter and $2u_i$ is the beam divergence angle. For a $\Gamma$-times expansion where

$$\Gamma = \frac{y_1}{y_0} \tag{48}$$

the beam divergence angle $2u_1$ of the expanded beam becomes

$$2u_1 \approx \frac{2u_0}{\Gamma} \tag{49}$$

where $2u_0$ stands for the divergence angle before the expansion; $2y_0$ is the beam diameter before and $2y_1$ after expansion; $2y_0$ is For example, a 10-fold expansion reduces the divergence angle to about 20 arcsec. Galilean and keplerian telescopes may be used as beam expanders (Fig. 14).

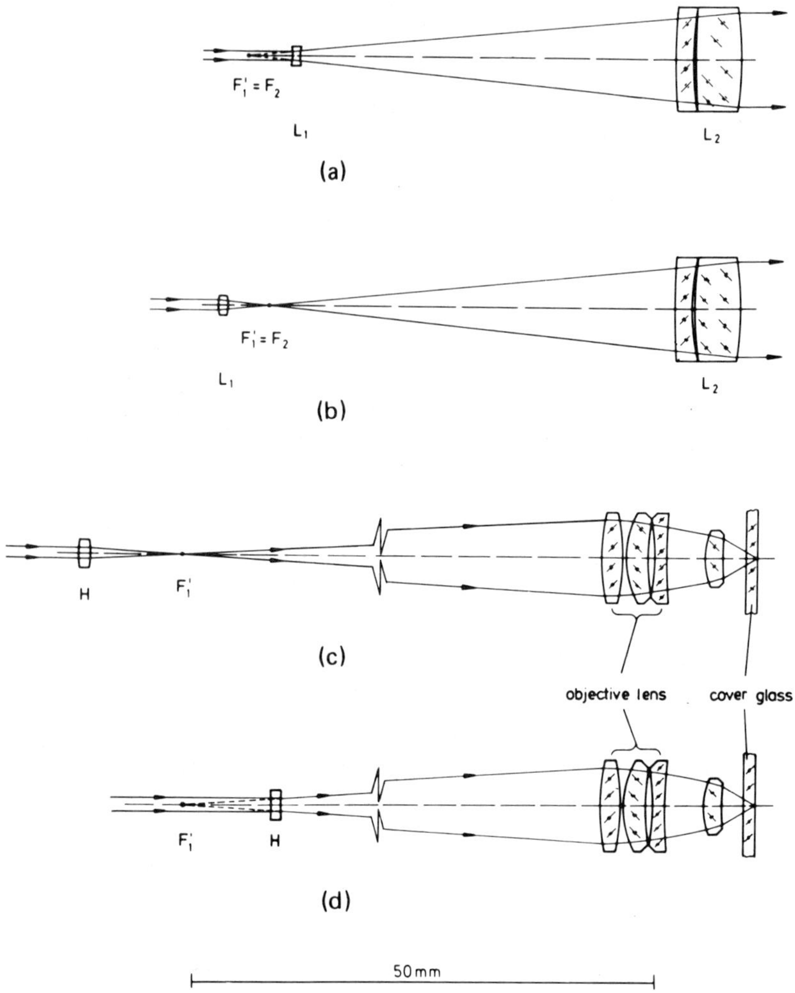


**Figure 14** Schematic of beam expanders, 12×; (a) Galilean beam expander, (b) Keplerian beam expander; $F_1'$ = back focus of lens $L_1$, $F_2$ = front focus of lens $L_2$. (c) Focusing lens H matched to a finite conjugate objective lens, 1:12; $F_1'$ = real back focus of lens H. (d) Focusing lens H matched to a finite conjugate objective lens, 1:12; $F_1'$ = virtual back focus of lens H.

The wave aberrations of the beam expander and the objective lens are additive. To obtain the best image quality already in the lens design the wave aberrations of the beam expander and the objective lens should cancel. The image quality of the beam expander alone should be comparable with those given in Figure 13 for objective lenses.

When gas lasers are used, objective lenses may be designed for finite conjugates with ratios of 1:12 to 1:15. In such cases an auxiliary lens H is required to focus the laser beam (Fig. 14c and d). This focus represents the object point for the objective lens and will be imaged into the disk plane. As far as the optical system design is concerned, it is irrelevant whether the auxiliary lens has a positive or a negative refractive power. In the first case the focus will be real (fig. 14c), and in the second it will be virtual (Fig. 14d). In these systems the polarizing beam splitter has to be placed in the divergent or convergent optical light path. Allowance for this needs to be made in the design of the lens and the polarizing beam splitter.

### 4.2. Collimator

The angular radiation distribution of a semiconductor laser is determined by the layer geometry and its CDH structure. The active layer is approximately 0.2 μm thick and 1.5–5 μm wide. For a semiconductor laser with a 0.2-μm-thick active layer, the far-field intensity distribution in the plane normal to the active layer as function of the angle from the optical axis is gaussian. The beam divergence $2u_1$ out to the half-maximum intensity level is approximately 60°. [See, e.g., Philips-type CQL10 (Fink et al., 1980–1981).] In the plane of the active layer the intensity distribution is determined by the structure of the gain-guided AlGaAs laser and in the far field has a beam divergence at the half-maximum intensity level of $2u_2$ approximately 34°. The divergence points of these two waves are not coincident. Therefore, semiconductor lasers are light sources with astigmatism in the emitted wavefront. Typically, the astigmatic difference $d_A$ in air for semiconductor lasers ranges between 18 and 25 μm [see, e.g., Finck et al. (1980-81) and Botez et al. (1982)]. These two intensity distributions as a function of the angle from the optical axis are shown schematically in Figure 15.

Recent developments in semiconductor lasers provide an active layer of 1.4 μm × 5 μm, which is well suited for applications that need high optical imaging quality. The bam has almost no

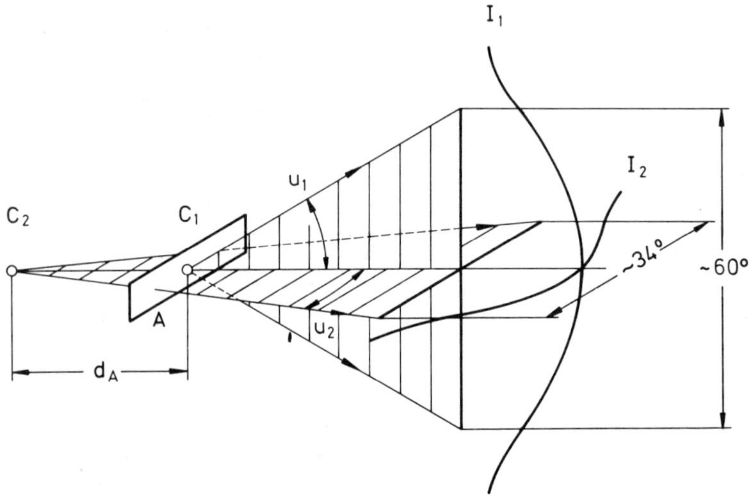


Figure 15 Angular intensity distribution in the far field of a semiconductor laser with a 0.2-μm-thick active layer. $d_A$ = astigmatic difference, $I_1$, $I_2$ = intensity distribution ($I_2$ simplified), A = active layer.

astigmatism. [See, e.g., RCA-type CHD-LOC (RCA, 1982).] Due to the greater thickness of the active layer, the beam divergence $2u_1$ is approximately 25°.

It is the task of the optical components following the light source to convert this radiation distribution into a parallel beam with an acceptable wave aberration. The collimator converts the radiation emitted by an object point into a parallel beam. Collimators are generally corrected for an object field diameter of 160 μm. This size is required to meet the mechanical manufacturing tolerances between the position of the collimator and the laser. A collimator with a NA of 0.5 has a collecting cone angle of 60°. To collect up to 50% of the divergent gaussian radiant flux from a laser diode, designs similar to that of the objective lenses of type b in Figure 12 may be used in reverse as collimators. For small collecting cone angles up to 20 or 25°, two-element collimators are adequate. A schematic representation of these collimators is given in Figure 16. Collimators with a focal length of nearly 20 mm are required if the polarizing beam splitter is placed between a laser

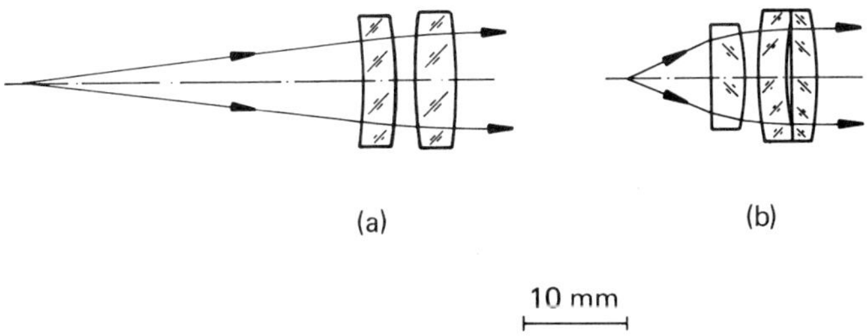


**Figure 16** Schematic of collimators for optical data storage technology. (a) Two-element collimator; f = 20 mm, NA = 0.12. (b) Three-element collimator; f = 4.5 mm, NA = 0.42.

diode and the collimator. In this case the collimator is also focusing the reflected beam onto the detector.

Incomplete collimation of the emitted radiation, even with automatic focusing, causes an image-quality degradation, which may be expressed as the difference between the rms wave aberration for a wave coming from a light source at a distance g and from infinity.

$$\Delta \sqrt{\overline{W^2}} = \left(\sqrt{\overline{W^2}}_g\right) - \left(\sqrt{\overline{W^2}}\right)_{g=\infty} \tag{50}$$

This degradation is influenced by the type of lens used. For a three-element lens of type b (Fig. 12), typical values are given in Table 15 for the image-quality degradation caused by a bundle of rays emerging from an object at a distance g. The values given are approximate.

### 4.3 Anamorphic Imaging

Because of the different radiation distributions in the two normal sections, the collimated light beam has an elliptical cross section whose major diameter-to-minor diameter ratio is given by

$$\gamma = \frac{\sin u_1}{\sin u_2} \tag{51}$$

Table 15 Typical Values for the Image-Quality Degradation by Using Nonparallel Incident Light Beams for Two NA Values

| Object distance of the light source g (m) | Image degradation, $\Delta\sqrt{\overline{W^2}}$ | |
|---|---|---|
| | For NA = 0.4 | For NA = 0.5 |
| 1.5 | Negligible | $\lambda/80$ |
| 1 | Negligible | $\lambda/50$ |
| 0.75 | $\lambda/100$ | $\lambda/30$ |
| 0.5 | $\lambda/65$ | $\lambda/20$ |

For commercially available semiconductor lasers, this ratio is nearly 1:0.6. The beam expansion in the direction of the minor diameter of the ellipse may be realized by anamorphic imaging. The needed anamorphic imaging leaves the image scale unchanged in the section containing the angle $2u_1$ but changes the image scale by 1:0.6 in the section containing the angle $2u_2$.

### 4.3.1. Prism

On passing through a prism the beam diameter lying in a principal section of the prism may be changed while the beam diameter normal to this section remains invariant. Figure 17 shows that after refraction of the incident beam at the prism surface (1), the beam diameter is changed in the plane of drawing. The expansion ratio is given by

$$\Gamma = \frac{d_1}{d_0} \tag{52}$$

where $d_0$ is the diameter of the incident beam and $d_1$ is the diameter after refraction. Hence with the law of refraction one obtains

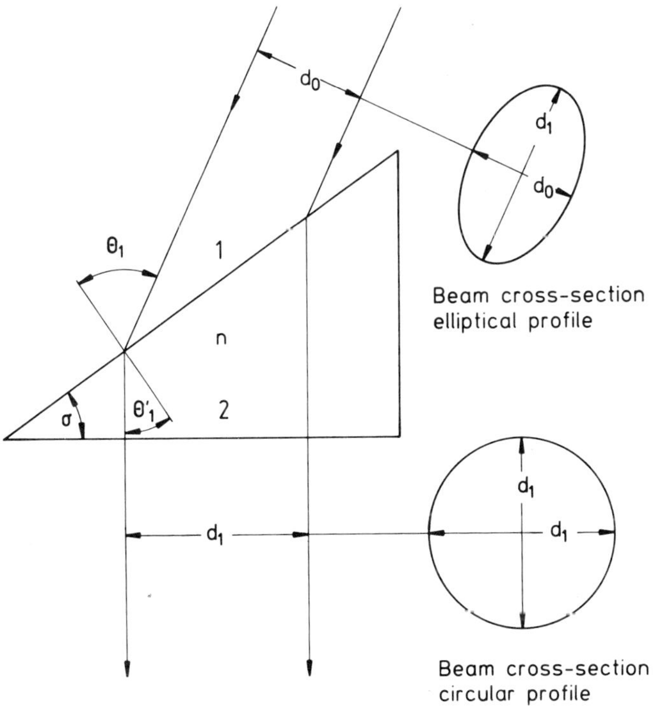


**Figure 17** Anamorphic beam expansion by a prism. $\sigma$ = prism angle, $\theta_1$ = angle of incidence at surface 1, $\theta_1'$ = angle of refraction at surface 1.

$$\Gamma = \frac{\cos \theta_1'}{\cos \theta_1} \tag{53}$$

where $\theta_1$ stands for the angle of incidence and $\theta_1'$ for the angle of refraction. This equation applies to the prism when the beam leaves the prism without refraction at surface 2.

Let $\sigma$ be the prism angle; then $\theta_1' = \sigma$. To achieve a required expansion factor $\Gamma$, the angle of incidence $\theta_1$ is determined by

$$\sin \theta_1 = n\sqrt{\frac{\Gamma^2 - 1}{\Gamma^2 n^2 - 1}} \tag{54}$$

where n is the refractive index of the prism material. Since $\theta_1' = \sigma$, the prism angle $\sigma$ is also determined.

For a polychromatic incident beam a glass type BK or PK of low dispersion must be selected for the manufacture of the prism. For a beam expansion ratio of 1.66, a BK7 glass type requires a prism apex angle of 35.2° and an incidence angle of 60.4°. This prism is represented in Figure 17. Due to dispersion by the glass, a semiconductor laser beam with a 5-nm bandwidth refracts at surface 1 into a colored fan of rays. For BK7 this fan of rays has a divergence angle $2\Delta\theta_1'$ of approximately 12 arcsec and is increased to $2\Delta\theta_2'$ about ±18 arcsec on leaving surface 2 of the prism. Therefore, in the focal plane of a lens colored fringes appear with an extension of

$$\Delta y \sim 2f\Delta\theta_2' \tag{55}$$

Thus, with a lens of 5-mm focal length, the diameter of the point image is increased by about 0.4 μm in the direction of the beam expansion. If the resolution of the image is unacceptably degraded by these colored fringes, the anamorphic imaging should be realized by a Brewster telescope.

#### 4.3.2. Brewster Telescope

Brewster telescopes are composed of two prisms with oppositely oriented prism apex angles, so that achromatization is achievable. A schematic representation of a Brewster telescope is shown in Figure 18. The telescope shown has an expansion ratio of 1.66 in the plane of the drawing and no expansion in the plane normal to the drawing. If $\theta_\nu$ is the incidence angle and $\theta_\nu'$ the angle of refraction at the $\nu$th prism surface, the expansion ratio is given by a generalization of Eq. (53), namely,

$$\Gamma = \left( \prod_{\nu=1}^{4} \frac{\cos\theta_\nu'}{\cos\theta_\nu} \right) \tag{56}$$

The expansion ratio may be adjusted to the required value by turning the prisms on pivots about the axis parallel to their refracting edges. For the 1.66× anamorphic beam expander shown, the emergent beam is displaced axially by 2 mm.

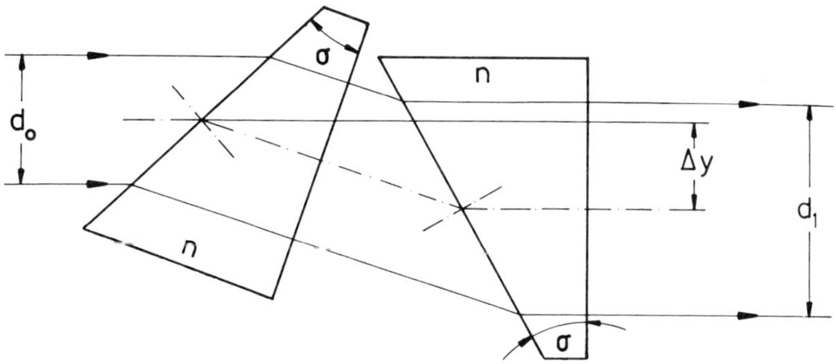


Figure 18 Schematic of a Brewster telescope. $\sigma$ = prism angle, $\sigma = 28.8°$; n = 1.52.

## 4.4. Astigmatism

Since the wavefront is orthogonal to the light rays emitted, the astigmatic difference in the far field of a semiconductor laser is the phase difference between the wavefront in the two sections in and normal to the plane of the active layer. As Figure 19 shows, the wavefront $W_1$ is emanating from the center $C_1$. The wavefront

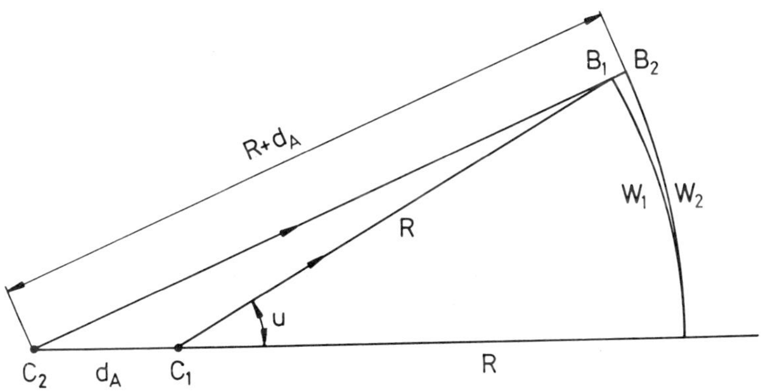


Figure 19 Wavefront connected with astigmatic differences of semiconductor lasers. $W_1$ = wavefront emanating from the center $C_1$, $W_2$ = wavefront emanating from the center $C_2$ is rotated about $C_2C_1$ by 90° into the plane of the drawing.

$W_2$ emanating from $C_2$ is rotated about $C_2C_1$ by 90° into the plane of the drawing. The optical path difference is given by the distance $B_1B_2$. If $C_1B_1$ is the radius R of a reference sphere such that $R \gg d_A$, the expression

$$C_2B_1 \approx R + d_A \cos u \tag{57}$$

can be derived from the geometry of the triangle $C_2C_1B$, where $C_2B_2 = R + d_A$. Then the difference $\Delta W$ between the wavefronts of the two sections in the far field is given by

$$\Delta W = B_1B_2 \approx d_A(1 - \cos u) \tag{58}$$

If the astigmatic difference $d_L$ is related to a medium with the refractive index $n_L$, the astigmatic difference in air is given by

$$d_A = \frac{d_L}{n_L} \tag{59}$$

and Eq. (59) may be written in the general form

$$\Delta W \approx \frac{d_L}{n_L}(1 - \cos u) \tag{60}$$

Values for this wavefront difference $\Delta W$ in the far field for a AlGa-As semiconductor laser with $n_L = 3.5$ and an astigmatic difference $d_A$ of 25 μm are given in Table 16. Like the wavefront aberration of lenses according to Eq. (27), this wavefront difference can be expressed as $\Delta W = \lambda/N$. The retardation between wavefronts $W_1$ and $W_2$ must be compensated for by a cylindrical lens to obtain a plane wavefront. Two possible solutions are given in connection with the collimator.

The first solution is for the case when $W_1$ emerges as a plane wavefront from the collimator. The cylindrical lens then has to present a negative refracting power in the section containing wavefront $W_2$ and a zero refracting power in the plane containing wavefront $W_1$. The design of the cylindrical lens should be such

**Table 16** Typical Values of Wavefront Difference ΔW Due to Astigmatism of Semiconductor Lasers as Function of the Semiaperture Angle Measured From the Optical Axis

| Semiaperture angle (deg) | Wavefront difference, ΔW | |
|---|---|---|
| | In μm | In λ (for λ = 0.8 μm) |
| 0 | 0 | 0 |
| 3 | 0.04 | λ/20 |
| 5 | 0.10 | λ/8 |
| 10 | 0.38 | λ/2 |
| 20 | 1.50 | λ/5 |
| 30 | 3.33 | λ/2 |

that in the plane containing $W_2$, the optical path difference between the edge and the center of the lens equals the wavefront difference caused by the astigmatism of the semiconductor laser. Hence this relationship may be written as

$$\frac{\Delta h}{n} = d_A(1 - \cos u) \tag{61}$$

where n is the refractive index of the cylindrical lens and Δh the difference in thickness between the edge and the center.

The second solution is for the case when $W_2$ emerges as a plane wavefront from the collimator. The cylindrical lens then has to present a positive refracting power in the section containing wavefront $W_1$ and a zero refracting power in the plane containing wavefront $W_2$.

Figure 20 shows a schematic representation of a cylindrical lens with negative refracting power, as required in the first case. It can be derived from lens geometry that for a planoconcave lens, the radius $r_1$ of the cylindrical surface is given by

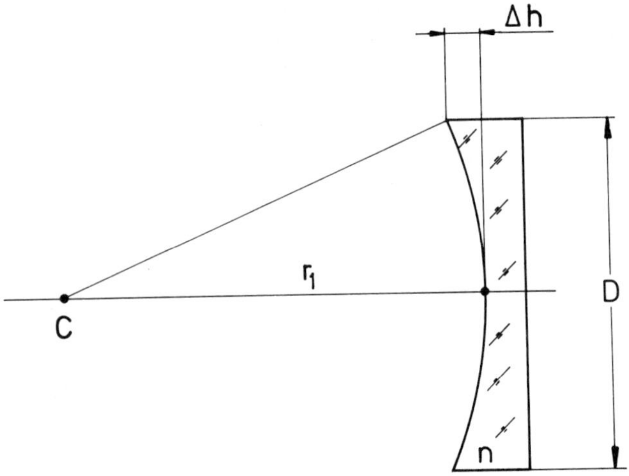


**Figure 20** Cylindrical lens with negative refracting power used to compensate for astigmatism. Δh = difference in thickness between edge and center, $r_1$ = radius of cylindrical surface.

$$r_1 \approx \frac{D^2}{8\,\Delta h} \tag{62}$$

where D is the lens diameter and $\Delta h \ll r_1$. Hence with Eq. (61) we obtain

$$r_1 \approx \frac{D^2}{8nd_A(1 - \cos u)} \tag{63}$$

Substituting D = 5 mm, $d_A$ = 25 μm, u = 30°, and n = 1.51, we obtain $r_1$ = 618 mm.

The focal length f of a thin cylindrical lens is given by the general formula

$$\frac{1}{f} = (n - 1)\left(\frac{1}{r_1} - \frac{1}{r_2}\right) \tag{64}$$

where $r_1$ and $r_2$ are the radii of curvature for the first and second surfaces, respectively. When for a planoconcave lens, $r_1 = -618$ mm and $r_2 = \infty$, the focal length $f = -1212$ mm.

## 5. QUARTER-WAVE PLATE

### 5.1. Optical Properties of Birefringent Crystals

In a uniaxial birefringent crystal electromagnetic radiation is resolved into two component waves, an extraordinary wave vibrating in the direction parallel to the optic axis and an ordinary wave vibrating normal to it. It shall be understood that the direction of vibration corresponds to the direction of the electric vector. From a point source the ordinary wave propagates as a spherical wave, while the extraordinary wave propagates as an ellipsoidal wave for which the velocity of propagation in the direction of the optic axis equals that of the ordinary wave. The refractive index n of a medium is given as the ratio of the velocity of light in vacuo $c_0$ to its velocity c in the medium:

$$n = \frac{c_0}{c} \tag{65}$$

The wavelength $\lambda$ is defined as

$$\lambda = \frac{c}{\nu} \tag{66}$$

where $\nu$ is the frequency of the light.

In a medium with refractive index n, the wavelength $\lambda$ is given by

$$\lambda = \frac{\lambda_0}{\nu} \tag{67}$$

where $\lambda_0$ is the wavelength in vacuo. Hence the frequency

$$\nu = \frac{c}{\lambda} \tag{68}$$

remains invariant in all media. We define an indicatrix as the surface generated by a spatial vector expressing the refractive index in polar coordinates. In a birefringent crystal the indicatrix for the refractive index $n_o$ of the ordinary wave is spherical. The refractive index $n_e$ of the extraordinary wave depends on the direction of propagation. For a uniaxial crystal the indicatrix is an ellipse of revolution. Along the optic axis, $n_e = n_o$. In a plane normal to the optic axis, however, we have $n_e > n_o$ for a crystal with a positive indicatrix and $n_e < n_o$ for a negative crystal. Among the crystals frequently used in optics, quartz has a positive indicatrix and calcite a negative indicatrix. A schematic representation of these two indicatrices is shown in Figure 21. The refractive indices of quartz and calcite are given in Table 17. The birefringence, that is, the difference of the refractive indices for the ordinary and extraordinary wave $(n_o - n_e)$ is nearly 17 times more for calcite than for quartz.

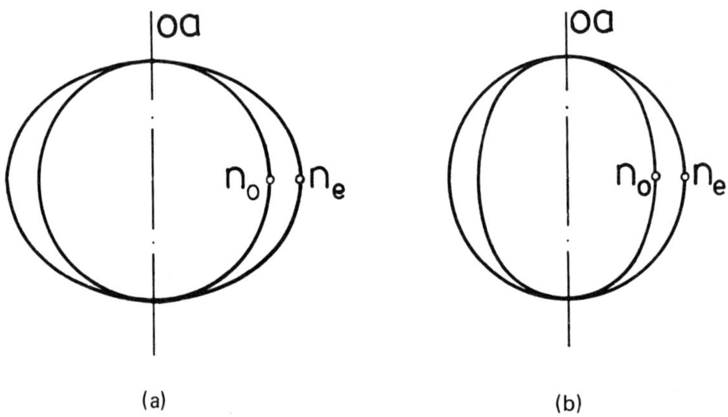


Figure 21 Schematic cross section of indicatrices of birefringent crystals. (a) Positive indicatrix. (b) negative indicatrix; oa = optic axis of the crystal, $n_o$ = refractive index for the ordinary wave, $n_e$ = refractive index for the extraordinary wave.

Table 17 Refractive Indices of Birefringent Crystals[a]

| | Quartz | | | Calcite | | |
|---|---|---|---|---|---|---|
| λ (nm) | $n_o$ | $n_e$ | Birefringence, $n_o - n_e$ | $n_o$ | $n_e$ | Birefringence, $n_o - n_e$ |
| 632.8 | 1.5426 | 1.5516 | −0.0090 | 1.6556 | 1.4849 | 0.1707 |
| 800 | 1.5383 | 1.5472 | −0.0089 | 1.6486 | 1.4821 | 0.1665 |

[a] $n_o$ = refractive index for the ordinary beam; $n_e$ = refractive index for the extraordinary beam.

### 5.2. Definition of a Quarter-Wave Plate

On traversing a birefringent plate, cut parallel to the optic axis of the crystal, the two component waves will progress with different changes in phase. The phase difference introduced by such a plate of thickness h is

$$\varphi = \frac{2\pi}{\lambda} h(n_o - n_e) \tag{69}$$

A quarter-wave plate has a phase difference of $\varphi = \pm\pi/2$, so that the plate must have a thickness

$$h = \frac{\lambda}{4} \left| \frac{2m + 1}{n_o - n_e} \right| \tag{70}$$

where m is any integer.

### 5.3. Theory of Quarter-Wave Plates

Consider a cartesian coordinate system in which a birefringent crystal plate, cut parallel to its optic axis, is oriented so that the optic axis is parallel to the y axis. Now a plane-polarized beam with the polarization plane in the azimuth angle $\alpha = \pi/4$ is incident to such a plate and enters into the crystal at point O (see Fig. 22). Then two component beams, ordinary and extraordinary, of equal amplitude propagate through the crystal. The azimuth angle indicates the angle between the vibration plane and the xz plane. The ordinary component vibrates in the x direction and the extraordinary component in the y direction.

Because of the different refractive indices $n_o$ and $n_e$, the two beams will progress with a phase difference $\varphi$ [Eq. (69)]. Thus the two waves may be described by

$$\begin{aligned} x &= a \sin \omega t \\ y &= a \sin (\omega t - \rho) \end{aligned} \tag{71}$$

If a is the amplitude of the two components, $a_0 = \sqrt{2}\, a$ is the amplitude of the incident beam; $\omega = 2\pi\nu$ is called the angular frequency and t is the time. By eliminating t, we obtain the equation of the vibration track:

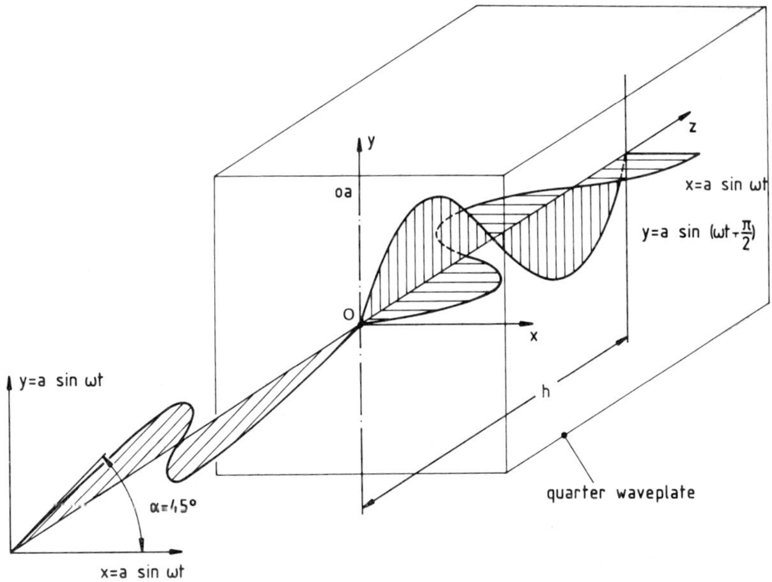


Figure 22 Schematic of a plane-polarized incident beam, oriented at 45° to the optic axis of the crystal, passing through a quarter-wave plate.

$$x^2 + y^2 - 2xy \cos \varphi - a^2 \sin^2 \varphi = 0 \tag{72}$$

The vibration track is the curve of the vibration vector when observed in the direction of propagation of the light. This equation is of the form $Ax^2 + 2Bxy + Cy^2 + D = 0$ and represents a conic section. As the discriminant $AC - B^2 \quad 0$, the conic section profile is an ellipse [see, e.g., Burri (1950)].

For a quarter-wave plate, $\varphi = \pi/2$, and Eq. (72) reduces to

$$x^2 + y^2 = a^2 \tag{73}$$

Equation (73) represents a circle; thus the resulting wave leaving the plate is circularly polarized. Figure 22 shows a schematic representation of the traversal of a plane-polarized beam, entering with an azimuth angle of 45°, within a quarter-wave plate.

If two quarter-wave plates are set one after another, the phase difference will be $\varphi = \pi$ and the vibration track becomes $(x + y)^2 = 0$. Hence it follows that

$$x = -y \tag{74}$$

The resulting wave after traversing through two quarter-wave plates, or twice through the same plate, is again a plane-polarized wave oriented such that the azimuth angle $\alpha = 3\pi/4$. The vibration plane of this wave is turned by 90° with respect to the plane of vibration of the incident beam. This feature is applied in the optical layout for data storage systems.

Let a plane-polarized beam for which in the azimuth $\alpha = \pi/4$ be incident on a quarter-wave plate with the orientation of the optic axis parallel to the yz plane. Further, let the emergent beam be reflected back from an optical disk and retraverse the same plate. Then the resulting emergent wave is plane-polarized, for which the azimuth $\alpha = 3\pi/4$. Thus the reflected beam may be distinguished from the incident beam and can be separated by modules reacting to the orientation of the plane of polarization. The general pattern of the vibration track according to Eq. (72) is an ellipse which may be transformed into the simpler principal axes equation by turning the coordinate system in accordance with the transformation equations

$$\begin{aligned} x &= x_1 \cos\alpha - y_1 \sin\alpha \\ y &= x_1 \sin\alpha - y_1 \cos\alpha \end{aligned} \tag{75}$$

in which the coordinate system $(x_1,y_1)$ is rotated by the angle $\alpha$ with respect to the coordinate system $(x,y)$. In the system $(x_1,y_1)$ the coefficient of the $x_1y_1$ term has to be equated to zero to obtain the principal axes equation. Hence to determine the rotation angle $\alpha$, the expression

$$\cos\varphi(\sin^2\alpha - \cos^2\alpha) = 0$$

is obtained.

Further, when $\cos\varphi \neq 0$ the solutions for $\alpha$ are

$$\alpha = \frac{\pi}{4}, \frac{3\pi}{4}, \ldots \tag{76}$$

When $\cos \varphi = 0$, $\varphi = \pi/2$, the resulting vibration track is circular such that the rotational angle $\alpha$ is irrelevant. For the value $\alpha = 3\pi/4$, Eq. (72) for the vibration track transforms to

$$\frac{x_1^2}{a_1^2} + \frac{y_1^2}{b_1^2} = 1 \tag{77}$$

in which the semimajor diameter $a_1$ and the semiminor diameter $b_1$ are given by

$$a_1 = \sqrt{2}\, a \sin \frac{\varphi}{2}$$
$$b_1 = \sqrt{2}\, a \cos \frac{\varphi}{2} \tag{78}$$

Figure 23 shows tracks of the vibration vector when a plane-polarized wave passes through a crystal plate. Equation (77) represents the main axes equation of an ellipse. The major diameter $2a_1$ is aligned to the azimuth $\alpha = 3\pi/4$ and thus rotated by 90° to the plane of polarization of the incident beam. From Eq. (78) the ratio of the principal diameters, that is, the ellipticity, is given by

$$\frac{b_1}{a_1} = \cot \frac{\varphi}{2} \tag{79}$$

The ellipticity of the vibration track is a function of the phase difference only. For a phase difference $\varphi = \pi$, $b_1 = 0$, and $a_1 = \sqrt{2}a$. In general, the emergent beam from such a crystal plate is elliptically polarized. For the special case when $\phi = \pi$, the elliptically polarized beam becomes a plane-polarized beam with an amplitude $\sqrt{2}\, a$, which equals the amplitude of the incident beam.

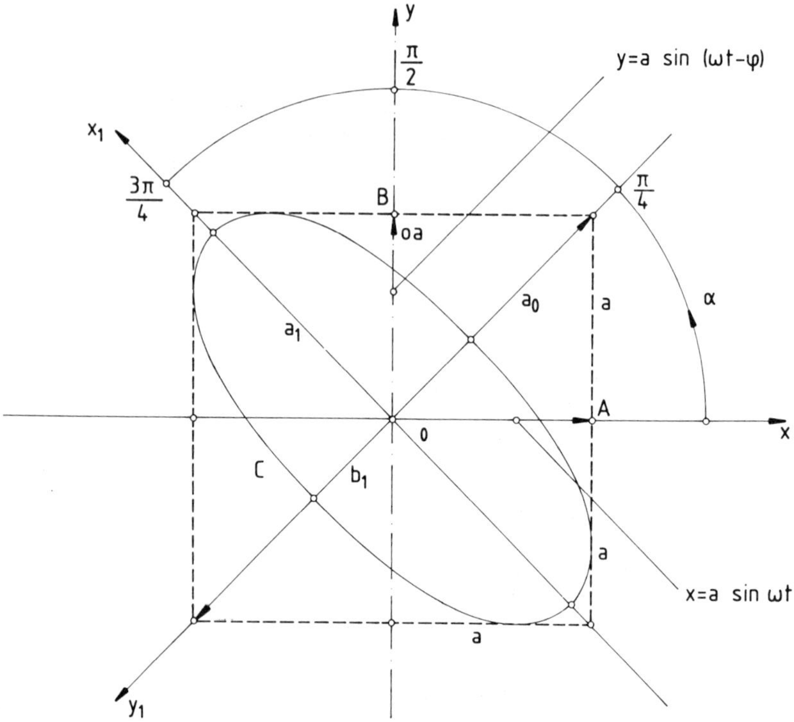


Figure 23 Tracks of the vibrating electrical vector when a plane-polarized beam passes through a crystal plate, observed in the direction of propagation. oa = optic axis of the crystal aligned to y axis; OA,OB = Incident plane-polarized beam components at right angles to each other, C = emergent elliptically polarized beam, (x,y) = initial coordinate system, ($x_1$,y1) = rotated coordinate system.

## 5.4. Engineering Tolerances of Quarter-Wave Plates

For the phase difference $\varphi = \pi + \Delta\varphi$, the ellipticity is given by

$$\frac{b_1}{a_1} = \tan\frac{\Delta\varphi}{2} = \frac{\Delta\varphi}{2} + \frac{1}{3}\left(\frac{\Delta\varphi}{2}\right)^3 + \cdots$$

and for $\Delta\varphi << \pi$, this series will be reduced to

$$\frac{b_1}{a_1} \approx \frac{\Delta\varphi}{2} \tag{80}$$

The ellipticity of the vibration track is a measure of the portion of the reflected beam that is not vibrating in the azimuth for which $\alpha = 3\pi/4$. The portion returns to the laser and may affect its light emission. The allowable portion depends on the type of laser and its operational mode. Approximately 1% is acceptable for semiconductor lasers.

If we differentiate Eq. (69), we obtain the expression

$$\Delta\varphi = \frac{2\pi}{\lambda}(n_o - n_e)\left(\Delta h - h\frac{\Delta\lambda}{\lambda}\right)$$

Hence for a constant wavelength $\lambda$, the thickness tolerance $\Delta h$ of a quarter-wave plate is given by

$$\Delta h = \frac{\lambda\Delta\varphi}{2\pi(n_o - n_e)} \tag{81}$$

Assuming an allowable phase variation $\Delta\varphi$ of approximately 0.02 for a wavelength of 0.8 μm, the thickness tolerance for a quarter-wave plate, cut from quartz parallel to the optic axis, is $\pm\Delta h <$ 0.6 μm. Inspection of Eq. (81) shows that this small tolerance will be increased only by materials having a smaller birefringence $(n_o - n_e)$. Because of its hardness quartz is a preferred optical material. With quartz a smaller effective birefringence may be achieved by cutting the plate so that its surfaces form an angle $\varepsilon$ to the optic axis (Fig. 24). For this cut the effective birefringence is given by

$$n_e(\varepsilon) - n_o = n_o\left(\frac{n_e\sqrt{1 + k^2}}{\sqrt{n_o^2 + k^2 n_e^2}} - 1\right) \tag{82}$$

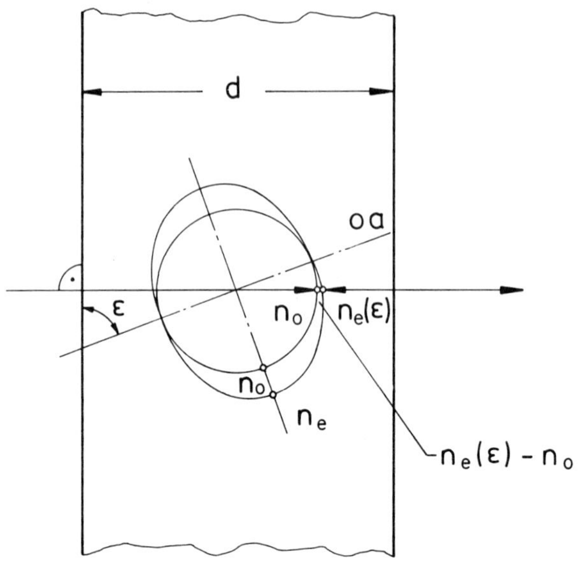


Figure 24 Orientation of the optic axis for a larger thickness tolerance for a quarter-wave plate cut from quartz. oa = optic axis, $\varepsilon$ = angle between crystal surface and optic axis; $n_o$ = refractive index of ordinary wave, indicatrix cross-section circular profile; $n_e(\varepsilon)$ = refractive index of extraordinary wave, indicatrix cross section elliptical profile.

where $k = \tan \varepsilon$. For $\varepsilon \sim 80°$ an effective birefringence of $n_e(\varepsilon) - n_o = 0.0002$ is obtained. Thus the thickness tolerance for such quarter-wave plates may be increased to $\pm\Delta h < 25\ \mu m$.

## 5.5. Adjustment

While adjusting an optical polarizing assembly of this type, special regard has to be paid to ensure that the vibration plane of the incident beam is inclined at an angle of 45° to the plane yz formed by the optic axis and the beam direction. In failing to do this, the two waves propagating through the crystal will have different amplitudes and Eq. (71) describing the two waves has to be replaced by

$$
\begin{aligned}
x &= a \sin \omega t \\
y &= b \sin (\omega t - \varphi)
\end{aligned}
\tag{83}
$$

where a and b are the amplitudes of the waves. Then the rotational angle $\alpha$ is given by

$$
\tan 2\alpha = \frac{2ab \cos \varphi}{a^2 - b^2} \tag{84}
$$

When $a \neq b$, the orientation and the length of the diameters of the resulting elliptical track are dependent on the amplitudes of the two waves in the crystal as well as on their phase difference.

## 6. POLARIZING BEAM SPLITTER

In optical data storage systems polarizing beam splitters are utilized to separate the beam reflected at the optical disk from the incident beam. The reflected beam is directed onto a detector, where the optically read digital information is converted into electrical signals.

A simple beam splitter is composed of a substrate on which there is deposited a semireflecting and semitransmitting thin film, reflecting 50% of the incident radiation and transmitting the other 50%. After the beam is reflected at the optical disk, it has to pass back through the same beam splitter. The maximum proportion of the incident radiation that can impinge onto the detector is 25% even if the absorption loss is negligible (Fig. 25). The other 25% goes back into the light source, while the remaining 50% is reflected on the first pass. Because of this low efficiency and because of the portion of light going back into the light source, this type of beam splitter should be used only with light sources which have a high enough output and with that semiconductor lasers which are not influenced by the 25% of the light energy that is reflected back into the laser. In the future suitable semiconductor lasers should become available.

After a beam has made a double pass through a quarter-wave plate, the plane of polarization of the reflected beam is at right angles to that of the incident beam. Therefore, the reflected beam, which is directed back along the incident beam, may be isolated efficiently either with beam splitter polarizers or with polarizing beam splitter layers.

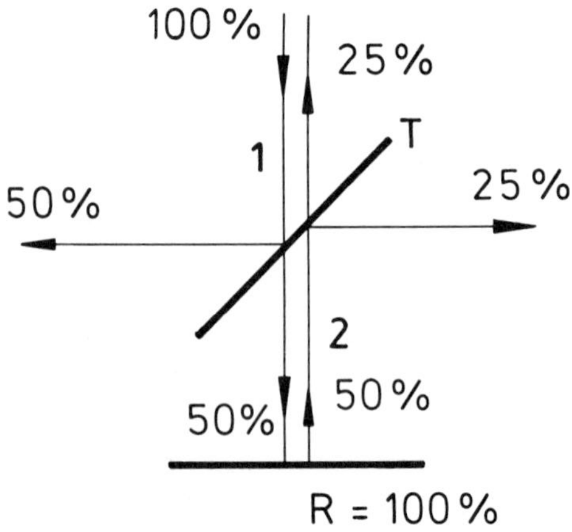


Figure 25 Schematic of a simple beam splitter. T, beam splitter; R, 100% reflector; 1, incident beam; 2, reflected beam.

### 6.1. Beam Splitter Polarizers

Beam splitter polarizers are based on the properties of birefringent crystals in which the ordinary and extraordinary waves are polarized at right angles to each other. For a beam directed normally to the optic axis of the crystal, the critical angles $\omega_o$ and $\varepsilon_o$ for total internal reflection are given by

$$\sin \omega_o = \frac{1}{n_o} \tag{85}$$

for the ordinary ray, and by

$$\sin \varepsilon_o = \frac{1}{n_e} \tag{86}$$

for the extraordinary ray. Values for these critical angles at a wavelength of 800 nm are given in Table 18 for calcite and quartz.

**Table 18** Critical Angles of Total Internal Reflection for Calcite and Quartz at a Wavelength of 800 nm

| Material | Critical angle for: Ordinary ray | Extraordinary ray |
|---|---|---|
| Calcite | 37°21' | 42°26' |
| Quartz | 40°33' | 40°16' |

### 6.1.1 Glan Polarizer

If a surface of a crystal prism is tilted so that the angle of incidence $\theta_t$ is between $\omega_o$ and $\varepsilon_o$, the ray with the higher refractive index suffers total internal reflection, while the other is transmitted. This property is used in Glan polarizers.

The Glan polarizer is composed of two calcite prisms separated at the hypotenuse face by a thin air gap. The mode of operating as a beam splitter polarizer in combination with a quarter-wave plate is shown in Figure 26. The prism's angle $\sigma$ is determined by $\sigma = \theta_t = (\omega_o + \varepsilon_o)/2$.

The returning beam reentering the Glan prism is polarized at right angles to the incident beam. It now vibrates in the crystal like an ordinary ray and will be reflected at the hypotenuse face. These beam splitter polarizers are suitable for beam divergence or convergence angles up to 8° and within a spectral region of $\Delta\lambda$, approximately 200 nm.

For light with the electric vector vibrating in the plane of incidence, the reflective $R_p$ at the boundary surface is given by Fresnel's formulas:

$$R_p = \frac{\tan^2(\theta_i - \theta_t)}{\tan^2(\theta_i + \theta_t)} \tag{87}$$

Similarly, for light with the electric vector vibrating at right angles to the plane of incidence, the reflectance $R_s$ is given by

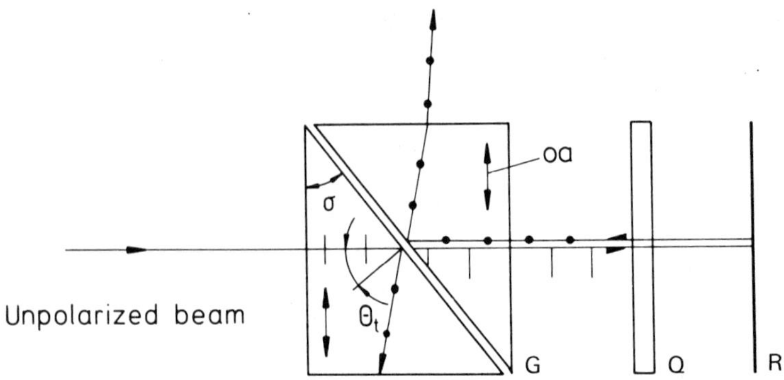


**Figure 26** Beam splitter polarizer composed of a Glan prism, a quarter-wave plate and a full reflector. G, Glan prism; Q, quarter-wave plate; R, full reflector; $\sigma$ = prism angle ($\sigma = 39°54'$ for $\lambda$ = 800 nm), ↕ = optic axis in the plane of the drawing.

$$R_s = \frac{\sin^2(\theta_i - \theta_t)}{\sin^2(\theta_i + \theta_t)} \tag{88}$$

The angles $\theta_i$ and $\theta_t$ are connected by Snell's law of refraction,

$$\sin\theta_i = n \sin\theta_t \tag{89}$$

where $\theta_i$ is the angle between the light ray and the normal to the refractive surface in air, and $\theta_t$ is the corresponding angle in the material. At near-normal incidence, Eq. (89) approximates $\theta_i = n\theta_t$ such that the difference between the parallel (p) and normal (s) reflectance components are insignificant and Eqs. (87) and (88) reduce to

$$R = \left(\frac{n-1}{n+1}\right)^2 \tag{90}$$

Semiconductor lasers are emitting plane-polarized light in which the electrical vector vibrates parallel to the p-n junction. Gain-guided lasers have typical polarization ratios above 10:1, 20:1, and 40:1 at 5, 10, and 20 mW total laser output power, respectively, as shown by Wagner et al. (1983). For a Glan polarizing beam splitter an efficiency in transmission of some 80% can be achieved for incident light that is plane-polarized in the plane of the drawing (Fig. 26). This efficiency may be improved with antireflection coatings.

### 6.1.2 Wollaston Prism

Two beams plane-polarized at right angles to each other can be separated by means of a Wollaston prism, which is often used in optical data storage technology. A Wollaston prism is composed of two rectangular prisms cut from quartz and cemented together at their hypotenuse faces (Fig. 27). To avoid total internal reflection at the hypotenuse surfaces, the refractive index of the cement used is close to the refractive indices of the quartz. In the first prism (1) the optic axis is oriented parallel to edge AB in the plane of the drawing. In the second prism (2) the optic axis is normal to the plane of the drawing.

In the first prism an unpolarized incident beam is resolved into two component beams: the extraordinary ray and the ordinary ray. The extraordinary ray's electric vector vibrates parallel

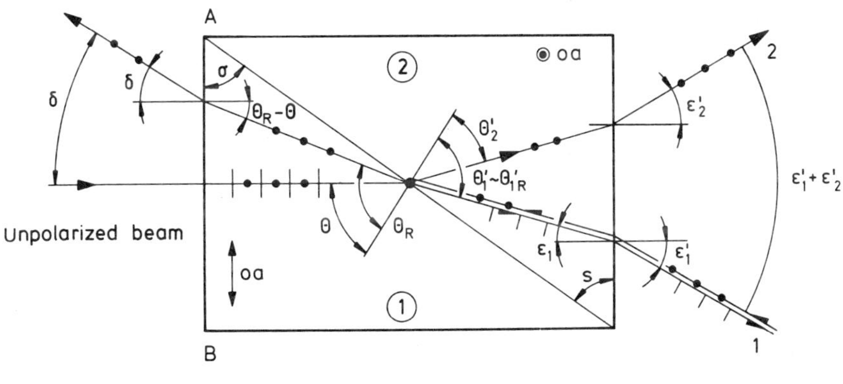


Figure 27 Wollaston prism. (1) first prism; (2) second prism. $\sigma$ = prism angle, ↕ = optic axis in the plane of the drawing, ⊙ = optic axis normal to the plane of the drawing.

to the optic axis and is subject to a refractive index $n_e$. In the second prism the vibration direction of this ray's electric vector is maintained but presents itself to the optic axis as an ordinary ray subject to a refractive index $n_o$. Thus at the hypotenuse surface there is refraction that is described by Snell's law as

$$n_e \sin \theta = n_o \sin \theta_1' \tag{91}$$

The electric vector of the ordinary ray in the first prism vibrates normally to the optic axis and will present itself to the optic axis as an extraordinary ray in the second prism with the refractive index $n_e$. For this ray the refraction at the hypotenuse surface by Snell's law is given by

$$n_o \sin \theta = n_e \sin \theta_2' \tag{92}$$

In Figure 27, by geometry the incidence angle $\theta$ equals the prism angle $\sigma$, and $\theta_1'$ and $\theta_2'$, are the angles of refraction for the first and second ray, respectively. Since $n_e > n_o$ for quartz, $\theta_1' > \theta$ for the first ray and $\theta_2' < \theta$ for the second ray. The angle of divergence $(\theta_1' - \theta_2')$ of the two beams in the crystal is symmetrical about the direction of the incident beam. This angle of divergence increases as the prism angle increases. At the boundary surface to air, $(\varepsilon_1' + \varepsilon_2')$ represents the angle of divergence of the two emergent beams and is described by

$$\varepsilon_1' + \varepsilon_2' \approx (\theta_1' - \theta_2')\left(\frac{n_o + n_e}{2}\right) \tag{93}$$

Because $\varepsilon_1' \neq \varepsilon_2'$, this angle of divergence is not quite symmetrical about the direction of the incident beam. Table 19 gives a couple of practical angles of divergence for the separation of the two emergent beams. Suppose that beam 1 reads the stored information. Then exit angle $\varepsilon_1'$ for beam 1 is determined by

$$n_o \sin \varepsilon_1 = \sin \varepsilon_1' \tag{94a}$$

where $\varepsilon_1 = \sigma - \theta_1'$.

**Table 19** Angles of Divergence of Two Beams Emerging From a Wollaston Prism Made of Quartz for Different Prism Angles at a Wavelength of 800 nm

| | Angles of divergence | |
|---|---|---|
| Prism angle, $\sigma$ | At the hypotenuse surface, $\theta_1' - \theta_2'$ | At the emergent face, $\varepsilon_1' + \varepsilon_2'$ |
| 45° | 40' | 1°2' |
| 55° | 58' | 1°30' |

The plane of polarization of the returning read beam is rotated by 90° after a double pass through a quarter-wave plate. This beam is incident at the Wollaston prism at the angle $\varepsilon_1'$ at which it emerges and is now an extraordinary ray in prism (2). The refraction at this entrance is determined by

$$\sin \varepsilon_1' = n_e \sin \varepsilon_{1R} \tag{94b}$$

where the index R indicates angles for the reflected beam. Since for quartz $n_e = n_o + 0.0089$ (Table 17), the angle $\varepsilon_{1R}$ differs from $\varepsilon_1$ by a similar portion. Thus at the hypotenuse face the angle of incidence $\theta_{1R}$, for the reflected beam is similar to $\theta_1'$ by the same proportion; hence $\theta_{1R} = \theta_1'$. Then from the refraction of the reflected beam at the hypotenuse face we obtain

$$n_e \sin \theta_{1R}' = n_o \sin \theta_R \tag{95}$$

By geometry the angle of incidence of the reflected beam at the front surface AB is $(\theta_R - \theta)$. Finally, the deflection angle $\delta$ in air to the direction of the incident beam is obtained by Snell's law of refraction:

$$\sin \delta = n_o \sin (\theta_R - \theta) \tag{96}$$

**Table 20** Deflection Angle for a Wollaston Prism Made of Quartz at a Wavelength of 800 nm

| Prism angle, $\sigma$ | Angle of deflection, $\delta$ |
|---|---|
| 45° | 1°2' |
| 55° | ∿1°30' |

Values for this deflection angle $\delta$ for different prism angles $\sigma$ are given in Table 20. If two Wollaston prisms are set one after another, the deflection angle is increased twofold. A technical design of a Wollaston prism with a larger beam deflection is shown in Figure 28. For an appropriately oriented plane-polarized incident light beam the transmittance of a Wollaston-type beam splitter polarizer is about 83%. Since there are two crystal-air interfaces, it is easy to improve the transmittance of the beam splitter by depositing antireflection coatings on these two interfaces.

## 6.2. Polarizing Beam Splitter Layers

Polarizing beam splitter layers are based on the interference at thin film layers and on the behavior of optical boundary surfaces near the Brewster angle.

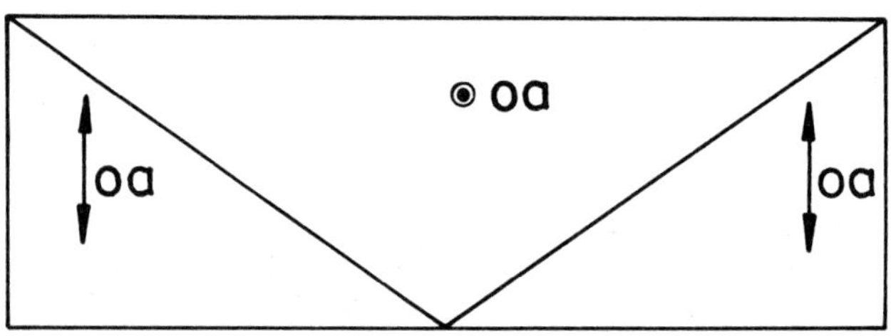


**Figure 28** Wollaston-type beam splitter polarizer with a greater beam deflection. ↕ = optic axis in the plane of the drawing, ⊙ = optic axis normal to the plane of the drawing.

6.2.1. Interference at Thin Layers

Let a light beam passing through a medium with a refractive index $n_o$ be incident on a thin layer with a refractive index $n_1$, which is deposited on a substrate with refractive index $n_2$. To determine reflectance and transmittance, the phase behavior in the thin layer has to be considered. As shown in Figure 29, the optical path difference $\Delta s$ between the reflected rays 1 and 2 is given by

$$\Delta s = 2n_1 h_1 \cos\theta_1 \qquad (97)$$

where $h_1$ is the layer thickness and $\theta_1$ the angle of refraction in the thin layer. Thus the phase difference $\varphi$ existing between the two rays is

$$\varphi = \frac{2\pi}{\lambda}(2h_1 n_1)\cos\theta_1 \qquad (98)$$

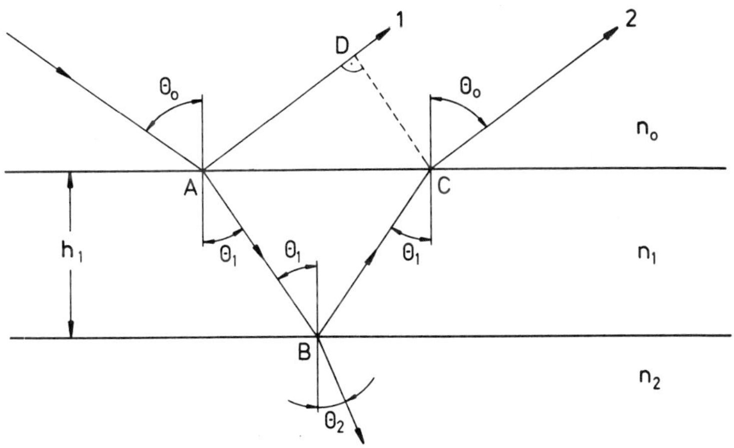


Figure 29 Path of the reflected and the transmitted rays for a beam incident obliquely at a thin layer. Optical path difference between rays 1 and 2: $[n_1 (AB + BC) - n_o AD]$.

With a phase difference of $\phi = \pi$, for the superimpose of two beams with equal amplitude described by

$$a \sin \omega t + a \sin (\omega t + \pi) = 0$$

no reflected wave arises. For this from Eq. (98) the condition for the thickness of the optical layer can be obtained as

$$n_1 h_1 = \frac{\lambda}{4 \cos \theta_1} \tag{99}$$

For $\theta_1 = 0$ we obtain $n_1 h_1 = \lambda/4$, the condition at normal incidence for the cancellation of two interfering reflected beams with a single layer coating.

To achieve a path difference of $\lambda/2$ and cancellation at oblique incidence, the thickness of the optical layer is, by Eq. (99), larger by the factor $(\cos \theta_1)^{-1}$ than for normal incidence. Including the multiple reflections at the two boundaries of the layer, for complete cancellation the intensities of light reflected at the two boundaries must be equal. This implies that at normal incidence the ratios of the refractive indices at each boundary are equal: namely, $n_o/n_1 = n_1/n_2$, or

$$n_1 = \sqrt{n_o n_2} \tag{100}$$

If for air as the first medium, $n_o = 1$, the condition reduces to $n_1 = \sqrt{n_2}$. Therefore, for a single layer a coating material with a low refractive index is required to cancel the reflected beams. For this purpose $MgF_2$, for which $n_1 = 1.38$, is normally used. For oblique incidence we need to distinguish between the two components $R_p$ and $R_s$ as given in Eqs. (87) and (88), respectively. For dielectric thin layers free of absorption, the transmittances for the two components are given by

$$T_p = 1 - R_p \tag{101}$$

and

$$T_s = 1 - R_s \tag{102}$$

6.2.2. Brewster's Angle

The denominators in Eqs. (87) and (88) are finite except when $(\theta_i + \theta_t) = \pi/2$. Then $\tan(\theta_i + \theta_t) = \infty$ and consequently $R_p = 0$. In this case the reflected and transmitted rays are perpendicular to each other (Fig. 30). Since $\sin\theta_t = \sin[(\pi/2) - \theta_i] = \cos\theta_i$, it follows from the law of refraction, that

$$\tan\theta_B = \frac{n_t}{n_i} = n_{it} \tag{103}$$

where $\theta_B$ is called the polarizing angle or Brewster's angle. $n_{it}$ is the ratio of the refractive indices of media i and t. Values for the Brewster's angle and the associated reflectances $R_p$ and $R_s$ as well as the associated transmittances $T_p$ and $T_s$ are given in Table

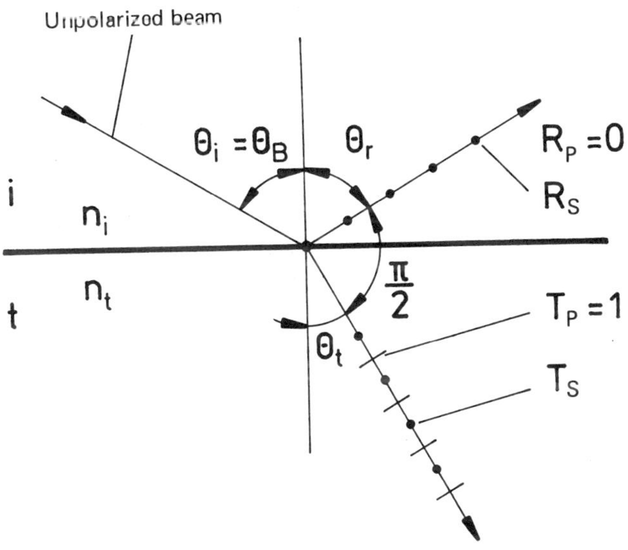


**Figure 30** Illustrating the polarizing angle (Brewster's angle). $(\theta_i = \theta_B)$ = angle of incidence = Brewster's angle; $\theta_t$ = angle after refraction, $\theta_r$ = angle of refraction $(\theta_r = \theta_i)$; $(\theta_i + \theta_t = \pi/2$ = reflected beam normal to the transmitted beam.

Table 21 Brewster's Angle and Values for the Reflectance and the Transmittance for Several Refractive Indices $n_{it}$[a]

| $n_{it}$ | $\theta_B$ | $R_p$ | $R_s$ | $T_p$ | $T_s$ |
|---|---|---|---|---|---|
| 1.4 | 54°28' | 0 | 0.11 | 1 | 0.89 |
| 1.6 | 58° | 0 | 0.19 | 1 | 0.81 |
| 1.8 | 60°57' | 0 | 0.28 | 1 | 0.72 |
| 2.0 | 63°27' | 0 | 0.36 | 1 | 0.64 |

[a] $n_{it}$ = ratio of the refractive indices of medium t and medium i, $\theta_B$ = Brewster's angle, $R_p$ = reflectance of the p wave, $R_s$ = reflectance of the s wave, $T_p$ = transmittance of the p wave, $T_s$ = transmittance of the s wave, p = direction of the electric vector vibrating in the plane of incidence, s = direction of the electric vector vibrating in a plane normal to the plane of incidence.

21. The transmittance $T_p = 1 - R_p$ is the proportion of light for which the electric vector vibrates in the plane of incidence and which is transmitted at a boundary surface. $T_s = 1 - R_s$ is the transmitted portion whose electric vector vibrates at right angles to the plane of incidence. If unpolarized light is incident at the Brewster's angle on an interface of two optical media for which the ratio of the refractive indices is $n_{it}$, the reflectance $R_p$ vanishes.

The degree of polarization $P_R$ of the reflected beam is defined by

$$P_R = \frac{R_p - R_s}{R_p + R_s} \tag{104}$$

and the degree of polarization $P_T$ of the transmitted beam is given by

$$P_T = \frac{T_p - T_s}{T_p + T_s} \tag{105}$$

For the Brewster's angle the reflected wave component is plane-polarized and from Eq. (104) the degree of polarization is $P_R = -1$. For the transmitted beam the degree of polarization is $P_T = 0.16$, $n_{it} = 1.8$. Because there is no reflected wave component $R_p$, the transmitted wave component $T_p$ is larger than the component $T_s$. For this reason there is also no loss of light by reflection when a plane-polarized light beam whose electric vector vibrates in the plane of incidence is transmitted through and is incident at Brewster's angle at the interface of two optical media.

### 6.2.3. Theory of Polarizing Beam Splitter Layers

The first polarizing beam splitter layer was designed by Banning (1947) using the principle of Brewster's angle. The s-polarized light is partially reflected and transmitted. To increase the s reflectance, retaining the p transmittance near unity, two or more materials are added to form a multilayer stack. Each layer thickness of the stack should be made one quarter-wave for the appropriate angle of incidence. In this stack alternating layers should be of high and low refractive index. Light that is reflected within a high-index layer will not suffer any phase shift on reflection, whereas at the reflection within a low-index layer a phase shift of $\pi$ will occur. Accordingly, the various components of the incident light produced by reflection at the successive boundaries will reappear at the entering surface all in phase, and hence they will recombine constructively. If a beam splitter is placed within a divergent or convergent optical beam, the asymmetric aberrations in the point image can be avoided only by a use of a beam splitter in the form of a planoparallel plate oriented perpendicular to the optical axis of the incident beam. The spherical aberration caused by this arrangement must be corrected for in the design of the entire imaging system to obtain a nearly diffraction-limited image. Therefore, polarizing beam splitters are built of two 45° right-angle prisms assembled together at their hypotenuse faces to form a cube. The multilayer stack is sandwiched between the hypotenuse faces.

Because it is shown in Figure 30 that at Brewster's angle $\theta_i + \theta_t = \pi/2$, the condition for the polarizing angle may be written as

$$\frac{n_H}{\cos \theta_H} = \frac{n_L}{\cos \theta_L} \qquad (106)$$

where n represents the refractive indices and θ the angles of incidence. The subscripts H and L refer to the high- and low-refractive-index materials, respectively.

Furthermore, for a ray traveling from one optical medium to another, the law of refraction may be written as

$$n_G \sin \theta_G = n_H \sin \theta_H = n_L \sin \theta_L \qquad (107)$$

Here the subscript G refers to the glass substrate. A polarizing beam splitter built in accordance with this principle is a cube composed of the two glass prisms with a stack of quarter-wave thin-film layers HLH ··· in between. There are an odd number of layers.

A schematic representation of the assembly of such a polarizing cube beam splitter is shown in Figure 31. If Eq. (106) is satisfied, it is guaranteed that within the layer stack the p-reflection component vanishes. To arrange that the beam is incident at the multilayer stack at Brewster's angle, the refractive index of the glass for the prisms of a polarizing cube beam splitter may be derived from Eqs. (106) and (107) to give

$$n_G = \frac{\sqrt{2}\, n_H n_L}{\sqrt{n_H^2 + n_L^2}} \qquad (108)$$

However, in this derivation Brewster's law is not fulfilled rigorously at the glass boundary to the first layer. This could be achieved only if the refractive indices of the glass and the second layer were equal.

Nonhygroscopic materials, which can be deposited by vacuum evaporation, are needed to manufacture durable multilayers which consist of alternating quarter-wave layers of high and low refractive index. These materials must have a high transmittance in the given spectral region and must be mechanically strong. Table 22 lists optical properties for some materials and an appropriate deposition technique.

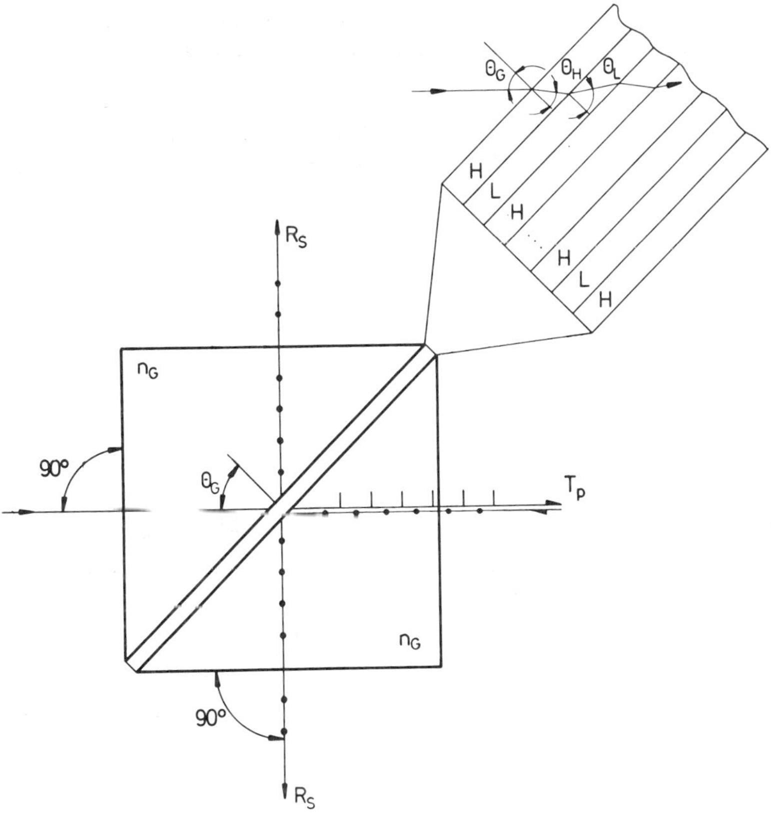


**Figure 31** Schematic of a polarizing beam splitter cube. H, layer with a high refractive index $n_H$; L, layer with a low refractive index $n_L$. $n_G$ = refractive index of a glass prism.

According to Eq. (108), for alternative thin-film depositions of ZnS and $MgF_2$, a refractive index of 1.68 is needed for the glass prism. The reflectance and the transmittance of such a polarizing beam splitter may be calculated. The reflectance is given by

$$R = \left( \frac{\eta_G^2 - \eta_H^2 (\eta_H / \eta_L)^{\nu - 1}}{\eta_G^2 + \nu_H^2 (\eta_H / \eta_L)^{\nu - 1}} \right)^2 \qquad (109)$$

Table 22 Characteristics of Thin-Layer Materials, Some with a Low and Some with a High Refractive Index, Suitable for a Thin-Film Multilayer Stack

| Material | Refractive index (visible) | Useful wave-length region (μm) | Evaporation technique |
|---|---|---|---|
| Cryolite ($Na_3AlF_6$) | 1.35 | 0.2–14 | Tantalum boat or horwitzer |
| Lithium fluoride (LiF) | 1.36 | 0.11–7 | Tantalum boat |
| Magnesium fluoride ($MgF_2$) | 1.38 | 0.2–10 | Tantalum boat |
| Silicon dioxide ($SiO_2$) | 1.46 | 0.2–8 | Electron bombardment |
| Titanium dioxide ($TiO_2$) | 2.30 (2.2–2.7, depending on structure) | 0.4–12 | Electron bombardment |
| Cerid oxide ($CeO_2$) | 2.35 | 0.4–15 | Tungsten boat |
| Zinc sulfide (ZnS) | 2.35 | 0.4–15 | Tantalum boat or horwitzer |
| Zinc selenide (ZnSe) | 2.58 | 0.6–15 | Platinum boat |

where for the p wave the η values are given by

$$\eta_G = n_G/\cos\theta_G$$
$$\eta_H = n_H/\cos\theta_H \qquad (109a)$$
$$\eta_L = n_L/\cos\theta_L$$

and for the s wave by

$$\begin{aligned} \eta_G &= n_G \cos \theta_G \\ \eta_H &= n_H \cos \theta_H \\ \eta_L &= n_L \cos \theta_L \end{aligned} \tag{109b}$$

where $\nu$ is the number of layers as shown by Macleod (1969). Therefore, with the imposed condition for Brewster's angle within the multilayer stack, it follows from Eqs. (106) and (109a) that $\eta_H = \eta_L$. If we use this relationship and substitute the set of equations (109a) into Eq. (109), the reflectance $R_p$ of the beam splitter for the p wave is given by

$$R_p = \left(\frac{n_G^2 \cos^2 \theta_H}{n_H^2 \cos^2 \theta_G} - 1\right)^2 \div \left(\frac{n_G^2 \cos^2 \theta_H}{n_H^2 \cos^2 \theta_G} + 1\right)^2 \tag{110}$$

Similarly, by substituting the set of equations (109b) into Eq. (109), the reflectance $R_s$ for the s wave is given by

$$R_s = \left[\frac{n_G^2 \cos^2 \theta_G}{n_H^2 \cos^2 \theta_H} - \left(\frac{n_H}{n_L}\right)^{2(\nu - 1)}\right]^2 \div \left[\frac{n_G^2 \cos^2 \theta_G}{n_H^2 \cos^2 \theta_H} + \left(\frac{n_H}{n_L}\right)^{2(\nu - 1)}\right]^2 \tag{111}$$

Since Brewster's condition is fulfilled in the multilayer stack, there are no p-wave reflections from the stack. Thus $R_p$ represents the reflectance of the p wave at the interface between glass and first layer with $n_H$. Therefore, the contribution to the reflectance from the multilayer stack is nonexistent in Eq. (110). The reflectance of the s wave increases as the number of layers increases and the higher the ratio of the refractive indices $n_H/n_L$ becomes. It has become the state of the art to deposit many layers.

In general, the separation of the p- and s-wave components with many layers offers some advantages. The angle of incidence is determined by the geometry of a cube, and hence, according to Eq. (108), also the refractive index of the prisms. To obtain a low reflectance for the p wave, the refractive indices $n_H$ and $n_L$ ought not to be too high or too low, respectively. This requirement is the opposite of the previous demand for a high $n_H/n_L$ ratio. The effects of different $n_H$ and $n_L$ values, and hence the value of $n_G$ for achievable reflectances of the p wave, are given in Table 23.

For given refractive indices of the layer stack, the reflectance of the s-wave may be increased, especially by increasing the number of layers deposited. This effect is shown in Table 24 for a multilayer stack with $n_H = 2.35$ (for ZnS), $n_L = 1.38$ (for $MgF_2$), and $n_G = 1.68$. Using the values given in Tables 23 and 24 together with Eq. (104) it can be calculated that a stack of seven or nine layers produces a degree of polarization $P_T$ of 0.996 or 0.999, respectively.

Such a polarizing beam splitter is used in optical data storage systems in combination with a quarter-wave plate. Consider a p-wave transmitted through a polarizing beam splitter. The vibration plane of the returning beam from the quarter-wave plate is rotated in azimuth by 90°. Therefore, the beam reenters the cube as an s wave and will be reflected from the multilayer stack due to the high reflectance $R_s$.

### 6.2.4. Performance of Polarizing Beam Splitter Layers

Figure 32 shows typical transmittance and reflectance characteristics for the p wave and the s wave of a polarizing beam splitter with 13 layers as a function of the angle of incidence $\theta_G$ in glass. These transmittance and reflectance characteristics allow such polarizing beam splitters to be used for beam divergence angles in glass up to a $\Delta\theta_G$ value of approximately ±5°, which corresponds to a beam divergence angle in air of about ±8°. In optical data storage systems the light emitted by a semiconductor laser may cover a spectral region of ±20 nm under working conditions.

To achieve high degrees of polarization $P_R$ and $P_T$ in the spectral region used, multilayer stacks with steep slopes in the transmittance curve $T_p(\lambda)$ and $T_s(\lambda)$ are required. To achieve

Table 23 Effects of Different Refractive Indices $n_H$ and $n_L$ on the Reflectance and Transmittance of the p Wave (Calculated Values)

| Low refractive index, $n_H$ | High refractive index, $n_L$ | $\frac{n_H}{n_L}$ | Refractive index of substrate, $n_G$ | Reflectance,[a] $R_p$ | Transmittance[a] $T_p$ |
|---|---|---|---|---|---|
| 2.35 | 1.38 | 1.70 | 1.68 | 0.02 | 0.98 |
| 2.30 | 1.38 | 1.67 | 1.67 | 0.015 | 0.985 |
| 2.30 | 1.46 | 1.58 | 1.74 | 0.01 | 0.99 |

[a] According to Eq. (110), $R_p$ and $T_p$ are independent of the number of layers $\nu$.

Table 24 Calculated Values for the Reflectance and Transmittance of the s Wave from a Polarizing Beam Splitter as a Function of the Number of Layers[a]

| Number of layers, $\nu$ | Reflectance, $R_s$ | Transmittance, $T_s$ |
|---|---|---|
| 1 | 0.24 | 0.76 |
| 3 | 0.85 | 0.15 |
| 5 | 0.98 | 0.02 |
| 7 | 0.998 | 0.002 |
| 9 | 0.9997 | 0.0003 |

[a] $n_H = 2.35$, $n_L = 1.38$, $n_G = 1.68$/

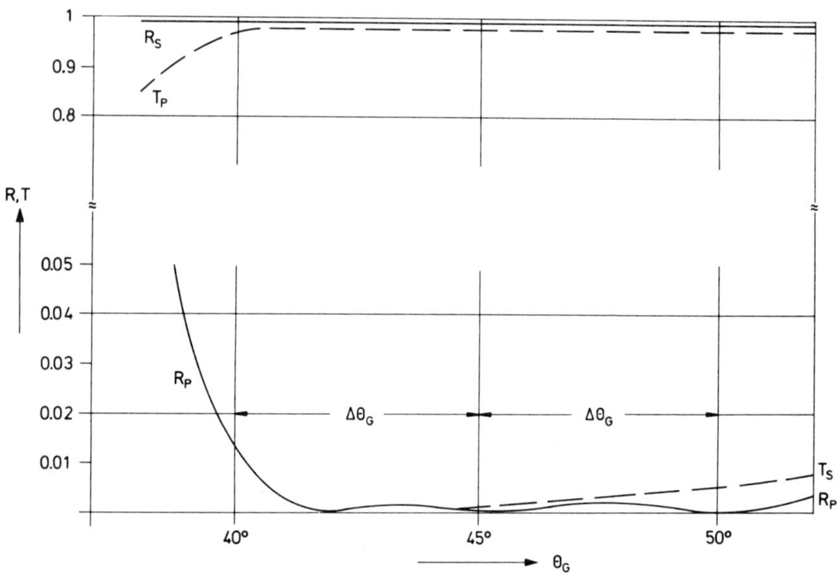


Figure 32 Transmittance and reflectance characteristics of a cubic polarizing beam splitter with thirteen layers as function of angle of incidence $\theta_G$ in glass. $T_p$ = transmittance of the p wave, $R_p$ = reflectance of p wave, $T_s$ = transmittance of the s wave, $R_s$ = reflectance of the s wave, p = direction of the electric vector vibrating in the plane of incidence, s = direction of the electric vector vibrating normal to the plane of incidence.

steep slopes in an extended spectral region, such polarizing beam splitter layers are made with some 30 layers. Figure 33 gives the characteristics for the p wave and s waves of a polarizing beam splitter as a function of wavelength.

If beam splitters are used in a nearly parallel light beam, the polarizing beam splitter multilayer stack may be deposited at the entrance surface of a single prism. As shown in Figure 17, prisms so coated are used simultaneously as anamorphic beam expanders. During the design of the prism, a compromise for the incident angle $\theta_1$ is made between the demand to fulfill Brewster's law and the required anamorphic expansion ratio according to Eq. (53). A schematic representation of such a prism is shown in Figure 34. The multilayer stack is deposited on surface 1. All

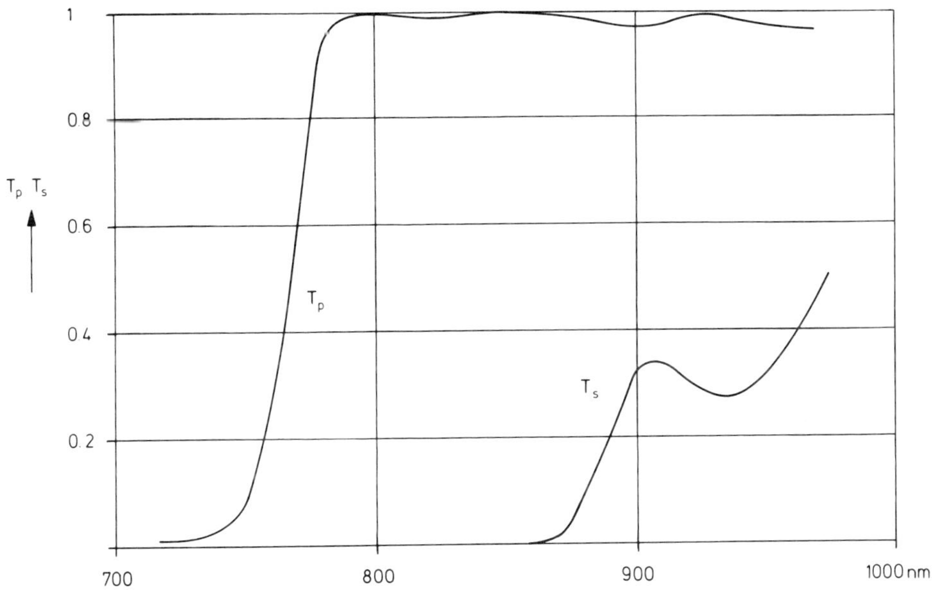


**Figure 33** Transmittance characteristics of a multilayer polarizing beam splitter as function of wavelength λ for the spectral region 780–860 nm. $T_p$ = transmittance of the p wave, $T_s$ = transmittance of the s wave. (Courtesy of OCLI Europe.)

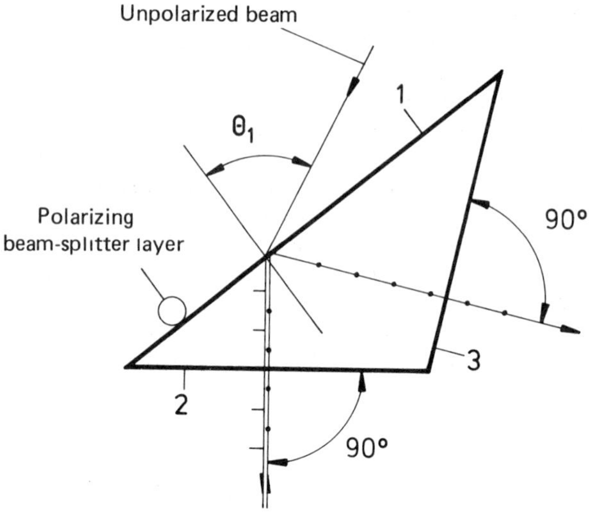


Figure 34 Polarizing beam splitter prism used simultaneously as an anamorphic beam expander. ○ = polarizing beam splitter layer, $\theta_1$ = angle of incidence.

the other prism surfaces are oriented orthogonally to the directions of the refracted and reflected beam. A returning beam through surface 2 with s-wave polarization is reflected at surface 1 and emerges through surface 3. Hence, for a beam in this direction, the prism acts a a planoparallel plate and may be used in divergent or convergent optical beams, provided that it is included in the design of the entire optical imaging system.

## 7. AUTOMATIC TRACE TRACKING

### 7.1. Automatic Focusing

For writing as well as for reading, the data storage film plane has to remain within the depth of focus of the objective lens when the optical disk is rotating. The specification range 1–2 μm for defocusing is so small that automatic focusing is necessary. The servo control system requires a signal that changes sign when the data storage film plane crosses the equilibrium position. In general, it is desirable that there be a nearly linear relationship between the signal amplitude and the displacement of the data storage film plane from the position of best focus and equilibrium. As shown in Figure 1, the beam reflected from the optical disk OD is deviated at the beam splitter T and then focused onto detector D. If the beam splitter is placed in an uncollimated light beam, the objective lens also may be used for focusing; however, if it is placed in a collimated light beam, an additional lens is required. Several methods exist for automatic focusing. The optical principles in the methods frequently used are now described.

### 7.1.1. Astigmatic Method

To obtain a signal response, for best focus a cylindrical lens is added to the focusing lens in the reflected beam. Adding a cylindrical lens produces anamorphic imaging, where the original point image is replaced by two axially separated focal lines. The first line lies in the focal plane of the spherical system; the second line, which is rotated in azimuth by 90°, lies in the focal plane of the combination formed by the spherical system and the cylindrical lens. If $f_1$ is the focal length of a spherical lens and $f_2$ the focal length of a cylindrical lens, the focal length f of the pair is given by

$$\frac{1}{f} = \frac{1}{f_1} + \frac{1}{f_2} - \frac{e}{f_1 f_2} \qquad (112)$$

where e is the principal-point separation between the two lenses. For two thin lenses close together, e becomes negligibly small and Eq. (112) reduces to

$$\frac{1}{f} = \frac{1}{f_1} + \frac{1}{f_2} \qquad (113)$$

The image profile changes uniformly between the saggital and the tangential focal lines. Therefore, midway between the lines the image formed is a uniform spot which normally has a circular profile. The distance between these two focal lines is the range for automatic focusing and has to be chosen so that it covers the defocusing range of the optical disk under dynamic conditions. The focal length $f_2$ of the cylindrical lens needed for an astigmatic difference $\Delta z$

$$\Delta z = f_1 - f \qquad (114)$$

is given in Table 25 for a 5-mm-focal-length spherical lens.

To distinguish the shape of the image spot a quadrupole position sensor is used. This sensor is composed of four quadrants, A, B, C, and D. The circular spot in the middle of the astigmatic difference is adjusted so that all four quadrants receive equal portions of the light. This is shown in Figure 35a. The two focal lines in a real system are elliptically shaped. The first one covers preponderantly quadrants B and C (Fig. 35b), and the second one covers quadrants A and D (Fig. 35c). The control signal for automatic focusing is obtained from a difference circuit connection formed between the sum of the diametrically opposing quadrants. The sign of this signal for different focal positions is given in Table 26.

The output signal of the quadrupole sensor is almost proportional to small displacements of the data storage film plane. The focusing process is continued until the sensor will deliver no more error signal. During this focusing process high accelerations may occur. Consideration must be given in the mechanical design of the actuator for coping with these accelerations. The sum of the

Table 25 Focal Length of a Cylindrical Lens to Compensate for the Astigmatic Difference in Combination with a Spherical Lens of 5-mm Focal Length

| Astigmatic difference, $\Delta z$ (μm) | Focal length of cylindrical lens, $f_2$ (mm) |
|---|---|
| 25 | 1000 |
| 50 | 500 |
| 100 | 250 |
| 200 | 125 |

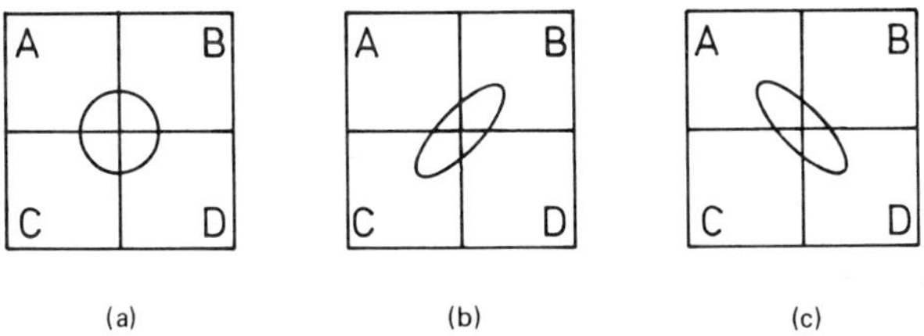


**Figure 35** Image profiles on a quadrupole position sensor composed of four quadrants A, B, C and D when (a) the optical disk is at the correct distance; (b) the position of the optical disk is too close; and (c) the position of the optical disk is too far.

four signals (A + B + C + D) emitted by the four quadrants represent the information being read. Figure 36 gives the circuit connections of a quadrupole position sensor with the output signals for automatic focusing and for the information being read.

### 7.1.2. Wobble Method

Another method for automatic focusing is by giving an axial wobble to the objective lens. The cross-sectional area of the beam varies as a function of the displacement $\Delta z$ from the focal position. Hence the ratio of the size of the holes, called pits, in the data storage film, to the beam cross sectional area varies as a function of $\Delta z$. This variation is in phase with the wobble

**Table 26** Output Signal of a Quadrupole Position Sensor Used for Automatic Astigmatic Focusing

| Circuit connection of a quadrupole position sensor | Signal | Shape of image spot |
|---|---|---|
| | <0 |  |
| (A + D) − (B + C) | =0 |  |
| | >0 |  |

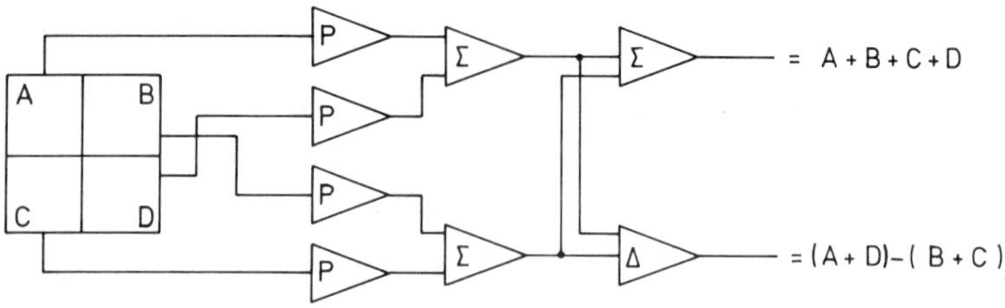


Figure 36 Circuit connections for a quadrupole position sensor. Σ = sum, Δ = difference, P = preamplifier, A + B + C + D = signal with the information being read, (A + D) − (B + C) = signal for automatic focusing.

on the one side of the focus and in phase opposition on the other side. The normalized intensity along the optical axis is given by Eq. (20). Hence it follows that because of defocusing, the intensity variation for small values of $\Delta z$ is a quadratic function in $\Delta z$ described by

$$I(z_0) - I(z) = \Delta I = \frac{1}{12} \frac{\pi^2}{\lambda^2} (NA)^4 (\Delta z)^2 \tag{115}$$

where z is the coordinate in the direction of the optical axis, $z_0$ the coordinate for best focus. The derivative $d(\Delta I)/d(\Delta z)$ of Eq. (115) is linear to the defocusing displacement $\Delta z$. Thus a corresponding signal may be obtained for servo loop control. A typical value for the amplitude of the wobble is 0.1 μm which is much smaller than the specification acquisition range 1–2 μm.

A wobble at high frequency may also be obtained by introducing vertical periodic relief along the progressing sequence of pits in the disk, called a depth-modulated track. Normally, the wobble is used for smaller ranges of defocusing.

### 7.1.3. Skew Beam Method

Another method for automatic focusing is the skew beam method, applied by Bouwhuis and Braat (1978) for the automatic focusing of optical disks. A schematic representation of the principle of this method is shown in Figure 37. Let us assume an off-axis

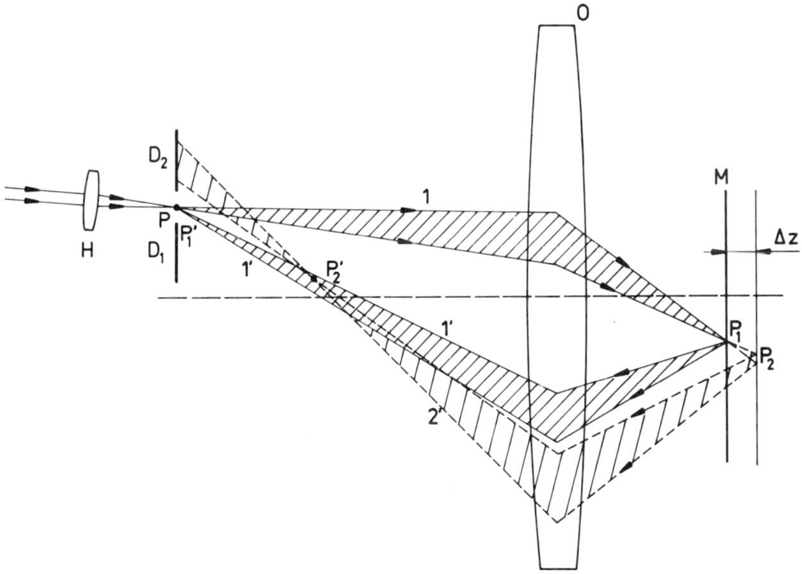


Figure 37 Schematic of the skew beam method for automatic focusing. H, auxiliary lens; $D_1$ and $D_2$, semidetectors; O, lens; M, mirror; 1, incident beam; 1', beam reflected at mirror M; 2', beam reflected at plane defocused by a value $\Delta z$.

object point P, illuminated by an auxiliary lens H, which is located between two semidetectors. This point is imaged by the lens O at the image point $P_1$ on the mirror M. In optical data storage technology the mirror M represents the data storage film plane. The reflected beam 1' from the mirror reimages the point $P_1$ at the image point $P_1'$, which is coincident with the object point P. Its spot covers equally semidetectors $D_1$ and $D_2$.

If, however, the mirror is displaced by an amount $\Delta z$ from the image plane of $P_1$, beam 1 strikes the mirror around $P_2$ and is reflected. The reflected beam 2' forms an image at $P_2$ of the object point $P_1$, and a defocused spot finally strikes the two detectors asymmetrically. Now the two detector signals are different and may be used to control the servo loop for automatic focusing. It should be mentioned that for automatic focusing by the skew beam method, the objective lens is required to have a flat field image surface.

### 7.1.4. Modified Foucault's Knife Edge Method

A further method for automatic focusing is based on a modification of Foucault's knife edge test. Instead of a knife edge, a biprism with a large apex angle is used. The beam reflected at the data storage film plane is divided by the biprism. Then, with either one of these two beams a control signal for automatic focusing may be obtained.

## 7.2. Radial Tracking

The information on an optical disk is contained in a spiral with an average track pitch of 1.6 μm. For disks rotating at speeds of 200 rpm up to as high as 1500 rpm, the reading light beam has to remain focused on the track with a radial accuracy of 0.1 μm. This accuracy cannot be achieved with mechanical guidance systems. Feedback systems using either a pivoting mirror in front of the objective lens (Fig. 12a) or a radial reset of the objective lens (Fig. 12b) are required for radial tracking. The deviation of the beam from the center of the track is read optically to obtain a signal for controlling the servo loop. Some of the methods used for radial tracking are related to those used for automatic focusing.

### 7.2.1. Method with Auxiliary Beams

In this method of automatic tracking two auxiliary beams are used which are slightly displaced from the center-line of the track: one to the left side of the track and the other to the right side so that each is partly on and partly alongside the track (Fig. 40). A diffraction grating may be used to split the incoming beam into a main beam for reading and two auxiliary beams for radial tracking. The theory of a diffraction grating will be described first.

A light beam incident on a grating is diffracted (Fig. 38). The maxima of the intensity distribution in the diffracted beams lie in the directions for which the optical path difference of the rays striking two neighboring grooves [AD−CB] is an integer number of the wavelength λ. The direction of the Fraunhofer diffraction maxima may be derived from Figure 38 and are determined by

$$\sin \Theta_1 - \sin \theta_0 = \frac{m\lambda}{d} \qquad m = 0, \pm 1, \pm 2, \ldots \qquad (116)$$

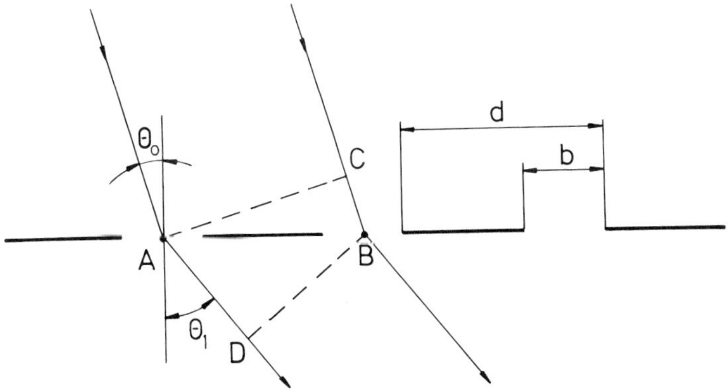


**Figure 38** Diffraction grating. d = grating spacing, b = width of a transparent groove, $\theta_0$ = angle of incidence, $\theta_1$ = angle of diffraction.

where d represents the grating spacing, $\theta_0$ the angle of maximum incidence, $\theta_1$ the angle of maximum diffraction, and m is called the order of interference. For m = 0 we obtain the position of the central maximum, which lies in the direction of the incident beam. The intensity of the secondary maxima reduces as the order of interference increases. The envelope for these maxima is given by the normalized diffraction pattern of a single slit of width b and may be written as

$$I_0(p) = \left[\frac{\sin(\pi bp/\lambda)}{\pi bp/\lambda}\right]^2 \tag{117}$$

where $p = \sin\theta_1 - \sin\theta_0$.

$I_0(p)$ of Eq. (117) first reaches zero when $p = \lambda/b$. For a grating with $b = d/2$, a grating in which the width of the transparent grooves is half the grating spacing, only the diffraction maxima of the first order have an intensity worth mentioning. Their direction is given by

$$p = \pm\frac{1}{2}\frac{\lambda}{b} \tag{118}$$

and the relative intensity is 64% of the intensity for the central maximum. The energy lost in this type of diffraction grating is 50% because the width of the transparent and opaque portions are equal. A suitably designed phase grating is almost lossless. The same type of beam splitting as already described is obtained by a phase grating that has a groove depth of $\lambda/n$, where n represents the refractive index of the substrate material. Figure 39 shows a cross-sectional profile of a phase grating. The application of a phase grating for generating two auxiliary beams for radial tracking in optical disk technology was described by Braat and Bouwhuis (1978). Almost the entire energy of the incident beam is concentrated in the zero- and first-order diffraction maxima; thus there is no more unwanted radiant background noise. When using a grating, the linear deflection $\Delta y$ in the focal plane of an objective lens of the first-order diffracted beam maxima with respect to the central on-axis zero-order beam maximum is given by

$$\Delta y \approx \pm f\frac{\lambda}{d} \tag{119}$$

where f represents the focal length of the objective lens.

The principle of radial tracking with two auxiliary beams is shown in Figure 40. After reflection at the optical disk, the two auxiliary beams, E' and F', each strike an additional photodiode, one arranged on the left side and one the right side of the

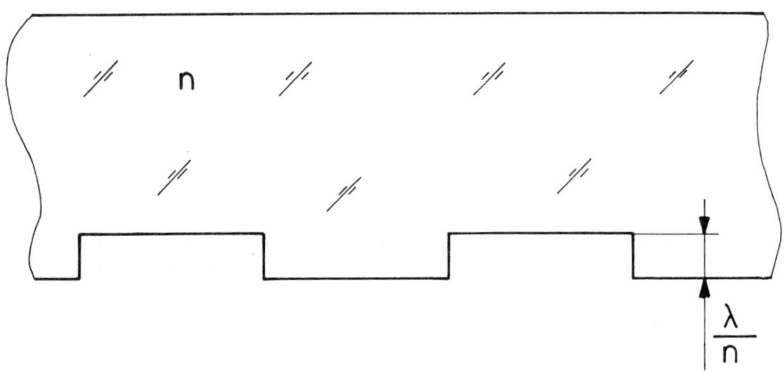


Figure 39 Cross-sectional profile of a phase grating.

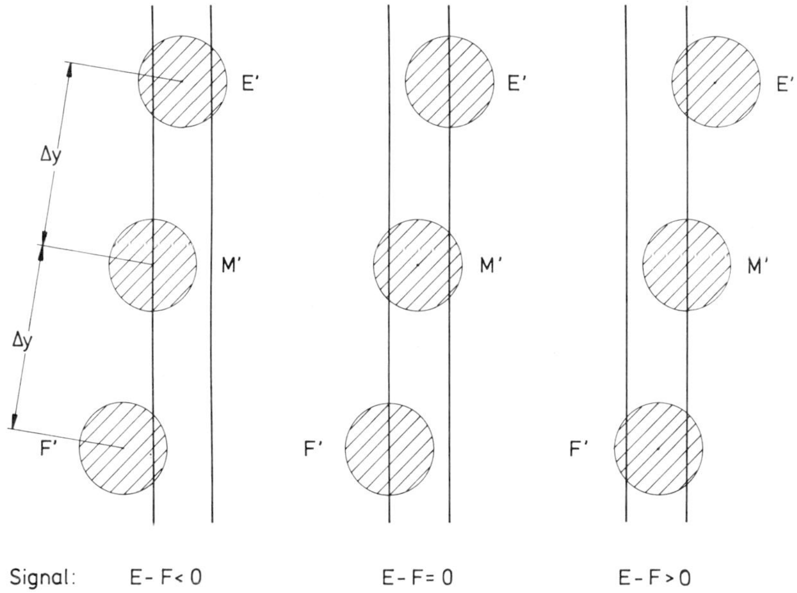


Figure 40 Principle of radial tracking with two auxiliary beams. M', main beam; E' and F', auxiliary beams; E and F, detector signals.

quadrupole sensor, which detects stored data information and the focusing signal. Only when the two beams E' and F' are in the same position relative to the track are the detector signals E and F equal. With a deviation in radial tracking the signal E becomes different from the signal F. The difference signal (E − F) changes sign depending on the direction of the deviation and may be used to control the servo loop for radial tracking. There are three possible cases:

$$E - F \begin{cases} < 0 \\ = 0 \\ > 0 \end{cases}$$

shown schematically in Figure 40.

The two auxiliary beams may also be formed by appropriate prism wedges or by a biprism. These elements are placed in the incident beam to form three beams. The central one for reading stored information falls on the track, and the other two, which are for radial tracking, fall partly on and partly alongside the track, one to the left and the other to the right.

### 7.2.2. Wobble Method

A second method for radial tracking is by introducing a radial wobble with an amplitude of 0.1 μm and which is synchronized to the scanning frequency of the information being read. The reflected radiant flux from the optical disk depends on the position of the spot formed by the objective lens with respect to the track. The wobble may be introduced either by the objective lens or by the track. The first case is called the wobble spot method, the second, the wobble track method. The detector signal is modulated with the wobble frequency. The radial tracking is optimized for a minimum of this signal. The signal to control the servo loop for radial tracking is obtained from the variations in this signal. According to the configuration of the overall system, the detector signal may be designed to be a quadratic function of the position of the spot with respect to the track. Then the first derivative of the detector signal is generated to provide a signal for servo loop control and is linear to the deviation of the spot from the track.

## 8. MEASURING TECHNIQUES FOR THE PERFORMANCE OF OBJECTIVE LENSES

Lenses intended for optical data storage technology need to be manufactured with the utmost care. The manufacturing procedure has to be supervised with the required level of quality control. It should be remembered that we are dealing with nearly diffraction-limited objective lenses with a resolving power of about 1200 c/mm. These lenses must be very light in weight. During automatic focusing and radial tracking they are subject to oscillation frequencies of up to 4 kHz as well as accelerations of 50 G. with peaks up to 300 G. Under these conditions the objective lenses must perform fully and provide long-term stability of image quality. An evaluation of the image quality of optical disk objective lenses is made by measuring the intensity distribution in the point image, by interferometric measurement and processing of the image forming wavefront, or by measurement of the modulation transfer function (MTF) of the lens.

### 8.1. Assessment Based on the Quality of a Point Source Image

Generally, objective lenses for optical disk technology are optimized for one infinite conjugate. Good agreement between the measured and the calculated image quality can be attained only if all the parameters of practical use, such as image scale, NA, spectral region, and cover glass thickness, are taken into account. In general, for noncritical alignment during the image assessment process, the objective lens is illuminated with a collimated laser beam so as to provide a uniform intensity distribution over the clear aperture of the lens. For semiconductor lasers the attainment of a uniform intensity distribution is done with a collimator, and for gas lasers with a beam expander. As shown in Figure 3, with this type of illumination a maximum of image quality is obtained in the central core of the point source image spot.

In evaluating the influence of manufacturing defects on image quality, the effect of beam truncation of the expanded laser beam may be neglected, even if this effect exists in practice. However, allowance has to be made for the effects of truncation in the description of image quality under the conditions of practical use.

To simulate the cover glass of the optical disk, a representative glass plate is inserted following the objective lens. The image spot formed in the focal plane of the objective lens should be magnified for visual judgment with, for example, a 100× microscope objective and a 16× eyepiece. Coherent illumination exists between the objective lens under test and the microscope. Therefore, the wave aberrations of the objective lens and the microscope objective are additive. For large numerical apertures and magnifications, special care has to be paid to the microscope objective. The degree of correction in the microscope lens should be better than that of the lens under test. Good workmanship in rotational symmetry of the microscope objective is necessary. Suitable microscope objective lenses are hard to find when the image quality specified for an optical disk objective lens with an NA or 0.5 or higher is better than $\lambda/25$.

Because of this difficulty, it is considered better to reflect the light from a concave spherical mirror whose center of curvature coincides with the focal point of the lens under test. Such mirrors can be manufactured with a wave aberration accuracy of up to $\lambda/100$ rms. In this setup the optical disk objective lens acts, with respect to the reflected beam from the mirror, as a lens for magnifying the image spot. A schematic representation of this optical arrangement is shown in Figure 41. On the return path the beam becomes recollimated and is reflected at the beam splitter $T_1$. Since this beam splitter is placed in a parallel light beam, no degradation of image quality occurs, provided that the surface quality and the homogeneity of the beam splitter are good.

The magnification of the intermediate image I is given by

$$m = \left|\frac{f_2}{f_1}\right| \tag{120}$$

where $f_1$ is the focal length of the optical disk objective lens $L_1$ and $f_2$ is the focal length of the auxiliary lens $L_2$. If, for example, $f_1 = 4.5$ mm and $f_2 = 112.5$ mm, the magnification m is 25. Because of the reduced NA needed for lens $L_2$, it is easier to realize an almost-diffraction-limited lens.

The intermediate real image I may be further scaled up and imaged on a detector. To determine the quality criterion Q, which according to Eq. (19), is represented by the ratio of two encircled energies, two diaphragms, with apertures of radii $r_1$ and $r_2$ respectively, may be swung successively in front of the detector. For testing lenses in high-volume production it is recommended that two detectors $D_1$ and $D_2$, with a fixed diaphragm $P_1$ and $P_2$ in front of each, be used. Beam splitters $T_2$ and $T_3$ are both located in convergent light beams, and to avoid image degradation, it is recommended that pellicles be used with $T_2$ and $T_3$. A pellicle is a stretched semitransparent thin-foil beam splitter.

For visual evaluation of the point source image spot, the intermediate real image has to be magnified. The lens equation valid for a thin lens element is given by

$$\frac{1}{a} + \frac{1}{a'} = \frac{1}{f} \tag{121}$$

where a and a' are the object and image distances, respectively, and f is the focal length. Hence it follows that the object distance a as a function of the image magnification m may be written as

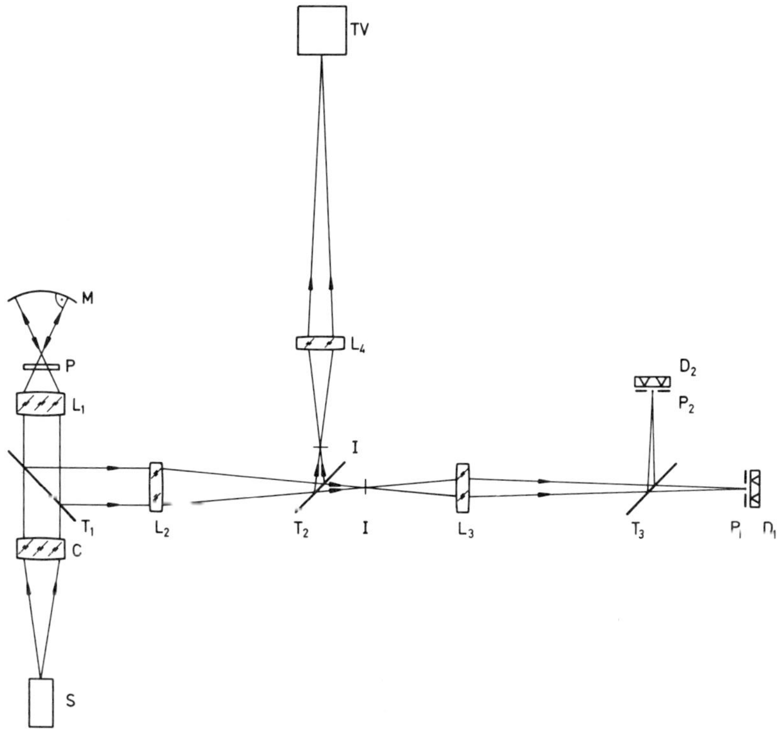


Figure 41 Instrumentation for measuring the point source image spot of objective lenses for optical disk technology. S, laser; C, collimator; $T_1$, beam splitter in a parallel light path; $L_1$, lens under test; P, cover glass plate; M, spherical concave mirror; $L_2$, lens; I, intermediate image; $T_2$ and $T_3$, beam splitters in convergent light path; $L_3$ and $L_4$, lenses for magnifying the real intermediate image I; $P_1$ and $P_2$, diaphragms with diameters $2r_1$ and $2r_2$, respectively; $D_1$ and $D_2$, detectors; TV, television camera with a monitor.

$$a = f\left(1 + \frac{1}{m}\right) \tag{122}$$

and the image distance a' as

$$a' = f(m + 1) \tag{123}$$

Thus the object-to-image distance is given by

$$a + a' = f\frac{(1 + m)^2}{m} \tag{124}$$

Therefore, to obtain sufficient magnification in a small space, a lens with a short focal length must be chosen for lens $L_3$. If the wavelength used in practice is in the visible spectral region, as for example, by using HeNe laser light, the intermediate image may be projected onto a ground-glass screen.

However, if the wavelength used is in the near infrared, such as that emitted by semiconductor lasers, the intermediate image has to be projected for viewing by a television (TV) camera and the magnified image observed on a TV monitor. The projected point source image spot may be judged visually, photographed, and compared with point source image spots of a known type having a known amount of aberration. Point source image spots with different types of aberrations have been compiled by Cagnet et al. (1962).

In Figure 42 are shown typical aberrations in the point source image spot of optical disk objectives as demonstrated by Kuttner (1983). These photographs were taken with LV-type objective lenses, as shown in Figure 12a, with an NA of 0.4 and a focal length of 8.3 mm for a wavelength of 632.8 nm. Figure 42(a) shows the point source image spot of a nearly symmetrical objective lens. Figure 42(b) shows a point source image spot with astigmatism. Figure 42(c) shows the point source image spot of an objective lens with coma as the predominant aberration. Finally, Figure 42(d) shows an image spot with a more pronounced display of coma. These aberrations are essentially caused by tilted elements in the objective lens assembly, but also by errors in curvature and by strain due to pressure on the elements.

The tilt of a lens surface can be measured as follows: A laser beam is directed along the axis of the test lens and focused by an auxiliary lens at the center of curvature of the surface under test

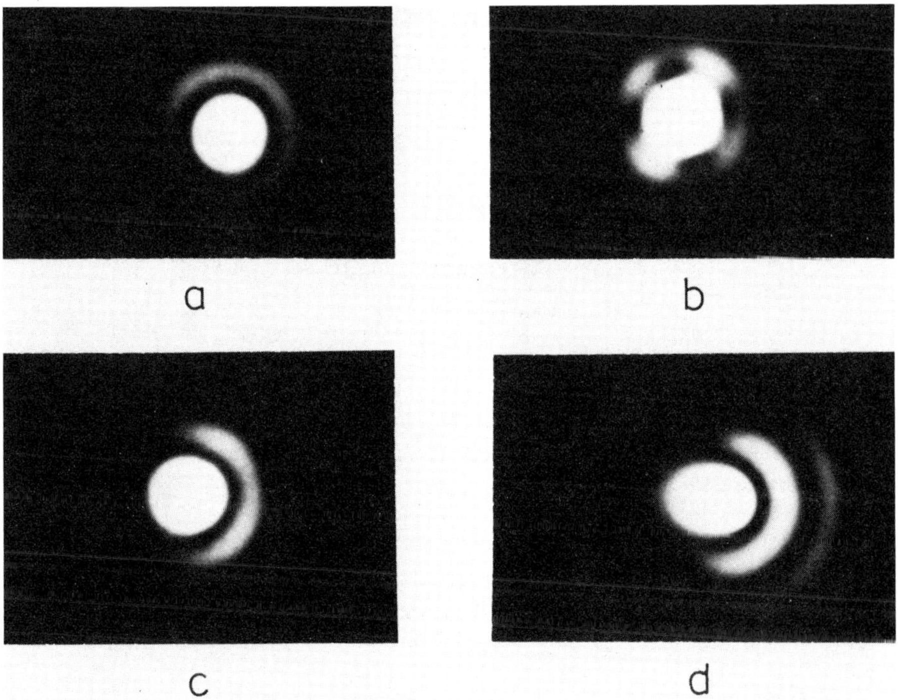


Figure 42 Point source image spots of optical disk objective lenses for which NA = 0.40, focal length = 8.3 mm, $\lambda$ = 632.8 nm. (a) Nearly symmetrical objective lens. (b) Objective lens with astigmatism. (c,d) Objective lenses with coma.

and then reflected from it. Only a nontilted surface reflects the beam back along the way of the incident beam. For a tilted surface the reflected beam is deviated. This deviation is measured with a position sensitive detector. Of course, allowance for the refractive power of the lens elements in front of this surface has to be made. Once these tilt angles are known, the wave aberration for the corresponding objective lens may be computed. Further, the rms value of the wave aberration and the intensity distribution in the point source image spot may also be determined and these results compared with the experimental measurements.

For the objective lenses that produced the point source image spots shown in Figure 42, the rms values of their wave aberration, including the effect of tilted elements, are given in Tablo 27.

**Table 27** RMS Values of the Wave Aberrations for Objective Lenses with an NA of 0.40, a Focal Length of 8.3 mm, at the Wavelength of $\lambda = 632.8$ nm

| Objective lens of Figure: | Prime aberration | RMS value[a] of the wave aberration, $\sqrt{\overline{W^2}}$ |
|---|---|---|
| 42a | Nearly diffraction limited | $\sim\lambda/40$ |
| 42b | Astigmatism | $\sim\lambda/28$ |
| 42c | Coma | $\sim\lambda/20$ |
| 42d | Coma | $\sim\lambda/10$ |

[a]Values given have been calculated with an allowance made for measured surface tilt angles.

The image quality of the objective lens with astigmatism is also affected by additional strain and errors in the surface curvatures of the lens elements.

## 8.2. Interferometric Measurement of a Wavefront Aberration

The wave aberration can be determined metrologically by comparing the wavefront leaving the objective lens with a reference sphere. The instrumentation for this measurement is called an interferometer. The suitable configurations for measuring the wave aberration of objective lenses are the Michelson-type interferometer in the modified form made by Twyman (1921) and the Fizeau-type interferometer in a modified form shown by Forman (1979). In either version a plane wavefront is used as a reference. The reference wave originates from the collimated beam, which is reflected from the reference surface. The surface figure of the reference plane must be better than the aspired accuracy of the interferometric measurement. To measure the wave aberration of optical disk objective lenses, a glass plate window has to be added to simulate the cover glass of the optical disk. For the infinite conjugate of an optical disk objective lens, an aligned collimated beam illuminates the clear aperture of the lens. For diffraction-limited objective lenses, the emerging wavefront is spherical. A spherical mirror with the center of curvature at the focus of the lens reflects the spherical wave, forming a convergent spherical wave. In principle it is not important whether the reference mirror is concave or convex. On its way back this wave passes through the lens again and reemerges as a plane wave that interferes with the plane reference wave. However, the wave aberrations of the objective lens will first form a nonspherical emergent wavefront that on the way back reemerges from the objective lens as a nonplanar wavefront that interferes with the plane reference wavefront.

### 8.2.1. Michelson-Type Interferometer

The Michelson-type interferometer is a two-beam interferometer. A schematic representation of a Michelson interferometer in the modified form after Twyman and Green (1916) is shown in Figure 43. A beam splitter T is inclined at 45° to an incident collimated light beam and divides it into a measuring beam and a reference beam. The reference beam is reflected at a plane reference mirror R. The measuring beam passes through the objective lens L and is reflected at the auxiliary mirror M. To avoid wavefront degradation, a beam splitter cube should be used in the parallel light beams.

If the interferometer is adjusted for a small tilt angle θ between the reflected measuring wavefront and the reflected reference wavefront, W' and W", respectively, a phase difference between these two waves, using Eq. (98), is given by

$$\varphi = \frac{2\pi}{\lambda} h \cos \theta \tag{125}$$

where h is the reduced distance between the two wavefronts. This distance is reduced in such a manner that we obtain a common point for the two waves at one rim of the beam cross section. For nearly normal incidence, cos θ = 1, and this leads us to the condition for dark fringes:

$$h = (2m + 1) \frac{\lambda}{2} ; \quad m = 0, 1, 2, \ldots \tag{126}$$

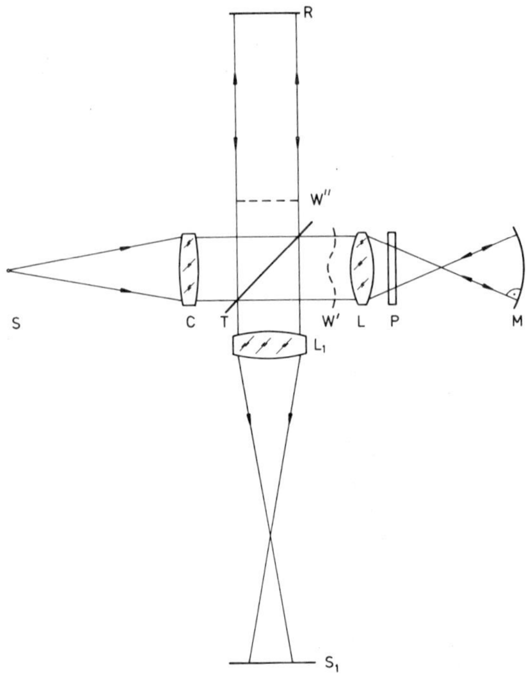


Figure 43 Michelson interferometer in the modified form after Twyman and Green, as used to test lenses. S, light source; C, collimator; T, beam splitter; L, optical disk lens under test; P, cover glass plate; M, spherical auxiliary mirror; R, plane reference mirror; W', reflected wavefront of the measuring beam; W'', reflected wavefront of the reference beam; $L_1$, imaging system; $S_1$, projection screen.

The separation s of the dark fringes in the plane of the interferogram is given by

$$s = \frac{\lambda}{2\theta} \tag{127}$$

where θ is in radians. The dark interference fringes are loci for constant distances h between the two interfering wavefronts. As a wave aberration is a deviation from planeness, it exhibits itself in the interferogram fringes as deviations from their straightness.

### 8.2.2. Fizeau-Type Interferometer

A schematic representation of a Fizeau-type interferometer in the modified form after Forman is shown in Figure 44. The version depicted in Figure 44 uses a gas laser as a light source. The incident light beam passes through the beam expander $B_1$ and $B_2$ and is divided at the semitransparent reference plane R into a reflected reference beam with a plane wavefront W" and a transmitted measuring beam. A plane disk with a semitransparent thin film deposited on its front surface is used as a reference plane. The rear surface of this disk is antireflection coated. The measuring beam passes through the objective lens and is reflected at the auxiliary mirror M; again it passes through the objective lens and is incident as a collimated beam with the wavefront W' at the reference plane, where it now interferes with the reference wavefront W".

### 8.2.3. Coherence Length

To obtain high-contrast interference fringes, the two interfering beams must be of equal intensity. In addition, the coherence

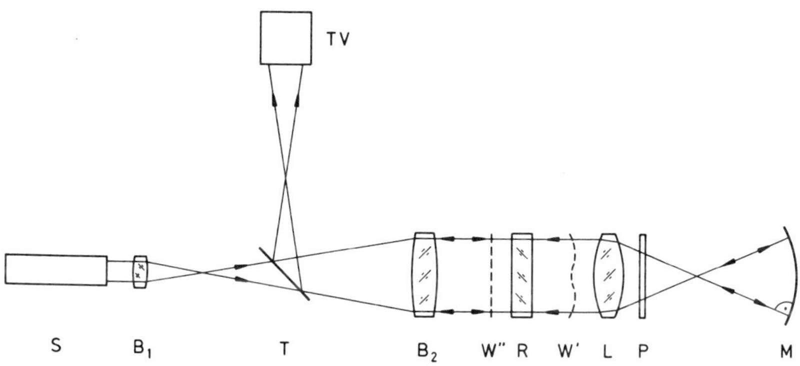


**Figure 44** Fizeau-type interferometer, as used to test lenses. S, laser; $B_1$ and $B_2$, beam expander; T, pellicle beam splitter; L optical disk lens under test; P, cover glass plate; M, spherical auxiliary mirror; R, plane disk with a semitransparent thin film deposited on its front surface, used as reference plane; W', reflected wavefront of the measuring beam; W", reflected wavefront of the reference beam; TV, television camera with a monitor.

length of the light source must be appropriate. Since all the wave trains emitted from a light source are damped waves, the coherence length is a measure of the capability of a wave train to interfere. The more sharply a light source becomes monochromatic, the longer the coherence length. To obtain high-contrast interference fringes using a light source that emits within a spectral interval $\Delta\lambda$ at a mean wavelength $\lambda$, the optical path difference $\delta$ of the two interfering waves must fulfill the condition

$$\delta \ll \frac{\lambda^2}{\Delta\lambda} \tag{128}$$

The coherence length $l$ of the emitted radiation is then given by

$$l = \frac{\lambda^2}{\pi\,\Delta\lambda} \tag{129}$$

as described by Françon and Slansky (1965). For example, a conventional mercury vapor lamp with $\Delta\lambda = 30$ nm at $\lambda = 546$ nm has a coherence length of 3 μm. If such a light source is used in an interferometer, special attention has to be paid to ensure that the measuring beam and the reference beam have equal optical path lengths. A long coherence length is obtained only from lasers with sufficient monochromatism. Because of the difference between the optical path lengths for the two interfering beams in the version of the Fizeau interferometer shown in Figure 44, only lasers with an adequate coherence length are suitable.

### 8.2.4. Processing

The interference fringes may be observed either directly or on a television monitor after suitable projection into a TV camera. The interference fringes may be evaluated automatically and the deduced rms value of the wave aberration used as a criterion for quality control. See, for example, the ZAPP processor by Forman (1979). The accuracy that is achievable by automatic interference fringe evaluation is $\lambda/100$ rms.

## 8.3. Measurement of the MTF of Optical Disk Lenses

### 8.3.1. MTF Instrumentation

Objective lenses for optical data storage technology have a high resolving power. The limiting spatial frequency for nearly diffraction-limited high-aperture lenses is more than 1000 c/mm. A system that measures the incoherent MTF up to this spatial frequency has been described by Kuttner (1968). The basic definition of MTF, according to Eq. (30), allows one to choose a point object or a line object with an intensity distribution $I_0(x)$. In the system mentioned, an incoherently illuminated narrow slit is used as a test target. If the test target slit is narrow enough, it corresponds in theory to a $\delta$ function such that the denominator of Eq. (30) degenerates into a constant value.

A fine-line image formed by the objective lens is scanned with a slit 0.5 μm wide, and from the signal registered the MTF is obtained by Fourier transformation. Let the transmittance of the scanning slit be described by the rectangular function S(x) as

$$S(x) = \begin{cases} 1 & \text{for } -\frac{b}{2} \leqslant 0 \leqslant \frac{b}{2} \\ 0 & \text{outside} \end{cases}$$

where b is the slit width. Now, if H(x) is the fine-line image, we obtain by scanning the convolution

$$H_1(x) = \int_{-\infty}^{+\infty} S(x - x_1)H(x_1)dx_1 \tag{130}$$

The Fourier transform of the slit function S is given by

$$\sigma(R) = \text{const.}\ \frac{\sin(\pi Rb)}{\pi Rb} \tag{131}$$

The Fourier transform of a convolution is given by the product of the two Fourier transforms. Hence applying this convolution theorem to Eq. (130), it follows that the Fourier transform of the fine-line image H, which is the MTF, is given by

$$T(R) = \chi_1(R) \frac{\pi R b}{\sin(\pi R b)} \tag{132}$$

where T(R) is the MTF as function of the spatial frequency R and $\chi_1(R)$ is the Fourier transform of the signal $H_1$ obtained by scanning the fine-line image. The influence of the finite width of the scanning slit is eliminated in this way. The correction factor $(\pi Rb)/\sin(\pi Rb)$ for a slit width of 0.5 μm is given in Table 28 as function of the spatial frequency.

A schematic representation of this system for measuring MTF is shown in Figure 45. To provide incoherent illumination of the test target slit T, a ground glass G is placed in front of the slit. Incoherent illumination of the test target slit may be achieved without a ground glass if the aperture angle of the illuminating system is at least twice that of the aperture angle of the lens under test. The illuminating system has to be adapted to the light source used. The version shown in Figure 45 assumes the use of light sources that emit radiation into a large solid angle, such as semiconductor lasers and tungsten-halogen lamps. For gas laser sources, however, an appropriate collecting lens is used that sufficiently converges the parallel beam onto the test. To adapt the spectral response when using tungsten-halogen lamps, a suitable spectral filter is placed in the parallel light path of the condenser system, $C_1$. A collimator C is used to simulate the infinite conjugate of the objective lens L. Because of the coherent light path between the collimator and this test lens, an ideal on-axis performance is required for the clear aperture used in the collimator.

**Table 28** Correction Factors for a 0.5—μm-Wide Scanning Slit

| Spatial frequency, R (c/mm) | Correction factor, $\frac{\pi Rb}{\sin(\pi Rb)}$ |
|---|---|
| 300 | 1.04 |
| 500 | 1.11 |
| 1000 | 1.57 |

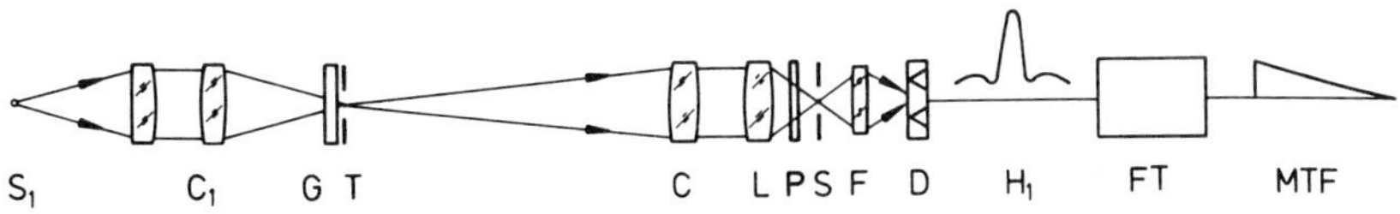


Figure 45 System for measuring the MTF. $S_1$, light source; $C_1$, condenser; G, ground glass; T, test target slit; C, collimator; L, optical disk lens under test; S, scanning slit; F, field lens; D, detector; H1, fine-line image signal characteristic obtained by scanning; FT, computer for the Fourier transformation; MTF, modulation transfer function.

The fine-line image is scanned by moving the scanning slit S at a constant speed across the image plane. Thus the signal obtained corresponds to Eq. (130).

### 8.3.2. Measuring Results

Figure 46 shows the measured incoherent MTFs of a uniformly illuminated optical disk objective lens with an NA of 0.5 and a focal

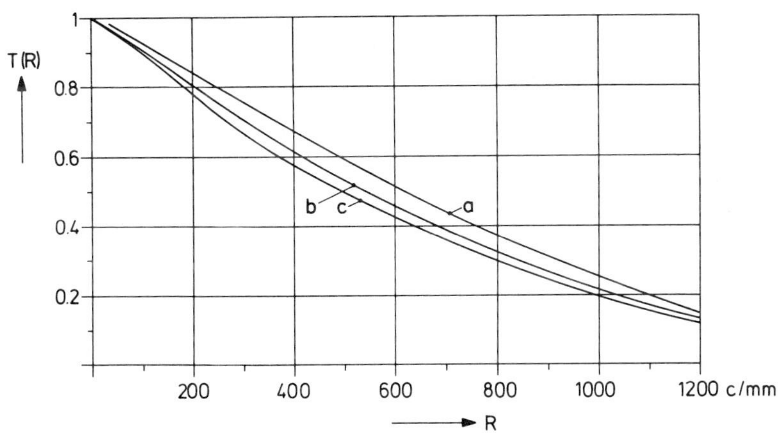


Figure 46 MTF data for an optical disk objective lens with NA = 0.50, focal length = 8.4 mm and $\lambda$ = 632.8 nm. a, Calculated MTF for an ideal lens; b, best MTF in azimuth; c, worst MTF in another azimuth.

length of 8.4 mm at the wavelength of 632.8 nm as a function of the spatial frequency R up to 1200 c/mm. For a more complete survey of the image quality of a real objective lens, the MTF is determined in azimuths at which the MTF is the best and worst (curves b and c, respectively, Fig. 46). Also shown in Figure 46 is the MTF of an ideal lens with the same NA (curve a). This allows a comparison of how close the MTFs of the real lens are to the theoretical limit.

The incoherent MTF of an ideal lens can be computed as follows. With $R_l$ as the limiting spatial frequency, a reduced spatial frequency r may be defined as

$$r = \frac{2R}{R_l} \tag{133}$$

Then the MTF T(r) of an ideal lens as a function of the reduced spatial frequency is given by

$$T(r) = \frac{2}{\pi} \arccos \frac{r}{2} - \frac{r}{2}\sqrt{1 - \left(\frac{r}{2}\right)^2} \tag{134}$$

Figure 46 shows that differences exist between the best MTF and the worst MTF measured in azimuth. These differences are attributed to residual manufacturing errors and are more pronounced in the medium-frequency region; values for these differences ($\Delta T$) may reach $\sim 0.05$. This difference is nearly equivalent to an rms wavefront aberration of $\lambda/20$. Furthermore, we can see that these measured MTFs have a $\Delta T$ about 0.05 lower than the MTF of an ideal lens.

## REFERENCES

Airy, G. B. (1835). *Trans. Camb. Philos. Soc.*, *5*, 283

Banning, M. (1947). *J. Opt. Soc. Am.*, *37*, 792.

Born, M., and Wolf, E. (1959). *Principles of Optics*, Pergamon Press, London.

Botex, D., Channin, D. J., and Ettenberg, M. (1982). *Opt. Eng.*, *21*, 1066.

Bouwhuis, G., and Braat, J. J. M. (1978). *Appl. Opt.* *17*, 1993.

Braat, J. J. M., and Bouwhuis, G. (1978). *Appl. Opt.*, *17*, 2013.
Burri, C. (1950). *Das Polarisationsmikroskop*, Birkhäuser, Basel, p. 29.
Cagnet, M., Françon, M., and Thier, J. C. (1962). *Atlas of Optical Phenomena*, Springer-Verlag, Berlin, p. 25.
Finck, J. C. J., van der Laak, H. J. M., and Schrama, J. T., (1980–81). *Philips Tech. Rundsch.*, *39*, 101.
Forman, P. (1979). *Proc. SPIE*, *192*, 41.
Françon, M., and Slansky, S. (1965). *Cohérence en optique*, Centre National de la Recherche Scientifique, Paris, p. 14.
Goodman, J. W. (1968). *Introduction to Fourier Optics*, McGraw-Hill, New York, p. 60.
Heemskerk, J. P. J. (1978). *Appl. Opt.*, *17*, 2007.
Hopkins, R. E., and Buzawa, M. J. (1976). *Opt. Eng.*, *15*, 90.
Jahnke, E., and Emde, F. (1945). *Tables of Functions with Formulae and Curves*, 4th ed., Dover, New York, p. 149.
Kogelnik, H. (1965). *Bell Syst. Tech. J.*, *44*, 455.
Kuttner, P. (1968). *Appl. Opt.*, *7*, 1029.
Kuttner, P. (1983). *Opt. Eng.*, *22*, 473.
MacLeod, H. A. (1969). *Thin-Film Optical Filters*, Adam Hilger, London, p. 303.
Maréchal, A., and Françon, M. (1960). *Diffraction: Structures des images*, Edition de la Revue d'Optique Théorique et Instrumentale, Paris, p. 152.
RCA (1982). *Handbook: Electro Optics and Devices*, RCA, New York.
Strehl, K. (1902). *Z. Instrumentenkd.*, *22*, 213.
Twyman, F. (1921). *Philos. Mag.* (6), *42*, 777.
Twyman, F., and Green, A. (1916). British Patent No. 103832.
Wagner, E. E., Angerstein, J., and Petermann, K. (1983). *Proc SPIE*, *396*, 189.
Zock, J. D., and Lee, T. C. (1972). *Appl. Opt.*, *11*, 2140.

# Index

## C

D



E

F

G



H

J



K



L

## Q



## R



## S

## T



## U



## V



## W